智能系统与技术丛书

TensorFlow 1.x Deep Learning Cookbook

TensorFlow 深度学习实战

［波兰］安东尼奥·古利（Antonio Gulli）
［印度］阿米塔·卡普尔（Amita Kapoor） 著

李飞 徐从安 顾佼佼 尹高扬 译
李健伟 王丽英 审校

机械工业出版社
China Machine Press

图书在版编目（CIP）数据

TensorFlow 深度学习实战 /（波）安东尼奥·古利（Antonio Gulli）等著；李飞等译 . — 北京：机械工业出版社，2019.1

（智能系统与技术丛书）

书名原文：TensorFlow 1.x Deep Learning Cookbook

ISBN 978-7-111-61575-0

I. T… II. ①安… ②李… III. 人工智能 - 算法 - 研究 IV. TP18

中国版本图书馆 CIP 数据核字（2018）第 275783 号

本书版权登记号：图字 01-2018-1455

Antonio Gulli, Amita Kapoor: TensorFlow 1.x Deep Learning Cookbook（ISBN: 1-78829-359-4）.

Copyright © 2017 Packt Publishing. First published in the English language under the title "TensorFlow 1.x Deep Learning Cookbook".

All rights reserved.

Chinese simplified language edition published by China Machine Press.

Copyright © 2019 by China Machine Press.

本书中文简体字版由 Packt Publishing 授权机械工业出版社独家出版。未经出版者书面许可，不得以任何方式复制或抄袭本书内容。

TensorFlow 深度学习实战

出版发行：机械工业出版社（北京市西城区百万庄大街22号 邮政编码：100037）	
责任编辑：陈佳媛	责任校对：殷 虹
印　　刷：北京市荣盛彩色印刷有限公司	版　　次：2019年1月第1版第1次印刷
开　　本：186mm×240mm　1/16	印　　张：22.75
书　　号：ISBN 978-7-111-61575-0	定　　价：99.00元

凡购本书，如有缺页、倒页、脱页，由本社发行部调换

客服热线：（010）88379426 88361066　　投稿热线：（010）88379604

购书热线：（010）68326294 88379649 68995259　　读者信箱：hzit@hzbook.com

版权所有·侵权必究
封底无防伪标均为盗版
本书法律顾问：北京大成律师事务所　韩光 / 邹晓东

译 者 序

凭借Google强大的开发实力，TensorFlow在2015年年底一出现就受到了极大的关注。与其他机器学习框架相比，TensorFlow在GitHub上的Fork数和Star数都是最多的，而且在图形分类、音频处理、推荐系统和自然语言处理等场景下都有丰富的应用。毫不夸张地说，TensorFlow已成为当下热门的深度学习框架。TensorFlow计算资源不仅包含CPU、GPU，还包括TPU，比如AlphaGo和AlphaZero就用到了TPU集群，大大提高了训练速度，当然还能够通过Google Cloud进行训练。

除了常见的网络结构外，TensorFlow还支持深度强化学习乃至其他计算密集的科学计算。用户能够将训练好的模型方便地部署到多种硬件、操作系统平台上，对Linux、Mac和Windows的支持也越来越成熟，同时在移动设备上表现得也很好。在工业生产环境中，硬件设备新旧不一，TensorFlow的异构性让它能够全面地支持各种硬件和操作系统。TensorFlow还集成了一个功能强大的可视化组件TensorBoard，能可视化网络结构和训练过程，对于观察复杂的网络结构和监控时间长、规模大的训练很有帮助。

本书对一些常见的模型进行了理论介绍，并给出了完整的实现代码，不仅能够对深度学习初学者进行理论与实践的指导，还能为开发人员提供程序设计借鉴。然而由于译者水平有限，特别担心未能完美传达原作者的意思，建议有条件的读者去阅读英文原著。

本书的翻译工作由海军航空大学的一些深度学习研究者完成，具体分工如下：第1～5章由徐从安翻译，第6～10、12章以及附录A由李飞完成，第11章由尹高扬完成，附录B由顾佼佼完成，李健伟、顾佼佼、王丽英对全书进行了审校。同时感谢刘锦涛、张梦玲、丁鹏程在本书翻译过程中提供了指导与帮助。

PREFACE

前　言

本书介绍如何有效地使用Google的开源框架TensorFlow进行深度学习。你将实现不同的深度学习网络，如卷积神经网络（CNN）、循环神经网络（RNN）、深度Q learning网络（DQN）和生成对抗网络（GAN），并将学习如何使用TensorFlow的高级封装工具Keras。

你将了解如何使用不同的深层神经架构来执行复杂的任务，并在一些常用数据集（如MNIST、CIFAR-10、Youtube8m等）上了解不同DNN的性能，不仅可以了解TensorFlow支持的不同移动和嵌入式平台，还可以了解如何为深度学习应用程序搭建云平台。你将深入了解TPU架构，以及它们将如何影响DNN的未来。

学完本书，你将理解深度学习实践技术，能够独立开发现实世界中的一些应用，开展如强化学习、GAN、自动编码机等领域的研究。

本书主要内容

第1章　讨论Google的开源框架TensorFlow，以及为什么它对深度学习有用。这一章将讨论如何在Mac、Windows和Ubuntu上为CPU和GPU安装TensorFlow，还将讨论在整本书中使用的其他Python包。我们将解释TensorFlow代码的两个组成部分、图的定义及其执行，学习使用TensorBoard来查看图表结构，了解TensorFlow常量、变量和占位符之间的区别，也会体验一下TensorFlow Estimator。

第2章　讨论回归及其应用。这一章将讨论回归中涉及的概念，了解它与聚类和分类有何不同，学习不同类型的损失函数，以及如何在TensorFlow中实现它们。我们将学习如何实现L1和L2正则化，并讨论梯度下降算法，学习如何优化它并在TensorFlow中实现它，还将简要介绍一下交叉熵函数及其实现。

第3章　涵盖人工神经网络基础知识，并解释为什么它可以完成DNN所要求的任务，还将学习如何选择不同的激活函数，使用它们来构建一个简单的感知机，并将其用于函数建模。在训练数据之前有必要了解数据的正则化，并将学习逐层构建多层感知机（MLP）。

第4章　讨论卷积过程以及如何提取特征，将学习CNN的三个重要的层：卷积层、池化层和全连接层。我们也将学习dropout正则化，以及为何其能提高性能，并学习不同的CNN架构，如LeNET和GoogleNET。

第5章　涵盖诸如面部识别CNN的一些成功案例。我们将写一个使用CNN进行情感分析的程序，并学习如何实现迁移学习。还将学习如何使用 VGG16 网络进行迁移学习，并用 VGGNet、ResNet、Inception 和 Xception 来学习图像的分类。我们将使用扩张 ConvNet、Wavenet 和 Nsynth 来生成一段音乐，还将学习如何做视觉问答、如何做视频分类。

第 6 章　将介绍 RNN 的基本单元、单词嵌入和时间排序。我们将简要讨论 LSTM 网络，并学习 seq2seq RNN，还学习如何使用 RNN 进行机器翻译、生成文本和预测未来值。

第 7 章　主要介绍无监督学习范式。我们将学习聚类和降维，学习像主成分分析（PCA）这样的技术，并了解它们如何用于降维。我们将学习均值聚类，了解地形图的概念，并学习如何训练自组织映射网络。我们将学习受限玻尔兹曼机（RBM），并讨论其架构和训练机制。我们将学习如何堆叠 RBM 来构建深度置信网络（DBN），并学习如何训练它们。我们还将使用预训练和微调情绪检测的概念来训练 DBN。

第 8 章　将揭开自动编码机的神秘面纱。我们将学习自动编码机及其应用程序，讨论各种可以使用自动编码机的真实案例，讨论编码和后续重建的过程，学习重建错误，学习稀疏自动编码机、KL 散度的概念，学习去噪自动编码机，并使用它们来根据被噪声污染的图像重建纯净的图像。我们将学习如何构建卷积自动编码机和堆叠自动编码机。

第 9 章　涵盖不同的强化学习算法。我们将学习 Q learning 算法、讨论 Bellman-Ford 方程，以及如何选择学习率、折扣因子，还将学习如何使用 OpenAI Gym 框架。我们将学习经验回放和缓存的概念来实现价值迭代 Q 网络，使用 Q learning 和策略梯度来构建游戏 agent。最后，我们将学习如何创建自己的 DQN，简要介绍 AlphaGo Zero 及其取得的胜利。

第 10 章　将介绍移动端深度学习的不同应用。我们将在 Windows 平台上学习如何在 Android studio 上使用 TensorFlow，学习如何使用 TensorFlow 和 XCode 来制作基于 ios 的应用程序，学习如何优化移动端的 TensorFlow 计算图，以及学习如何为移动设备转换 TensorFlow 计算图。

第 11 章　从生成对抗网络开始，首先探索不同的预测模型。我们将阐述 GAN 及其运行背后的动机，了解基本的 GAN 架构，并探索一些非常酷的 GAN 应用。我们将学习另一个生成网络——变分自动编码机。最后，我们将了解最近提出的胶囊网络。

第 12 章　解释云环境、Docker、容器的概念，以及如何使用它们，学习如何使用多个 CPU 或多个服务器运行分布式 TensorFlow，学习如何设置 AWS 进行深度学习，学习如何为深度学习应用设置谷歌云，学习如何为深度学习应用设置 Microsoft Azure 云。我们还将了解其他可用的云服务。

附录 A　简要介绍 AutoML 和孪生网络。

附录 B　包括张量处理单元和它的基本架构，以及它将如何影响 DNN 的未来。

阅读本书前的准备工作

为更好地学习本书，你需要安装 Python 3.5 版本（https://www.continuum.io/ downloads）

以及 TensorFlow（www.tensorflow.org）。建议使用以下硬件配置：
- CPU 架构：x86_64
- 系统内存：8 GB～32 GB
- CPU：4～8 核
- GPU：可选，最低 NVDIA GTX 650

本书读者对象

本书主要面向想要定期执行机器学习任务的数据科学家、机器学习从业者和深度学习爱好者。对深度神经网络已有了解并希望获得 CNN 和 RNN 等方面实践经验的人会发现本书很有用。

本书结构

在这本书中，你会发现几个频繁出现的小标题（准备工作、具体做法、解读分析、更多内容、拓展阅读），具体含义如下。

准备工作
这部分主要介绍需要做什么，并介绍如何安装所需的软件并进行初步设置。

具体做法
这部分包括实现相应功能的具体步骤。

解读分析
这部分通常包含对具体步骤的详细解释。

更多内容
这部分是扩充知识，以使读者对其有更多的了解。

拓展阅读
这部分将列出一些相关的网址。

下载示例代码

本书的示例代码，可以从 http://www.packtpub.com 通过个人账号下载，也可以访问华章图书官网 http://www.hzbook.com，通过注册并登录个人账号下载。

作者简介

  Antonio Gulli 是企业领导和软件部门高管，具备创新精神和执行力，并乐于发现和管理全球高科技人才。他是搜索引擎、在线服务、机器学习、信息检索、数据分析以及云计算等方面的专家。他已经在欧洲四个国家获得了从业经验，并管理过欧美六个国家的有关团队。目前，他在谷歌华沙担任网站主管和云计算主管，推动 Serverless、Kubernetes 和 Google Cloud UX 等项目在欧洲的发展。以前，Antonio 曾作为全球领先出版商 Elsevier 的副总裁帮助创新学术搜索任务，而在此之前，他曾作为微软的首席工程师开展查询建议和新闻搜索项目。他还曾担任 Ask.com 的首席技术官，推动多媒体和新闻搜索技术的发展。Antonio 已经申请了 20 多项专利，发表了多篇学术论文，并是多个国际会议的高级 PC 成员。他相信，要想成功必须把管理、研究技巧、执行力和销售态度统一起来。

  感谢每位读者对本书的关注和信任。在 LinkedIn 和 Facebook 上收到的评论数量让我感到诚惶诚恐：每一位读者都为本书的改进提供了巨大的帮助。也要感谢以下各位在写作过程中给予的支持：Susana、Ewa、Ignacy、Dawid、Max、Jarek、Jerzy、Nina、Laura、Antonella、Eric、Ettore、Francesco、Liubov、Marco、Fabio、Giacomo、Saskia、Christina、Wieland 和 Yossi。非常感谢我的合作者 Amita，以及她的宝贵意见和建议。特别感谢本书的审校者 Eric Brewer、Corrado Zoccolo 和 Sujit Pal 审校了整本书的内容。特别感谢我的经理 Eyal 在写作过程中对我的支持与信任。Charlotte Menora (http://bistrocharlotte.pl/) 是位于华沙的一家酒吧，工作之余我在那里写作了本书的部分章节。这是一个鼓舞人心的地方，如果你访问波兰，一定不要错过华沙这个现代化的酷炫城市。最后，非常感谢 Packt 的整个编辑团队，尤其是 Tushar Gupta 和 Tejas Limkar 的支持以及不断的督促和提醒。感谢你们的耐心。

  Amita Kapoor 是印度德里大学电子学系副教授。她在过去的 20 年里一直在教授神经网络课程。她于 1996 年取得电子学硕士学位，并于 2011 年获得博士学位。在攻读博士学位期间，她获得了著名的 DAAD 奖学金，这笔奖金资助了她在德国卡尔斯鲁厄理工学院的部分研究工作。她曾获得 2008 年度国际光电子大会颁发的最佳演讲奖。她是 OSA（美国光学学会）、IEEE（美国电气和电子工程师协会）、INNS（国际神经网络协会）和 ISBS（印度佛学研究协会）等专业机构的成员。Amita 在国际期刊和会议上发表了 40 多篇文章。她最近的研究领域包括机器学习、人工智能、神经网络、机器人学、佛学（偏心理学和哲学）以及 AI 伦理学。

我试图在这本书中总结我在深度神经网络领域学到的知识。我以一种让读者容易理解和运用的方式呈现，所以本书的主要动力来自于你——读者。感谢这本书的每一位读者，是你们激励着我不断前行，尤其是在我懒惰的时候。我还要感谢加尔各答大学的 Parongama Sen 教授在 1994 年向我介绍了这个主题，还有我的朋友 Nirjara Jain 和 Shubha Swaminathan 在大学图书馆里与我讨论阿西莫夫、他的故事以及未来神经网络将改变社会的预见。非常感谢我的合作者 Antonio Guili 的宝贵意见，以及本书的审校者 Narotam Singh 和 Nick McClure 对全书内容精心审阅并重新检查代码。最后，非常感谢 Packt 的整个编辑团队，尤其是 Tushar Gupta 和 Tejas Limkar 的支持以及不断的督促和提醒。

ABOUT THE REVIEWERS
审校者简介

Narotam Singh 自 1996 年以来一直在印度地球科学部气象部门工作。他一直积极参与 GoI 在信息技术和通信领域的各种技术项目及培训。他于 1996 年毕业于电子学专业,并分别于 1994 年及 1997 年获得计算机工程学士和硕士学位。他目前正在从事神经网络、机器学习和深度学习等领域的研究工作。

Nick McClure 目前是美国华盛顿州西雅图 PayScale 公司的高级数据科学家。此前,他曾任职于 Zillow 及 Caesar's Entertainment 公司。分别在蒙大拿大学、圣本笃学院和圣约翰大学攻读应用数学学位。Nick 还撰写了由 Packt 出版公司出版的《TensorFlow Machine Learning Cookbook》。

他热衷于学习,并在分析、机器学习和人工智能领域颇有建树。Nick 经常会把一些想法放在自己的博客 fromdata.org 或者 Twitter 账号 @nfmcclure 上。

Corrado Zoccolo 是 Google 的高级软件工程师,在分布式索引和信息检索系统方面拥有超过 10 年的从业经验。

我想感谢许多帮助我的人,特别是:我的妻子 Ermelinda,多年来一直支持我致力于计算机科学研究;Marco Vanneschi 教授,向我介绍美丽的分布式系统世界;我在 Google 的第一任经理 Peter Dickman,他让我走上了正确的职业生涯,我每天都在和同事们一起学习。

目 录

译者序
前言
作者简介
审校者简介

第1章 TensorFlow 简介 ………… 1
1.1 引言 ……………………………… 1
1.2 TensorFlow 安装 ……………… 2
1.3 Hello world ………………… 6
1.4 理解 TensorFlow 程序结构 …… 8
1.5 常量、变量和占位符 ………… 10
1.6 使用 TensorFlow 执行矩阵操作 … 15
1.7 使用数据流图 ………………… 17
1.8 从 0.x 迁移到 1.x ……………… 18
1.9 使用 XLA 提升运算性能 …… 19
1.10 调用 CPU/GPU 设备 ………… 21
1.11 TensorFlow 与深度学习 ……… 24
1.12 DNN 问题需要的 Python 包 … 28

第2章 回归 …………………………… 30
2.1 引言 ……………………………… 30
2.2 选择损失函数 ………………… 31
2.3 TensorFlow 中的优化器 ……… 33
2.4 读取 CSV 文件和数据预处理 … 36
2.5 房价估计——简单线性回归 … 39
2.6 房价估计——多元线性回归 … 42
2.7 MNIST 数据集的逻辑回归 …… 45

第3章 神经网络——感知机 ………… 50
3.1 引言 ……………………………… 50
3.2 激活函数 ……………………… 52
3.3 单层感知机 …………………… 58
3.4 计算反向传播算法的梯度 …… 60
3.5 使用 MLP 实现 MNIST 分类器 … 63
3.6 使用 MLP 逼近函数来预测波士顿房价 …………………… 66
3.7 调整超参数 …………………… 71
3.8 高级 API——Keras …………… 72

第4章 卷积神经网络 ………………… 75
4.1 引言 ……………………………… 75
4.2 创建一个 ConvNet 来分类手写 MNIST 数字 ………………… 79
4.3 创建一个 ConvNet 来分类 CIFAR-10 数据集 ……………… 84
4.4 用 VGG19 做风格迁移的图像重绘 ……………………………… 87

4.5 使用预训练的 VGG16 网络进行迁移学习 · 96
4.6 创建 DeepDream 网络 · · · · · · · · · · · · 100

第 5 章 高级卷积神经网络 · · · · · · · · · 105
5.1 引言 · 105
5.2 为情感分析创建一个 ConvNet · · · · 106
5.3 检验 VGG 预建网络学到的滤波器 · 109
5.4 使用 VGGNet、ResNet、Inception 和 Xception 分类图像 · · · · · · · · · · · · 113
5.5 重新利用预建深度学习模型进行特征提取 · 125
5.6 用于迁移学习的深层 InceptionV3 网络 · 126
5.7 使用扩张 ConvNet、WaveNet 和 NSynth 生成音乐 · · · · · · · · · · · · · · · · 129
5.8 关于图像的问答 · · · · · · · · · · · · · · · · 134
5.9 利用预训练网络进行视频分类的 6 种方法 · 140

第 6 章 循环神经网络 · · · · · · · · · · · · · · 144
6.1 引言 · 144
6.2 神经机器翻译——seq2seq RNN 训练 · 150
6.3 神经机器翻译——seq2seq RNN 推理 · 156
6.4 你所需要的是注意力—另一个 seq2seq RNN 例子 · · · · · · · · · · · · · · · · 157
6.5 使用 RNN 像莎士比亚一样写作 · 161
6.6 基于 RNN 学习预测比特币价格 · · · 165

6.7 多对一和多对多的 RNN 例子 · · · · 174

第 7 章 无监督学习 · · · · · · · · · · · · · · · · 176
7.1 引言 · 176
7.2 主成分分析 · 176
7.3 k 均值聚类 · 181
7.4 自组织映射 · 186
7.5 受限玻尔兹曼机 · · · · · · · · · · · · · · · · · · 191
7.6 基于 RBM 的推荐系统 · · · · · · · · · · · · 196
7.7 用 DBN 进行情绪检测 · · · · · · · · · · · · 198

第 8 章 自动编码机 · · · · · · · · · · · · · · · · 205
8.1 引言 · 205
8.2 标准自动编码机 · · · · · · · · · · · · · · · · · · 207
8.3 稀疏自动编码机 · · · · · · · · · · · · · · · · · · 212
8.4 去噪自动编码机 · · · · · · · · · · · · · · · · · · 217
8.5 卷积自动编码机 · · · · · · · · · · · · · · · · · · 221
8.6 堆叠自动编码机 · · · · · · · · · · · · · · · · · · 225

第 9 章 强化学习 · · · · · · · · · · · · · · · · · · 231
9.1 引言 · 231
9.2 学习 OpenAI Gym · · · · · · · · · · · · · · · 232
9.3 用神经网络智能体玩 Pac-Man 游戏 · 235
9.4 用 Q learning 玩 Cart-Pole 平衡游戏 · 238
9.5 用 DQN 玩 Atari 游戏 · · · · · · · · · · · · 244
9.6 用策略梯度网络玩 Pong 游戏 · · · · 252

第 10 章 移动端计算 · · · · · · · · · · · · · · · 259
10.1 引言 · 259

- 10.2 安装适用于 macOS 和 Android 的 TensorFlow mobile ············ 260
- 10.3 玩转 TensorFlow 和 Android 的示例 ····················· 265
- 10.4 安装适用于 macOS 和 iPhone 的 TensorFlow mobile ············ 268
- 10.5 为移动设备优化 TensorFlow 计算图 ····················· 271
- 10.6 为移动设备分析 TensorFlow 计算图 ····················· 273
- 10.7 为移动设备转换 TensorFlow 计算图 ····················· 275

第 11 章 生成式模型和 CapsNet ··· 278
- 11.1 引言 ······················ 278
- 11.2 学习使用简单 GAN 虚构 MNIST 图像 ····················· 284
- 11.3 学习使用 DCGAN 虚构 MNIST 图像 ····················· 289
- 11.4 学习使用 DCGAN 虚构名人面孔和其他数据集 ············ 294
- 11.5 实现变分自动编码机 ·········· 297
- 11.6 学习使用胶囊网络击败 MNIST 前期的最新成果 ············ 305

第 12 章 分布式 TensorFlow 和云深度学习 ················· 319
- 12.1 引言 ······················ 319
- 12.2 在 GPU 上使用 TensorFlow ····· 322
- 12.3 玩转分布式 TensorFlow：多个 GPU 和一个 CPU ············ 323
- 12.4 玩转分布式 TensorFlow：多服务器 ····················· 324
- 12.5 训练分布式 TensorFlow MNIST 分类器 ····················· 326
- 12.6 基于 Docker 使用 TensorFlow Serving ···················· 328
- 12.7 使用计算引擎在谷歌云平台上运行分布式 TensorFlow ········ 330
- 12.8 在谷歌 CloudML 上运行分布式 TensorFlow ················· 333
- 12.9 在 Microsoft Azure 上运行分布式 TensorFlow ················· 334
- 12.10 在 Amazon AWS 上运行分布式 TensorFlow ················· 337

附录 A 利用 AutoML 学会学习（元学习）··················· 342

附录 B TensorFlow 处理器 ········· 350

CHAPTER 1

第 1 章

TensorFlow 简介

任何曾经试图在 Python 中只利用 NumPy 编写神经网络代码的人都知道那是多么麻烦。编写一个简单的一层前馈网络的代码尚且需要 40 多行代码，当增加层数时，编写代码将会更加困难，执行时间也会更长。

TensorFlow 使这一切变得更加简单快捷，从而缩短了想法到部署之间的实现时间。在本书中，你将学习如何利用 TensorFlow 的功能来实现深度神经网络。

1.1 引言

TensorFlow 是由 Google Brain 团队为深度神经网络（DNN）开发的功能强大的开源软件库，于 2015 年 11 月首次发布，在 Apache 2.x 协议许可下可用。截至今天，短短的两年内，其 GitHub 库（https://github.com/tensorflow/tensorflow）大约 845 个贡献者共提交超过 17000 次，这本身就是衡量 TensorFlow 流行度和性能的一个指标。下图列出了当前流行的深度学习框架，从中能够清楚地看到 TensorFlow 的领先地位。

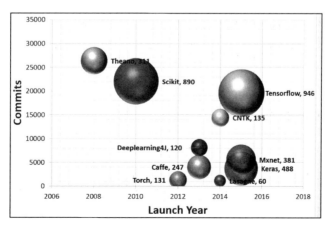

该图基于 2017 年 7 月 12 日 GitHub 库的数据；每个气泡都有一个注释（包括框架、贡献者）。

先来了解一下 TensorFlow 究竟是什么，以及它为什么在 DNN 研究人员和工程师中如此受欢迎。开源深度学习库 TensorFlow 允许将深度神经网络的计算部署到任意数量的 CPU 或 GPU 的服务器、PC 或移动设备上，且只利用一个 TensorFlow API。你可能会问，还有很多其他的深度学习库，如 Torch、Theano、Caffe 和 MxNet，那 TensorFlow 与其他深度学习库的区别在哪里呢？包括 TensorFlow 在内的大多数深度学习库能够自动求导、开源、支持多种 CPU/GPU、拥有预训练模型，并支持常用的 NN 架构，如递归神经网络（RNN）、卷积神经网络（CNN）和深度置信网络（DBN）。TensorFlow 则还有更多的特点，如下：

- 支持所有流行语言，如 Python、C++、Java、R 和 Go。
- 可以在多种平台上工作，甚至是移动平台和分布式平台。
- 它受到所有云服务（AWS、Google 和 Azure）的支持。
- Keras——高级神经网络 API，已经与 TensorFlow 整合。
- 与 Torch / Theano 比较，TensorFlow 拥有更好的计算图表可视化。
- 允许模型部署到工业生产中，并且容易使用。
- 有非常好的社区支持。
- TensorFlow 不仅仅是一个软件库，它是一套包括 TensorFlow，TensorBoard 和 TensorServing 的软件。

谷歌 research 博客（`https://research.googleblog.com/2016/11/celebrating-tensorflows-first-year-html`）列出了全球一些使用 TensorFlow 开发的有趣项目：

- Google 翻译运用了 TensorFlow 和 TPU（Tensor Processing Units）。
- Project Magenta 能够使用强化学习模型生成音乐，运用了 TensorFlow。
- 澳大利亚海洋生物学家使用了 TensorFlow 来发现和理解濒临灭绝的海牛。
- 一位日本农民运用 TensorFlow 开发了一个应用程序，使用大小和形状等物理特性对黄瓜进行分类。

使用 TensorFlow 的项目还有很多。本书旨在让读者理解 TensorFlow 在深度学习模型中的应用，使读者可以轻松地将模型用于数据集并开发有用的应用程序。每章包含一系列处理技术问题、依赖性、代码和解读的示例，在每章的最后，还有一个功能完善的深度学习模型。

1.2 TensorFlow 安装

本节将介绍在不同的操作系统（Linux、Mac 和 Windows）上如何全新安装 TensorFlow 1.3。首先了解安装 TensorFlow 的必要要求，TensorFlow 可以在 Ubuntu 和 macOS 上基于 native pip、Anaconda、virtualenv 和 Docker 进行安装，对于 Windows 操作系统，可以使用 native pip 或 Anaconda。

Anaconda 适用于这三种操作系统，安装简单，在同一个系统上维护不同的项目环境也很方便，因此本书将基于 Anaconda 安装 TensorFlow。有关 Anaconda 及其环境管理的更多详细信息，请参考 `https://conda.io/docs/user-guide/index.html`。

本书中的代码已经在以下平台上进行了测试：

- Windows 10，Anaconda 3，Python 3.5，TensorFlow GPU，CUDA toolkit 8.0，cuDNN v5.1，NVDIA GTX 1070
- Windows 10 / Ubuntu 14.04 / Ubuntu 16.04 / macOS Sierra，Anaconda 3，Python 3.5，TensorFlow（CPU）

准备工作

TensorFlow 安装的前提是系统安装了 Python 2.5 或更高版本，书中的例子是以 Python 3.5（Anaconda 3 版）为基础设计的。为了安装 TensorFlow，首先确保你已经安装了 Anaconda。可以从网址（`https://www.continuum.io/downloads`）中下载并安装适用于 Windows / macOS 或 Linux 的 Anaconda。

安装完成后，可以在窗口中使用以下命令进行安装验证：

```
conda --version
```

安装了 Anaconda，下一步决定是否安装 TensorFlow CPU 版本或 GPU 版本。几乎所有计算机都支持 TensorFlow CPU 版本，而 GPU 版本则要求计算机有一个 CUDA compute capability 3.0 及以上的 NVIDIA GPU 显卡（对于台式机而言最低配置为 NVIDIA GTX 650）。

> **CPU 与 GPU 的对比**：中央处理器（CPU）由对顺序串行处理优化的内核（4～8 个）组成。图形处理器（GPU）具有大规模并行架构，由数千个更小且更有效的核芯（大致以千计）组成，能够同时处理多个任务。

对于 TensorFlow GPU 版本，需要先安装 CUDA toolkit 7.0 及以上版本、NVDIA ® 驱动程序和 cuDNN v3 或以上版本。Windows 系统还另外需要一些 DLL 文件，读者可以下载所需的 DLL 文件或安装 Visual Studio C ++。还有一件事要记住，cuDNN 文件需安装在不同的目录中，并需要确保目录在系统路径中。当然也可以将 CUDA 库中的相关文件复制到相应的文件夹中。

具体做法

1. 在命令行中使用以下命令创建 conda 环境（如果使用 Windows，最好在命令行中以管理员身份执行）：

```
conda create -n tensorflow python=3.5
```

2. 激活 conda 环境：

```
# Windows
activate tensorflow
#Mac OS/ Ubuntu:
source activate tensorflow
```

3. 该命令应提示：

```
# Windows
(tensorflow)C:>
# Mac OS/Ubuntu
(tensorflow)$
```

4. 根据要在 conda 环境中安装的 TensorFlow 版本，输入以下命令：

```
## Windows
# CPU Version only
(tensorflow)C:>pip install --ignore-installed --upgrade
https://storage.googleapis.com/tensorflow/windows/cpu/tensorflow-1.
3.0cr2-cp35-cp35m-win_amd64.whl

# GPU Version
(tensorflow)C:>pip install --ignore-installed --upgrade
https://storage.googleapis.com/tensorflow/windows/gpu/tensorflow_gp
u-1.3.0cr2-cp35-cp35m-win_amd64.whl

## Mac OS
# CPU only Version
(tensorflow)$ pip install --ignore-installed --upgrade
https://storage.googleapis.com/tensorflow/mac/cpu/tensorflow-1.3.0c
r2-py3-none-any.whl

# GPU version
(tensorflow)$ pip install --ignore-installed --upgrade
https://storage.googleapis.com/tensorflow/mac/gpu/tensorflow_gpu-1.
3.0cr2-py3-none-any.whl

## Ubuntu
# CPU only Version
(tensorflow)$ pip install --ignore-installed --upgrade
https://storage.googleapis.com/tensorflow/linux/cpu/tensorflow-1.3.
0cr2-cp35-cp35m-linux_x86_64.whl

# GPU Version
(tensorflow)$ pip install --ignore-installed --upgrade
https://storage.googleapis.com/tensorflow/linux/gpu/tensorflow_gpu-
1.3.0cr2-cp35-cp35m-linux_x86_64.whl
```

5. 在命令行中输入 `python`。

6. 输入以下代码：

```
import tensorflow as tf
message = tf.constant('Welcome to the exciting world of Deep Neural
Networks!')
with tf.Session() as sess:
    print(sess.run(message).decode())
```

7. 输出如下：

```
C:\Users\am>activate tensorflow

(tensorflow) C:\Users\am>python
Python 3.5.3 |Continuum Analytics, Inc.| (default, Feb 22 2017, 21:28:42) [MSC v.1900 64 bit (AMD64)] on win32
Type "help", "copyright", "credits" or "license" for more information.
>>> import tensorflow as tf
>>> message = tf.constant('Welcome to the exciting world of Deep Neural Networks!')
>>> with tf.Session() as sess:
...     print(sess.run(message).decode())
...
2017-06-04 15:23:17.129659: W c:\tf_jenkins\home\workspace\release-win\device\gpu\os\windows\tensorflow\core\platfc
2017-06-04 15:23:17.129809: W c:\tf_jenkins\home\workspace\release-win\device\gpu\os\windows\tensorflow\core\platfc
2017-06-04 15:23:17.129961: W c:\tf_jenkins\home\workspace\release-win\device\gpu\os\windows\tensorflow\core\platfc
2017-06-04 15:23:17.130111: W c:\tf_jenkins\home\workspace\release-win\device\gpu\os\windows\tensorflow\core\platfc
2017-06-04 15:23:17.130264: W c:\tf_jenkins\home\workspace\release-win\device\gpu\os\windows\tensorflow\core\platfc
2017-06-04 15:23:17.130392: W c:\tf_jenkins\home\workspace\release-win\device\gpu\os\windows\tensorflow\core\platfc
2017-06-04 15:23:17.130504: W c:\tf_jenkins\home\workspace\release-win\device\gpu\os\windows\tensorflow\core\platfc
2017-06-04 15:23:17.130991: W c:\tf_jenkins\home\workspace\release-win\device\gpu\os\windows\tensorflow\core\platfc
2017-06-04 15:23:17.466909: I c:\tf_jenkins\home\workspace\release-win\device\gpu\os\windows\tensorflow\core\commor
name: GeForce GTX 1070
major: 6 minor: 1 memoryClockRate (GHz) 1.683
pciBusID 0000:01:00.0
Total memory: 8.00GiB
Free memory: 6.66GiB
2017-06-04 15:23:17.467072: I c:\tf_jenkins\home\workspace\release-win\device\gpu\os\windows\tensorflow\core\commor
2017-06-04 15:23:17.468448: I c:\tf_jenkins\home\workspace\release-win\device\gpu\os\windows\tensorflow\core\commor
2017-06-04 15:23:17.468824: I c:\tf_jenkins\home\workspace\release-win\device\gpu\os\windows\tensorflow\core\commor
Welcome to the exciting world of Deep Neural Networks!
>>>
```

8. 在命令行中禁用 conda 环境（Windows 调用 `deactivate` 命令，MAC / Ubuntu 调用 `source deactivate` 命令）。

解读分析

Google 使用 wheel 标准分发 TensorFlow，它是 `.whl` 后缀的 ZIP 格式文件。Python 3.6 是 Anaconda 3 默认的 Python 版本，且没有已安装的 wheel。在编写本书时，Python 3.6 支持的 wheel 仅针对 Linux / Ubuntu，因此，在创建 TensorFlow 环境时，这里指定 Python 3.5。接着新建 conda 环境，命名为 `tensorflow`，并安装 pip, python, wheel 及其他软件包。

conda 环境创建后，调用 `source activate/activate` 命令激活环境。在激活的环境中，使用 `pip install` 命令安装所需的 TensorFlow（从相应的 TensorFlow-API URL 下载）。尽管有利用 conda forge 安装 TensorFlow CPU 的 Anaconda 命令，但 TensorFlow 推荐使用 pip install。在 conda 环境中安装 TensorFlow 后，就可以禁用了。现在可以执行第一个 TensorFlow 程序了。

程序运行时，可能会看到一些警告（W）消息和提示（I）消息，最后是输出代码：

```
Welcome to the exciting world of Deep Neural Networks!
```

恭喜你已成功安装并执行了第一个 TensorFlow 代码，在下一节中更深入地通读代码。

拓展阅读

另外，你也可以安装 Jupyter notebook：

1. 安装 `ipython`：

```
conda install -c anaconda ipython
```

2. 安装 `nb_conda_kernels`：

```
conda install -channel=conda-forge nb_conda_kernels
```

3. 启动 `Jupyter notebook`：

```
jupyter notebook
```

 这将会打开一个新的浏览器窗口。

如果已安装了 TensorFlow，则可以调用 `pip install--upgrade tensorflow` 进行升级。

1.3 Hello world

在任何计算机语言中学习的第一个程序是都是 Hello world，本书中也将遵守这个惯例，从程序 Hello world 开始。上一节进行 TensorFlow 安装验证的代码如下：

```
import tensorflow as tf
message = tf.constant('Welcome to the exciting world of Deep Neural Networks!')
 with tf.Session() as sess:
    print(sess.run(message).decode())
```

下面一起看一下这段简单的代码。

具体做法

1. 导入 tensorflow，这将导入 TensorFlow 库，并允许使用其精彩的功能。

```
import tensorflow as tf
```

2. 由于要打印的信息是一个常量字符串，因此使用 `tf.constant`：

```
message = tf.constant('Welcome to the exciting world of Deep Neural Networks!')
```

3. 为了执行计算图，利用 `with` 语句定义 Session，并使用 `run` 来运行：

```
with tf.Session() as sess:
    print(sess.run(message).decode())
```

4. 输出中包含一系列警告消息（W），具体取决于所使用的计算机和操作系统，并声明如果针对所使用的计算机进行编译，代码运行速度可能会更快：

```
The TensorFlow library wasn't compiled to use SSE instructions, but
these are available on your machine and could speed up CPU
```

```
computations.
The TensorFlow library wasn't compiled to use SSE2 instructions,
but these are available on your machine and could speed up CPU
computations.
The TensorFlow library wasn't compiled to use SSE3 instructions,
but these are available on your machine and could speed up CPU
computations.
The TensorFlow library wasn't compiled to use SSE4.1 instructions,
but these are available on your machine and could speed up CPU
computations.
The TensorFlow library wasn't compiled to use SSE4.2 instructions,
but these are available on your machine and could speed up CPU
computations.
The TensorFlow library wasn't compiled to use AVX instructions, but
these are available on your machine and could speed up CPU
computations.
The TensorFlow library wasn't compiled to use AVX2 instructions,
but these are available on your machine and could speed up CPU
computations.
The TensorFlow library wasn't compiled to use FMA instructions, but
these are available on your machine and could speed up CPU
computations.
```

5. 如果使用 TensorFlow GPU 版本，则还会获得一系列介绍设备的提示消息（I）：

```
Found device 0 with properties:
name: GeForce GTX 1070
major: 6 minor: 1 memoryClockRate (GHz) 1.683
pciBusID 0000:01:00.0
Total memory: 8.00GiB
Free memory: 6.66GiB
DMA: 0
0:   Y
Creating TensorFlow device (/gpu:0) -> (device: 0, name: GeForce
GTX 1070, pci bus id: 0000:01:00.0)
```

6. 最后是在会话中打印的信息：

```
Welcome to the exciting world of Deep Neural Networks
```

解读分析

前面的代码分为三个主要部分。第一部分 import 模块包含代码将使用的所有库，在目前的代码中只使用 TensorFlow，其中语句 `import tensorflow as tf` 则允许 Python 访问 TensorFlow 所有的类、方法和符号。第二个模块包含图形定义部分，在这里，创建想要的计算图。在本例中计算图只有一个节点，tensor 常量消息由字符串"`Welcome to the exciting world of Deep Neural Networks`"构成。第三个模块是通过会话执行计算图，这部分使用 `with` 关键字创建了会话，最后在会话中执行以上计算图。

现在来解读输出。收到的警告消息提醒 TensorFlow 代码可以以更快的速度运行，这能够通过从 source 安装 TensorFlow 来实现（本章后面的章节中会提及）。收到的提示消息给出计算设备的信息。这两个消息都是无害的，如果不想看到它们，可以通过以下两行代码实现：

```
import os
os.environ['TF_CPP_MIN_LOG_LEVEL']='2'
```

以上代码用于忽略级别 2 及以下的消息（级别 1 是提示，级别 2 是警告，级别 3 是错误）。

该程序打印计算图执行的结果，计算图的执行则使用 `sess.run()` 语句，`sess.run` 求取 message 中所定义的 tensor 值；计算图执行结果输入到 `print` 函数，并使用 `decode` 方法改进，`print` 函数向 `stdout` 输出结果：

```
b'Welcome to the exciting world of Deep Neural Networks'
```

这里的输出结果是一个字节字符串。要删除字符串引号和"b"（表示字节，byte）只保留单引号内的内容，可以使用 `decode()` 方法。

1.4 理解 TensorFlow 程序结构

TensorFlow 与其他编程语言非常不同。首先通过将程序分为两个独立的部分，构建任何拟创建神经网络的蓝图，包括计算图的定义及其执行。起初这对于传统程序员来说看起来很麻烦，但是正是图定义和执行的分开设计让 TensorFlow 能够多平台工作以及并行执行，TensorFlow 也因此更加强大。

计算图：计算图是包含节点和边的网络。本节定义所有要使用的数据，也就是张量（tensor）对象（常量、变量和占位符），同时定义要执行的所有计算，即运算操作对象（Operation Object，简称 OP）。每个节点可以有零个或多个输入，但只有一个输出。网络中的节点表示对象（张量和运算操作），边表示运算操作之间流动的张量。计算图定义神经网络的蓝图，但其中的张量还没有相关的数值。

为了构建计算图，需要定义所有要执行的常量、变量和运算操作。常量、变量和占位符将在下一节中介绍，数学运算操作将在矩阵运算章节中详细讨论。本节将用一个简单的例子描述程序结构，例子中，通过定义并执行计算图来实现两个向量相加。

计算图的执行：使用会话对象来实现计算图的执行。会话对象封装了评估张量和操作对象的环境。这里真正实现了运算操作并将信息从网络的一层传递到另外一层。不同张量对象的值仅在会话对象中被初始化、访问和保存。在此之前张量对象只被抽象定义，在会话中才被赋予实际的意义。

具体做法

通过以下步骤定义一个计算图：

1. 在此以两个向量相加为例给出计算图。假设有两个向量 `v_1` 和 `v_2` 将作为输入提供给 Add 操作。建立的计算图如下：

2. 定义该图的相应代码如下所示：

```
v_1 = tf.constant([1,2,3,4])
v_2 = tf.constant([2,1,5,3])
v_add = tf.add(v_1,v_2)   # You can also write v_1 + v_2 instead
```

3. 然后在会话中执行这个图：

```
with tf.Session() as sess:
    prin(sess.run(v_add))
```

以上两行相当于下面的代码。上面的代码的优点是不必显式写出关闭会话的命令。

```
sess = tf.Session()
print(ses.run(tv_add))
sess.close()
```

4. 运行结果是显示两个向量的和：

[3 3 8 7]

> 请记住，每个会话都需要使用 `close()` 来明确关闭，而 `with` 格式可以在运行结束时隐式关闭会话。

解读分析

计算图的构建非常简单。添加变量和操作，并按照逐层建立神经网络的顺序传递它们（让张量流动）。TensorFlow 还允许使用 with `tf.device()` 命令来使用具有不同计算图形对象的特定设备（CPU / GPU）。在例子中，计算图由三个节点组成，`v_1` 和 `v_2` 表示这两个向量，Add 是要对它们执行的操作。

接下来，为了使这个图生效，首先需要使用 `tf.Session()` 定义一个会话对象 `sess`。然后使用 Session 类中定义的 `run` 方法运行它，如下所示：

```
run (fetches, feed_dict=None, options=None, run_metadata)
```

运算结果的值在 `fetches` 中提取；在示例中，提取的张量为 `v_add`。run 方法将导致在每次执行该计算图的时候，都将对与 `v_add` 相关的张量和操作进行赋值。如果抽取的不是 `v_add` 而是 `v_1`，那么最后给出的是向量 `v_1` 的运行结果。

[1,2,3,4]

此外，一次可以提取一个或多个张量或操作对象，例如，如果结果抽取的是 `[v_1, v_2, v_add]`，那么输出如下：

```
[array([1, 2, 3, 4]), array([2, 1, 5, 3]), array([3, 3, 8, 7])]
```

在同一段代码中，可以有多个会话对象。

拓展阅读

你一定会问为什么必须编写这么多行的代码来完成一个简单的向量加，或者显示一条简单的消息。其实你可以利用下面这一行代码非常方便地完成这个工作：

```
print(tf.Session().run(tf.add(tf.constant([1,2,3,4]),tf.constant([2,1,5,3]))))
```

编写这种类型的代码不仅影响计算图的表达，而且当在 for 循环中重复执行相同的操作（OP）时，可能会导致占用大量内存。养成显式定义所有张量和操作对象的习惯，不仅可使代码更具可读性，还可以帮助你以更清晰的方式可视化计算图。

> 使用 TensorBoard 可视化图形是 TensorFlow 最有用的功能之一，特别是在构建复杂的神经网络时。我们构建的计算图可以在图形对象的帮助菜单下进行查看。

如果你正在使用 Jupyter Notebook 或者 Python shell 进行编程，使用 `tf.InteractiveSession` 将比 `tf.Session` 更方便。`InteractiveSession` 使自己成为默认会话，因此你可以使用 `eval()` 直接调用运行张量对象而不用显式调用会话。下面给出一个例子：

```
sess = tf.InteractiveSession()

v_1 = tf.constant([1,2,3,4])
v_2 = tf.constant([2,1,5,3])

v_add = tf.add(v_1,v_2)

print(v_add.eval())

sess.close()
```

1.5 常量、变量和占位符

最基本的 TensorFlow 提供了一个库来定义和执行对张量的各种数学运算。张量可理解为一个 n 维矩阵。所有类型的数据，包括标量、矢量和矩阵等都是特殊类型的张量。

数据的类型	张量	形状	数据的类型	张量	形状
标量	0-D 张量	[]	矩阵	2-D 张量	$[D_0,D_1]$
向量	1-D 张量	[D0]	张量	N-D 张量	$[D_0,D_1,...D_{n-1}]$

TensorFlow 支持以下三种类型的张量：
- 常量
- 变量
- 占位符

常量：常量是其值不能改变的张量。

变量：当一个量在会话中的值需要更新时，使用变量来表示。例如，在神经网络中，权重需要在训练期间更新，可以通过将权重声明为变量来实现。变量在使用前需要被显示初始化。另外需要注意的是，常量存储在计算图的定义中，每次加载图时都会加载相关变量。换句话说，它们是占用内存的。另一方面，变量又是分开存储的。它们可以存储在参数服务器上。

占位符：用于将值输入 TensorFlow 图中。它们可以和 `feed_dict` 一起使用来输入数据。在训练神经网络时，它们通常用于提供新的训练样本。在会话中运行计算图时，可以为占位符赋值。这样在构建一个计算图时不需要真正地输入数据。需要注意的是，占位符不包含任何数据，因此不需要初始化它们。

具体做法

从常量开始说明：

1. 声明一个标量常量：

```
t_1 = tf.constant(4)
```

2. 一个形如 [1，3] 的常量向量可以用如下代码声明：

```
t_2 = tf.constant([4, 3, 2])
```

3. 要创建一个所有元素为零的张量，可以使用 tf.zeros() 函数。这个语句可以创建一个形如 [M，N] 的零元素矩阵，数据类型（dtype）可以是 int32、float32 等。

```
tf.zeros([M,N],tf.dtype)
```

例如：

```
zero_t = tf.zeros([2,3],tf.int32)
# Results in an 2×3 array of zeros: [[0 0 0], [0 0 0]]
```

4. 还可以创建与现有 Numpy 数组或张量常量具有相同形状的张量常量，如下所示：

```
tf.zeros_like(t_2)
# Create a zero matrix of same shape as t_2
tf.ones_like(t_2)
# Creates a ones matrix of same shape as t_2
```

5. 创建一个所有元素都设为 1 的张量。下面的语句即创建一个形如 [M，N]、元素均为 1 的矩阵：

```
tf.ones([M,N],tf.dtype)
```

例如:

```
ones_t = tf.ones([2,3],tf.int32)
# Results in an 2×3 array of ones:[[1 1 1], [1 1 1]]
```

更进一步,还有以下语句:

1. 在一定范围内生成一个从初值到终值等差排布的序列:

```
tf.linspace(start, stop, num)
```

2. 相应的值为 (stop-start)/(num-1)
3. 例如:

```
range_t = tf.linspace(2.0,5.0,5)
# We get: [ 2.   2.75  3.5  4.25  5. ]
```

4. 从开始(默认值 = 0)生成一个数字序列,增量为 delta(默认值 = 1),直到终值(但不包括终值):

```
tf.range(start,limit,delta)
```

下面给出实例:

```
range_t = tf.range(10)
# Result: [0 1 2 3 4 5 6 7 8 9]
```

TensorFlow 允许创建具有不同分布的随机张量:

1. 使用以下语句创建一个具有一定均值(默认值 = 0.0)和标准差(默认值 = 1.0)、形状为 [M,N] 的正态分布随机数组:

```
t_random = tf.random_normal([2,3], mean=2.0, stddev=4, seed=12)

# Result: [[ 0.25347459  5.37990952  1.95276058], [-1.53760314
1.2588985   2.84780669]]
```

2. 创建一个具有一定均值(默认值 = 0.0)和标准差(默认值 = 1.0)、形状为 [M,N] 的截尾正态分布随机数组:

```
t_random = tf.truncated_normal([1,5], stddev=2, seed=12)
# Result: [[-0.8732627 1.68995488 -0.02361972 -1.76880157
-3.87749004]]
```

3. 要在种子的 [minval(default = 0),maxval] 范围内创建形状为 [M,N] 的给定伽马分布随机数组,请执行如下语句:

```
t_random = tf.random_uniform([2,3], maxval=4, seed=12)

# Result: [[ 2.54461002  3.69636583  2.70510912], [ 2.00850058
3.84459829   3.54268885]]
```

4. 要将给定的张量随机裁剪为指定的大小,使用以下语句:

```
tf.random_crop(t_random, [2,5],seed=12)
```

这里，t_random 是一个已经定义好的张量。这将导致随机从张量 t_random 中裁剪出一个大小为 [2,5] 的张量。

很多时候需要以随机的顺序来呈现训练样本，可以使用 tf.random_shuffle() 来沿着它的第一维随机排列张量。如果 t_random 是想要重新排序的张量，使用下面的代码：

```
tf.random_shuffle(t_random)
```

5. 随机生成的张量受初始种子值的影响。要在多次运行或会话中获得相同的随机数，应该将种子设置为一个常数值。当使用大量的随机张量时，可以使用 tf.set_random_seed() 来为所有随机产生的张量设置种子。以下命令将所有会话的随机张量的种子设置为 54：

```
tf.set_random_seed(54)
```

> 种子只能有整数值。

现在来看看变量：

1. 它们通过使用变量类来创建。变量的定义还包括应该初始化的常量/随机值。下面的代码中创建了两个不同的张量变量 t_a 和 t_b。两者将被初始化为形状为 [50,50] 的随机均匀分布，最小值 = 0，最大值 = 10：

```
rand_t = tf.random_uniform([50,50], 0, 10, seed=0)
t_a = tf.Variable(rand_t)
t_b = tf.Variable(rand_t)
```

> 变量通常在神经网络中表示权重和偏置。

2. 下面的代码中定义了两个变量的权重和偏置。权重变量使用正态分布随机初始化，均值为 0，标准差为 2，权重大小为 100×100。偏置由 100 个元素组成，每个元素初始化为 0。在这里也使用了可选参数名以给计算图中定义的变量命名。

```
weights = tf.Variable(tf.random_normal([100,100],stddev=2))
bias = tf.Variable(tf.zeros[100], name = 'biases')
```

3. 在前面的例子中，都是利用一些常量来初始化变量，也可以指定一个变量来初始化另一个变量。下面的语句将利用前面定义的权重来初始化 weight2：

```
weight2=tf.Variable(weights.initialized_value(), name='w2')
```

4. 变量的定义将指定变量如何被初始化，但是必须显式初始化所有的声明变量。在计算图的定义中通过声明初始化操作对象来实现：

```
intial_op = tf.global_variables_initializer().
```

5. 每个变量也可以在运行图中单独使用 tf.Variable.initializer 来初始化：

```
bias = tf.Variable(tf.zeros([100,100]))
 with tf.Session() as sess:
     sess.run(bias.initializer)
```

6. 保存变量：使用 `Saver` 类来保存变量，定义一个 Saver 操作对象：

```
saver = tf.train.Saver()
```

7. 介绍完常量和变量之后，我们来讲解最重要的元素——占位符，它们用于将数据提供给计算图。可以使用以下方法定义一个占位符：

```
tf.placeholder(dtype, shape=None, name=None)
```

8. dtype 定占位符的数据类型，并且必须在声明占位符时指定。在这里，为 x 定义一个占位符并计算 y = 2 * x，使用 `feed_dict` 输入一个随机的 4×5 矩阵：

```
x = tf.placeholder("float")
y = 2 * x
data = tf.random_uniform([4,5],10)
with tf.Session() as sess:
    x_data = sess.run(data)
    print(sess.run(y, feed_dict = {x:x_data}))
```

解读分析

> 所有常量、变量和占位符将在代码的计算图部分中定义。如果在定义部分使用 print 语句，只会得到有关张量类型的信息，而不是它的值。

为了得到相关的值，需要创建会话图并对需要提取的张量显式使用运行命令，如下所示：

```
print(sess.run(t_1))
# Will print the value of t_1 defined in step 1
```

拓展阅读

很多时候需要大规模的常量张量对象；在这种情况下，为了优化内存，最好将它们声明为一个可训练标志设置为 `False` 的变量：

```
t_large = tf.Variable(large_array, trainable = False)
```

TensorFlow 被设计成与 Numpy 配合运行，因此所有的 TensorFlow 数据类型都是基于 Numpy 的。使用 `tf.convert_to_tensor()` 可以将给定的值转换为张量类型，并将其与 TensorFlow 函数和运算符一起使用。该函数接受 Numpy 数组、Python 列表和 Python 标量，并允许与张量对象互操作。

下表列出了 TensorFlow 支持的常见的数据类型（来自 `TensorFlow.org`）：

数据类型	TensorFlow 类型	数据类型	TensorFlow 类型
DT_FLOAT	tf.float32	DT_STRING	tf.string
DT_DOUBLE	tf.float64	DT_BOOL	tf.bool
DT_INT8	tf.int8	DT_COMPLEX64	tf.complex64
DT_UINT8	tf.uint8	DT_QINT32	tf.qint32

请注意，与 Python / Numpy 序列不同，TensorFlow 序列不可迭代。试试下面的代码：

```
for i in tf.range(10)
```

你会得到一个错误提示：

```
#TypeError("'Tensor' object is not iterable.")
```

1.6 使用 TensorFlow 执行矩阵操作

矩阵运算，例如执行乘法、加法和减法，是任何神经网络中信号传播的重要操作。通常在计算中需要随机矩阵、零矩阵、一矩阵或者单位矩阵。

本节将告诉你如何获得不同类型的矩阵，以及如何对它们进行不同的矩阵处理操作。

具体做法

1. 开始一个交互式会话，以便得到计算结果：

```
import tensorflow as tf

#Start an Interactive Session
sess = tf.InteractiveSession()

#Define a 5x5 Identity matrix
I_matrix = tf.eye(5)
print(I_matrix.eval())
# This will print a 5x5 Identity matrix

#Define a Variable initialized to a 10x10 identity matrix
X = tf.Variable(tf.eye(10))
X.initializer.run()  # Initialize the Variable
print(X.eval())
# Evaluate the Variable and print the result

#Create a random 5x10 matrix
A = tf.Variable(tf.random_normal([5,10]))
A.initializer.run()

#Multiply two matrices
product = tf.matmul(A, X)
print(product.eval())
```

```
#create a random matrix of 1s and 0s, size 5x10
b = tf.Variable(tf.random_uniform([5,10], 0, 2, dtype= tf.int32))
b.initializer.run()
print(b.eval())
b_new = tf.cast(b, dtype=tf.float32)
#Cast to float32 data type

# Add the two matrices
t_sum = tf.add(product, b_new)
t_sub = product - b_new
print("A*X _b\n", t_sum.eval())
print("A*X - b\n", t_sub.eval())
```

2. 一些其他有用的矩阵操作，如按元素相乘、乘以一个标量、按元素相除、按元素余数相除等，可以执行如下语句：

```
import tensorflow as tf

# Create two random matrices
a = tf.Variable(tf.random_normal([4,5], stddev=2))
b = tf.Variable(tf.random_normal([4,5], stddev=2))

#Element Wise Multiplication
A = a * b

#Multiplication with a scalar 2
B = tf.scalar_mul(2, A)

# Elementwise division, its result is
C = tf.div(a,b)

#Element Wise remainder of division
D = tf.mod(a,b)

init_op = tf.global_variables_initializer()
with tf.Session() as sess:
    sess.run(init_op)
    writer = tf.summary.FileWriter('graphs', sess.graph)
    a,b,A_R, B_R, C_R, D_R = sess.run([a , b, A, B, C, D])
    print("a\n",a,"\nb\n",b, "a*b\n", A_R, "\n2*a*b\n", B_R,
"\na/b\n", C_R, "\na%b\n", D_R)

writer.close()
```

`tf.div` 返回的张量的类型与第一个参数类型一致。

解读分析

所有加法、减、除、乘（按元素相乘）、取余等矩阵的算术运算都要求两个张量矩阵是相同的数据类型，否则就会产生错误。可以使用 `tf.cast()` 将张量从一种数据类型转换为另一种数据类型。

拓展阅读

如果在整数张量之间进行除法，最好使用 `tf.truediv(a,b)`，因为它首先将整数张量转换为浮点类，然后再执行按位相除。

1.7 使用数据流图

TensorFlow 使用 TensorBoard 来提供计算图形的图形图像。这使得理解、调试和优化复杂的神经网络程序变得很方便。TensorBoard 也可以提供有关网络执行的量化指标。它读取 TensorFlow 事件文件，其中包含运行 TensorFlow 会话期间生成的摘要数据。

具体做法

1. 使用 TensorBoard 的第一步是确定想要的 OP 摘要。以 DNN 为例，通常需要知道损失项（目标函数）如何随时间变化。在自适应学习率的优化中，学习率本身会随时间变化。可以在 `tf.summary.scalar` OP 的帮助下得到需要的术语摘要。假设损失变量定义了误差项，我们想知道它是如何随时间变化的：

```
loss = tf...
tf.summary.scalar('loss', loss)
```

2. 还可以使用 `tf.summary.histogram` 可视化梯度、权重或特定层的输出分布：

```
output_tensor  = tf.matmul(input_tensor, weights) + biases
tf.summary.histogram('output', output_tensor)
```

3. 摘要将在会话操作中生成。可以在计算图中定义 `tf.merge_all_summaries` OP 来通过一步操作得到摘要，而不需要单独执行每个摘要操作。

4. 生成的摘要需要用事件文件写入：

`tf.summary.Filewriter:`

```
writer = tf.summary.Filewriter('summary_dir', sess.graph)
```

5. 这会将所有摘要和图形写入 `summary_dir` 目录中。

6. 现在，为了可视化摘要，需要从命令行中调用 TensorBoard：

```
tensorboard --logdir=summary_dir
```

7. 接下来，打开浏览器并输入地址 `http://localhost:6006/`（或运行 TensorBoard 命令后收到的链接）。

8. 你会看到类似于下面的图，顶部有很多标签。Graphs（图表）选项卡能将运算图可视化：

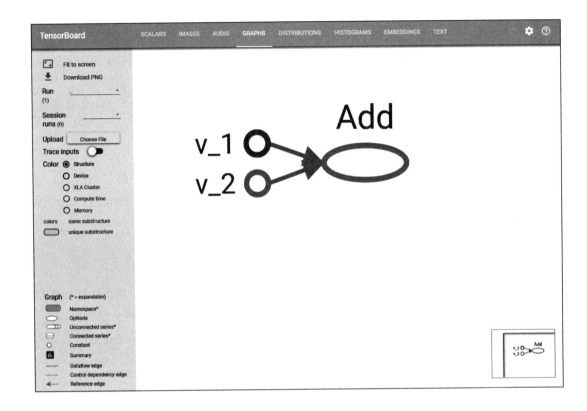

1.8 从 0.x 迁移到 1.x

TensorFlow 1.x 不提供向后兼容性。这意味着在 TensorFlow 0.x 上运行的代码可能无法在 TensorFlow 1.0 上运行。因此，如果代码是用 TensorFlow 0.x 框架编写的，你需要升级它们（旧的 GitHub 存储库或你自己的代码）。这一节将指出 TensorFlow 0.x 和 TensorFlow 1.0 之间的主要区别，并展示如何使用脚本 `tf_upgrade.py` 自动升级 TensorFlow 1.0 的代码。

具体做法

1. 从网址 https://github.com/tensorflow/tensorflow/tree/master/tensorflow/tools/compatibility 下载 `tf_upgrade.py`。

2. 如果要将一个文件从 TensorFlow 0.x 转换为 TensorFlow 1.0，请在命令行使用以下命令：

```
python tf_upgrade.py --infile old_file.py --outfile upgraded_file.py
```

3. 例如，如果有一个名为 test.py 的 TensorFlow 程序文件，可使用下述命令：

```
python tf_upgrade.py --infile test.py --outfile test_1.0.py
```

4. 这将创建一个名为 test_1.0.py 的新文件。
5. 如果要迁移目录中的所有文件，请在命令行中使用以下命令：

```
python tf_upgrade.py --intree InputDIr --outtree OutputDir
 # For example, if you have a directory located at
/home/user/my_dir you can migrate all the python files in the
directory located at /home/user/my-dir_1p0 using the above command
as:
python tf_upgrade.py --intree /home/user/my_dir --outtree
/home/user/my_dir_1p0
```

6. 在大多数情况下，该目录还包含数据集文件；可以使用以下命令确保非 Python 文件也会被复制到新目录（上例中的 my-dir_1p0）中：

```
python tf_upgrade.py --intree /home/user/my_dir --outtree
/home/user/my_dir_1p0 -copyotherfiles True
```

7. 在所有这些情况下，都会生成一个 `report.txt` 文件。该文件包含转换的细节和过程中的任何错误。
8. 对于无法更新的部分代码，需要阅读 `report.txt` 文件并手动升级脚本。

拓展阅读

`tf_upgrade.py` 有一些局限性：
- 它不能改变 `tf.reverse()` 的参数，因此必须手动修复。
- 对于参数列表重新排序的方法，如 `tf.split()` 和 `tf.reverse_split()`，它会尝试引入关键字参数，但实际上并不能重新排列参数。
- 有些结构必须手动替换，例如：

```
tf.get.variable_scope().reuse_variables()
```

替换为：

```
with tf.variable_scope(tf.get_variable_scope(), resuse=True):
```

1.9 使用 XLA 提升运算性能

加速线性代数（Accelerated linear algebra，XLA）是线性代数领域的专用编译器。根据 https://www.tensorflow.org/performance/xla/，它仍处于实验阶段，用于优化 TensorFlow 计算。它可以提高服务器和移动平台的执行速度、内存使用率和可移植性。它提供了双向 JIT（Just In Time）编译或 AoT（Ahead of Time）编译。使用 XLA，你可以生成平台相关的二进制文件（针对大量平台，如 x64、ARM 等），可以针对内存和速度进行优化。

准备工作

目前，XLA 并不包含在 TensorFlow 的二进制版本中。用时需要从源代码构建它。从源代码构建 TensorFlow，需要 TensorFlow 版的 LLVM 和 Bazel。`TensorFlow.org` 仅支持从 macOS 和 Ubuntu 的源代码构建。从源代码构建 TensorFlow 所需的步骤如下（参见 `https://www.tensorflow.org/install/install_sources`）：

1. 确定要安装哪个版本的 TensorFlow——仅支持 CPU 的 TensorFlow 或支持 GPU 的 TensorFlow。

2. 复制 TensorFlow 存储库：

```
git clone https://github.com/tensorflow/tensorflow
cd tensorflow
git checkout Branch #where Branch is the desired branch
```

3. 安装以下依赖：
- Bazel
- TensorFlow 的 Python 依赖项
- 对 GPU 版本，需要 NVIDIA 软件包以支持 TensorFlow

4. 配置安装。在这一步中，需要选择不同的选项，如 XLA、Cuda 支持、Verbs 等：

```
./configure
```

5. 使用 bazel-build。

6. 对于仅使用 CPU 的版本：

```
bazel build --config=opt
//tensorflow/tools/pip_package:build_pip_package
```

7. 如果有兼容的 GPU 设备，并且需要 GPU 支持，请使用：

```
bazel build --config=opt --config=cuda
//tensorflow/tools/pip_package:build_pip_package
```

8. 成功运行后，将获得一个脚本：`build_pip_package`。

9. 按如下所示运行这个脚本来构建 `whl` 文件：

```
bazel-bin/tensorflow/tools/pip_package/build_pip_package
/tmp/tensorflow_pkg
```

10. 安装 `pip` 包：

```
sudo pip install /tmp/tensorflow_pkg/tensorflow-1.1.0-py2-none-any.whl
```

现在你已经准备好了。

具体做法

TensorFlow 生成 TensorFlow 图表。在 XLA 的帮助下，可以在任何新类型的设备上运行 TensorFlow 图表。

1. JIT 编译：在会话级别中打开 JIT 编译：

```
# Config to turn on JIT compilation
config = tf.ConfigProto()
config.graph_options.optimizer_options.global_jit_level =
tf.OptimizerOptions.ON_1

sess = tf.Session(config=config)
```

2. 这是手动打开 JIT 编译：

```
jit_scope = tf.contrib.compiler.jit.experimental_jit_scope

x = tf.placeholder(np.float32)
with jit_scope():
    y = tf.add(x, x)  # The "add" will be compiled with XLA.
```

3. 还可以通过将操作指定在特定的 XLA 设备（XLA_CPU 或 XLA_GPU）上，通过 XLA 来运行计算：

```
with tf.device \
("/job:localhost/replica:0/task:0/device:XLA_GPU:0"):
    output = tf.add(input1, input2)
```

AoT 编译：独立使用 tfcompile 将 TensorFlow 图转换为不同设备（手机）的可执行代码。

TensorFlow.org 中关于 tfcompile 的论述：

tfcompile 采用一个由 TensorFlow 的 feed 和 fetch 概念所标识的子图，并生成一个实现该子图的函数。feed 是函数的输入参数，fetch 是函数的输出参数。所有的输入必须完全由 feed 指定；生成的剪枝子图不能包含占位符或变量节点。通常将所有占位符和变量指定值，这可确保生成的子图不再包含这些节点。生成的函数打包为一个 cc_library，带有导出函数签名的头文件和一个包含实现的对象文件。用户编写代码以适当地调用生成的函数。

要进行相同的高级步骤，可以参考 https://www.tensorflow.org/ performance/xla/tfcompile。

1.10 调用 CPU/GPU 设备

TensorFlow 支持 CPU 和 GPU。它也支持分布式计算。可以在一个或多个计算机系统的多个设备上使用 TensorFlow。TensorFlow 将支持的 CPU 设备命名为"/device:CPU:0"（或"/cpu: 0"），第 i 个 GPU 设备命名为"/device:GPU:I"（或"/gpu:I"）。

如前所述，GPU 比 CPU 要快得多，因为它们有许多小的内核。然而，在所有类型的计算中都使用 GPU 也并不一定都有速度上的优势。有时，比起使用 GPU 并行计算在速度上的优势收益，使用 GPU 的其他代价相对更为昂贵。为了解决这个问题，TensorFlow 可以选择将计算放在一个特定的设备上。默认情况下，如果 CPU 和 GPU 都存在，TensorFlow 会优先考虑 GPU。

具体做法

TensorFlow 将设备表示为字符串。本节展示如何在 TensorFlow 中指定某一设备用于矩阵乘法的计算。要验证 TensorFlow 是否确实在使用指定的设备（CPU 或 GPU），可以创建会话，并将 log_device_placement 标志设置为 True，即：

```
config=tf.ConfigProto(log_device_placement=True):
```

1. 如果你不确定设备，并希望 TensorFlow 选择现有和受支持的设备，则可以将 `allow_soft_placement` 标志设置为 `True`：

```
config=tf.ConfigProto(allow_soft_placement=True,
log_device_placement=True)
```

2. 手动选择 CPU 进行操作：

```
with tf.device('/cpu:0'):
    rand_t = tf.random_uniform([50,50], 0, 10, dtype=tf.float32, seed=0)
    a = tf.Variable(rand_t)
    b = tf.Variable(rand_t)
    c = tf.matmul(a,b)
    init = tf.global_variables_initializer()

sess = tf.Session(config)
sess.run(init)
print(sess.run(c))
```

3. 得到以下输出：

可以看到,在这种情况下,所有的设备都是'/ cpu:0'。

4. 手动选择一个 GPU 来操作:

```
with tf.device('/gpu:0'):
    rand_t = tf.random_uniform([50,50], 0, 10, dtype=tf.float32, seed=0)
    a = tf.Variable(rand_t)
    b = tf.Variable(rand_t)
    c = tf.matmul(a,b)
    init = tf.global_variables_initializer()

sess = tf.Session(config=tf.ConfigProto(log_device_placement=True))
sess.run(init)
print(sess.run(c))
```

5. 输出现在更改为以下内容:

6. 每个操作之后的 '/cpu:0' 现在被替换为 '/gpu:0'。

7. 手动选择多个 GPU:

```
c=[]
for d in ['/gpu:1','/gpu:2']:
    with tf.device(d):
        rand_t = tf.random_uniform([50, 50], 0, 10, dtype=tf.float32, seed=0)
        a = tf.Variable(rand_t)
        b = tf.Variable(rand_t)
        c.append(tf.matmul(a,b))
        init = tf.global_variables_initializer()

sess = tf.Session(config=tf.ConfigProto(allow_soft_placement=True,log_device_placement=True))
sess.run(init)
print(sess.run(c))
sess.close()
```

8. 在这种情况下,如果系统有 3 个 GPU 设备,那么第一组乘法将由 `'/:gpu:1'` 执行,第二组乘以 `'/gpu:2'` 执行。

解读分析

函数 `tf.device()` 选择设备(CPU 或 GPU)。`with` 块确保设备被选择并用于其操作。`with` 块中定义的所有变量、常量和操作将使用在 `tf.device()` 中选择的设备。会话配置使用 `tf.ConfigProto` 进行控制。通过设置 `allow_soft_placement` 和 `log_device_placement` 标志,告诉 TensorFlow 在指定的设备不可用时自动选择可用的设备,并在执行会话时给出日志消息作为描述设备分配的输出。

1.11 TensorFlow 与深度学习

DNN 现在是 AI 社区的流行词。最近,DNN 在许多数据科学竞赛 / Kaggle 竞赛中获得了多次冠军。自从 1962 年 Rosenblat 提出感知机(Perceptron)以来,DNN 的概念就已经出现了,而自 Rumelhart、Hinton 和 Williams 在 1986 年发现了梯度下降算法后,DNN 的概念就变得可行了。直到最近 DNN 才成为全世界 AI / ML 爱好者和工程师的最爱。

主要原因在于现代计算能力的可用性,如 GPU 和 TensorFlow 等工具,可以通过几行代码轻松访问 GPU 并构建复杂的神经网络。

作为一名机器学习爱好者,你必须熟悉神经网络和深度学习的概念,但为了完整起见,我们将在这里介绍基础知识,并探讨 TensorFlow 的哪些特性使其成为深度学习的热门选择。

神经网络是一个生物启发式的计算和学习模型。像生物神经元一样,它们从其他细胞(神经元或环境)获得加权输入。这个加权输入经过一个处理单元并产生可以是二进制或连续(概率,预测)的输出。**人工神经网络**(ANN)是这些神经元的网络,可以随机分布或排列成一个分层结构。这些神经元通过与它们相关的一组权重和偏置来学习。

下图对生物神经网络和人工神经网络的相似性给出了形象的对比:

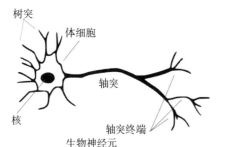

来源:https://commons.wikimedia.org/wiki/
File:Neuron_-_annotated.svg

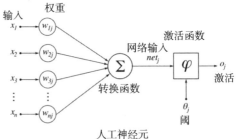

来源:https://commons.wikimedia.org/wiki/
File:Rosenblattperceptron.png

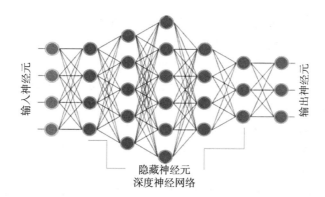

深度神经网络

根据 Hinton 等人的定义,深度学习(https://www.cs.toronto.edu/~hinton/absps/NatureDeepReview.pdf)是由多个处理层(隐藏层)组成的计算模型。层数的增加会导致学习时间的增加。由于数据量庞大,学习时间进一步增加,现今的 CNN 或**生成对抗网络**(GAN)的规范也是如此。因此,为了实际实现 DNN,需要高计算能力。NVDIA 公司 GPU 的问世使其变得可行,随后 Google 的 TensorFlow 使得实现复杂的 DNN 结构成为可能,而不需要深入复杂的数学细节,大数据集的可用性为 DNN 提供了必要的数据来源。TensorFlow 成为最受欢迎的深度学习库,原因如下:

- TensorFlow 是一个强大的库,用于执行大规模的数值计算,如矩阵乘法或自动微分。这两个计算是实现和训练 DNN 所必需的。
- TensorFlow 在后端使用 C / C ++,这使得计算速度更快。
- TensorFlow 有一个高级机器学习 API(tf.contrib.learn),可以更容易地配置、训练和评估大量的机器学习模型。
- 可以在 TensorFlow 上使用高级深度学习库 Keras。Keras 非常便于用户使用,并且可以轻松快速地进行原型设计。它支持各种 DNN,如 RNN、CNN,甚至是两者的组合。

具体做法

任何深度学习网络都由四个重要部分组成:数据集、定义模型(网络结构)、训练 / 学习和预测 / 评估。可以在 TensorFlow 中实现所有这些:

数据集:DNN 依赖于大量的数据。可以收集或生成数据,也可以使用可用的标准数据集。TensorFlow 支持三种主要的读取数据的方法,可以在不同的数据集中使用;本书中用来训练建立模型的一些数据集介绍如下。

MNIST:这是最大的手写数字(0~9)数据库。它由 60000 个示例的训练集和 10000 个示例的测试集组成。该数据集存放在 Yann LeCun 的主页(http://yann.lecun.com/exdb/mnist/)中。这个数据集已经包含在 tensorflow.examples.tutorials.mnist 的 TensorFlow 库中。

CIFAR10：这个数据集包含了 10 个类别的 60000 幅 32×32 彩色图像，每个类别有 6000 幅图像。其中训练集包含 50000 幅图像，测试数据集包含 10000 幅图像。数据集的 10 个类别分别是：飞机、汽车、鸟、猫、鹿、狗、青蛙、马、船和卡车。该数据由多伦多大学计算机科学系维护（https://www.cs.toronto.edu/kriz/cifar.html）。

WORDNET：这是一个英文的词汇数据库。它包含名词、动词、副词和形容词，被归为一组认知同义词（Synset），即代表相同概念的词语，例如 shut 和 close，car 和 automobile 被分组为无序集合。它包含 155287 个单词，组织在 117659 个同义词集合中，总共 206941 个单词对。该数据集由普林斯顿大学维护（https://wordnet.princeton.edu/）。

ImageNET：这是一个根据 WORDNET 层次组织的图像数据集（目前只有名词）。每个有意义的概念（synset）由多个单词或单词短语来描述。每个子空间平均由 1000 幅图像表示。目前共有 21841 个同义词，共有 14197122 幅图像。自 2010 年以来，每年举办一次 ImageNet 大规模视觉识别挑战赛（ILSVRC），将图像分类到 1000 个对象类别中。这项工作是由美国普林斯顿大学、斯坦福大学、A9 和谷歌赞助（http://www.image-net.org/）。

YouTube-8M：这是一个由数百万 YouTube 视频组成的大型标签视频数据集。它有大约 700 万个 YouTube 视频网址，分为 4716 个小类，并分为 24 个大类。它还提供预处理支持和框架功能。数据集由 Google Research（https://research.google.com/youtube8m/）维护。

读取数据：在 TensorFlow 中可以通过三种方式读取数据：通过 `feed_dict` 传递数据，从文件中读取数据以及使用预加载的数据。在整本书中都使用这三种方式来读取数据。接下来，你将依次学习每种数据读取方式。

1. **数据传递**：在这种情况下，运行每个步骤时都会使用 `run()` 或 `eval()` 函数调用中的 `feed_dict` 参数来提供数据。这是在占位符的帮助下完成的，这个方法允许传递 Numpy 数组数据。可以使用 TensorFlow 的以下代码：

```
...
y = tf.placeholder(tf.float32)
x = tf.placeholder(tf.float32).
...
with tf.Session as sess:
   X_Array = some Numpy Array
   Y_Array = other Numpy Array
   loss= ...
sess.run(loss,feed_dict = {x: X_Array, y: Y_Array}).
...
```

这里，`x` 和 `y` 是占位符；使用它们，在 `feed_dict` 的帮助下传递包含 X 值的数组和包含 Y 值的数组。

2. **从文件中读取**：当数据集非常大时，使用此方法可以确保不是所有数据都立即占用内存（例如 60 GB 的 YouTube-8m 数据集）。从文件读取的过程可以通过以下步骤完成：

- 使用字符串张量["file0","file1"]或者[("file%d"i)for in in range(2)]的方式创建文件命名列表，或者使用files = tf.train.match_filenames_once('*.JPG')函数创建。
- **文件名队列**：创建一个队列来保存文件名，此时需要使用tf.train.string_input_producer函数：

```
filename_queue = tf.train.string_input_producer(files)
# where files is the list of filenames created above
```

这个函数还提供了一个选项来排列和设置批次的最大数量。整个文件名列表被添加到每个批次的队列中。如果选择了shuffle=True，则在每个批次中都要重新排列文件名。

- **Reader** 用于从文件名队列中读取文件。根据输入文件格式选择相应的阅读器。read方法是标识文件和记录（调试时有用）以及标量字符串值的关键字。例如，文件格式为.csv时：

```
reader = tf.TextLineReader()
key, value = reader.read(filename_queue)
```

- **Decoder**：使用一个或多个解码器和转换操作来将值字符串解码为构成训练样本的张量：

```
record_defaults = [[1], [1], [1]]
col1, col2, col3 = tf.decode_csv(value,
record_defaults=record_defaults)
```

3. **预加载的数据**：当数据集很小时可以使用，可以在内存中完全加载。因此，可以将数据存储在常量或变量中。在使用变量时，需要将可训练标志设置为False，以便训练时数据不会改变。预加载数据为TensorFlow常量时：

```
# Preloaded data as constant
training_data = ...
training_labels = ...
with tf.Session as sess:
    x_data = tf.Constant(training_data)
    y_data = tf.Constant(training_labels)
...

# Preloaded data as Variables
training_data = ...
training_labels = ...
with tf.Session as sess:
    data_x = tf.placeholder(dtype=training_data.dtype,
shape=training_data.shape)
    data_y = tf.placeholder(dtype=training_label.dtype,
shape=training_label.shape)
    x_data = tf.Variable(data_x, trainable=False, collections[])
    y_data = tf.Variable(data_y, trainable=False, collections[])
...
```

一般来说，数据被分为三部分：训练数据、验证数据和测试数据。

4. **定义模型**：建立描述网络结构的计算图。它涉及指定信息从一组神经元到另一组神经元的超参数、变量和占位符序列以及损失／错误函数。你将在本章后面的章节中了解更多有关计算图的内容。

5.**训练／学习**：在 DNN 中的学习通常基于梯度下降算法（第 2 章将详细讨论），其目的是要找到训练变量（权重／偏置），将损失／错误函数（如步骤 2 中所定义的）最小化。这是通过初始化变量并使用 `run()` 来实现的：

```
with tf.Session as sess:
    ....
    sess.run(...)
    ...
```

6.**评估模型**：一旦网络被训练，通过 `predict()` 函数使用验证数据和测试数据来评估网络。这可以评价模型是否适合相应数据集，可以避免过拟合或欠拟合的问题。一旦模型取得让人满意的精度，就可以部署在生产环境中了。

拓展阅读

在 TensorFlow 1.3 中，增加了一个名为 TensorFlow Estimator 的新功能。TensorFlow Estimator 使创建神经网络模型的任务变得更加容易，它是一个封装了训练、评估、预测和服务过程的更高层次的 API。它提供了使用预先制作的估算器的选项，或者可以编写自己的定制估算器。通过预先制定的估算器，不再需要担心构建计算或创建会话，它会处理所有这些。

目前 TensorFlow Estimator 有 6 个预先制定的估算器。使用 TensorFlow 预制的 Estimator 的另一个优点是，它本身也可以在 TensorBoard 上创建可视化的摘要。有关 Estimator 的更多详细信息，请访问 `https://www.google.com/supporttensorflow.org/programmers_guide/estimators`。

1.12 DNN 问题需要的 Python 包

TensorFlow 能够实现大部分神经网络的功能。但是，这还是不够的。对于预处理任务、序列化甚至绘图任务，还需要更多的 Python 包。

具体做法

下面列出了一些常用的 Python 包。

1.**Numpy**：这是用 Python 进行科学计算的基础包。它支持 n 维数组和矩阵的计算，还拥有大量的高级数学函数。这是 TensorFlow 所需的必要软件包，因此，使用 pip install tensorflow 时，如果尚未安装 Numpy，它将被自动安装。

2.**Matplolib**：这是 Python 2D 绘图库。使用它可以只用几行代码创建各类图，包括直方、条形图、错误图、散点图和功率谱等。它可以使用 pip 进行安装：

```
pip install matplotlib
# or using Anaconda
conda install -c conda-forge matplotlib
```

3. OS：这包括在基本的 Python 安装中。它提供了一种使用操作系统相关功能（如读取、写入及更改文件和目录）的简单便携方式。

4. Pandas：这提供了各种数据结构和数据分析工具。使用 Pandas，您可以在内存数据结构和不同格式之间读取和写入数据。可以读取 .csv 和文本文件。可以使用 pip install 或 conda install 进行安装。

5. Seaborn：这是一个建立在 Matplotlib 上的专门的统计数据可视化工具。

6. H5fs：H5fs 是能够在 HDFS（分层数据格式文件系统）上运行的 Linux 文件系统（也包括其他带有 FUSE 实现的操作系统，如 macOS X）。

7. PythonMagick：这是 `ImageMagick` 库的 Python 绑定。它是一个显示、转换和编辑光栅图像及矢量图像文件的库。它支持超过 200 个图像文件格式。它可以使用 `ImageMagick` 提供的源代码来安装。某些 .whl 格式也可用 `pip install` (http://www.lfd.uci.edu/%7Egohlke/pythonlibs/#pythonmagick) 来安装。

8. TFlearn：TFlearn 是一个建立在 TensorFlow 之上的模块化和透明的深度学习库。它为 TensorFlow 提供更高级别的 API，以促进和加速实验。它目前支持最近的大多数深度学习模型，如卷积、LSTM、BatchNorm、BiRNN、PReLU、残差网络和生成网络。它只适用于 TensorFlow 1.0 或更高版本。请使用 `pip install tflearn` 安装。

9. Keras：Keras 也是神经网络的高级 API，它使用 TensorFlow 作为其后端。它可以运行在 Theano 和 CNTK 之上。添加图层只需要一行代码，非常用户友好，可以使用 pip install keras 来安装。

拓展阅读

你可以通过以下网址找到关于 TensorFlow 安装的更多资料：

- https://www.tensorflow.org/install/
- https://www.tensorflow.org/install/install_sources
- http://llvm.org/
- https://bazel.build/

CHAPTER 2

第 2 章

回 归

本章介绍如何使用 TensorFlow 完成回归。

2.1 引言

回归是数学建模、分类和预测中最古老但功能非常强大的工具之一。回归在工程、物理学、生物学、金融、社会科学等各个领域都有应用，是数据科学家常用的基本工具。

回归通常是机器学习中使用的第一个算法。通过学习因变量和自变量之间的关系实现对数据的预测。例如，对房价估计时，需要确定房屋面积（**自变量**）与其价格（**因变量**）之间的关系，可以利用这一关系来预测给定面积的房屋的价格。可以有多个影响因变量的自变量。因此，回归有两个重要组成部分：自变量和因变量之间的**关系**，以及不同自变量对因变量**影响的强度**。

以下是几种常用的回归方法：
- **线性回归**：使用最广泛的建模技术之一。已存在 200 多年，已经从几乎所有可能的角度进行了研究。线性回归假定输入变量（X）和单个输出变量（Y）之间呈线性关系。它旨在找到预测值 Y 的线性方程：

$$Y_{hat} = W^\mathsf{T} X + b$$

其中，$X = (x_1, x_2, ..., x_n)$ 为 n 个输入变量，$W = (w_1, w_2, ..., w_n)$ 为线性系数，b 是偏置项。目标是找到系数 W 的最佳估计，使得预测值 Y 的误差最小。使用最小二乘法估计线性系数 W，即使预测值 (Y_{hat}) 与观测值 (Y) 之间的差的平方和最小。因此，这里尽量最小化损失函数：

$$\mathrm{loss} = \sum_{i=1}^{p} Y_i - Y_{hat_i}$$

其中，需要对所有训练样本的误差求和。根据输入变量 X 的数量和类型，可划分出多种线性回归类型：简单线性回归（一个输入变量，一个输出变量），多元线性回归（多个输入变量，一个输出变量），多变量线性回归（多个输入变量，多个输出变量）。更多线性回归的相关内容，可参考 https://en.wikipedia.org/wiki/Linear_regression。
- **逻辑回归**：用来确定一个事件的概率。通常来说，事件可被表示为类别因变量。事件的概率用 logit 函数（Sigmoid 函数）表示：

$$P(Y_{hat}=1|X=x)=\frac{1}{1+e^{-(b+W^T+x)}}$$

现在的目标是估计权重 $W=(w_1, w_2,..., w_n)$ 和偏置项 b。在逻辑回归中，使用最大似然估计量或随机梯度下降来估计系数。损失函数通常被定义为交叉熵项：

$$loss=\sum_{i=1}^{P}Y_i\log(Y_{hat_i})+(1-Y_i)\log(1-Y_{hat_i})$$

逻辑回归用于分类问题，例如，对于给定的医疗数据，可以使用逻辑回归判断一个人是否患有癌症。如果输出类别变量具有两个或更多个层级，则可以使用多项式逻辑回归。另一种用于两个或更多输出变量的常见技术是 OneVsAll。对于多类型逻辑回归，交叉熵损失函数被修改为：

$$loss=\sum_{i=1}^{P}\sum_{j=1}^{K}Y_{ij}\log(Y_{hat_{ij}})$$

其中，K 是类别总数。更多逻辑回归的相关内容，可参考 https://en.wikipedia.org/wiki/Logistic_regression。
- **正则化**：当有大量的输入特征时，需要正则化来确保预测模型不会太复杂。正则化可以帮助防止数据过拟合。它也可以用来获得一个凸损失函数。有两种类型的正则化——L1 和 L2 正则化，其描述如下：
  - 当数据高度共线时，**L1 正则化**也可以工作。在 L1 正则化中，与所有系数的绝对值的和相关的附加惩罚项被添加到损失函数中。L1 正则化的正则化惩罚项如下：

  $$L1\_penalty=\lambda\sum_{i=1}^{n}|W_i|$$

  - **L2 正则化**提供了稀疏的解决方案。当输入特征的数量非常大时，非常有用。在这种情况下，惩罚项是所有系数的平方之和：

  $$L2\_penalty=\lambda\sum_{i=1}^{n}W_i^2$$

其中，λ 是正则化参数。

2.2 选择损失函数

正如前面所讨论的，在回归中定义了损失函数或目标函数，其目的是找到使损失最小化的系数。本节将介绍如何在 TensorFlow 中定义损失函数，并根据问题选择合适的损失函数。

准备工作

声明一个损失函数需要将系数定义为变量,将数据集定义为占位符。可以有一个常学习率或变化的学习率和正则化常数。在下面的代码中,设 m 是样本数量,n 是特征数量,P 是类别数量。这里应该在代码之前定义这些全局参数:

```
m = 1000
n = 15
P = 2
```

具体做法

1. 在标准线性回归的情况下,只有一个输入变量和一个输出变量:

```
# Placeholder for the Training Data
X = tf.placeholder(tf.float32, name='X')
Y = tf.placeholder(tf.float32, name='Y')

# Variables for coefficients initialized to 0
w0 = tf.Variable(0.0)
w1 = tf.Variable(0.0)

# The Linear Regression Model
Y_hat = X*w1 + w0

# Loss function
loss = tf.square(Y - Y_hat, name='loss')
```

2. 在多元线性回归的情况下,输入变量不止一个,而输出变量仍为一个。现在可以定义占位符 X 的大小为 [m, n],其中 m 是样本数量,n 是特征数量,代码如下:

```
# Placeholder for the Training Data
X = tf.placeholder(tf.float32, name='X', shape=[m,n])
Y = tf.placeholder(tf.float32, name='Y')

# Variables for coefficients initialized to 0
w0 = tf.Variable(0.0)
w1 = tf.Variable(tf.random_normal([n,1]))

# The Linear Regression Model
Y_hat = tf.matmul(X, w1) + w0

# Multiple linear regression loss function
loss = tf.reduce_mean(tf.square(Y - Y_hat, name='loss'))
```

3. 在逻辑回归的情况下,损失函数定义为交叉熵。输出 Y 的维数等于训练数据集中类别的数量,其中 P 为类别数量:

```
# Placeholder for the Training Data
X = tf.placeholder(tf.float32, name='X', shape=[m,n])
Y = tf.placeholder(tf.float32, name='Y', shape=[m,P])

# Variables for coefficients initialized to 0
```

```
w0 = tf.Variable(tf.zeros([1,P]), name='bias')
w1 = tf.Variable(tf.random_normal([n,1]), name='weights')

# The Linear Regression Model
Y_hat = tf.matmul(X, w1) + w0

# Loss function
entropy = tf.nn.softmax_cross_entropy_with_logits(Y_hat,Y)
loss = tf.reduce_mean(entropy)
```

4. 如果想把 L1 正则化加到损失上，那么代码如下：

```
lamda = tf.constant(0.8)  # regularization parameter
regularization_param = lamda*tf.reduce_sum(tf.abs(W1))

# New loss
loss += regularization_param
```

5. 对于 L2 正则化，代码如下：

```
lamda = tf.constant(0.8)  # regularization parameter
regularization_param = lamda*tf.nn.l2_loss(W1)

# New loss
loss += regularization_param
```

解读分析

你学会了如何实现不同类型的损失函数。那么根据手头的回归任务，你可以选择相应的损失函数或设计自己的损失函数。在损失项中也可以结合 L1 和 L2 正则化。

拓展阅读

为确保收敛，损失函数应为凸的。一个光滑的、可微分的凸损失函数可以提供更好的收敛性。随着学习的进行，损失函数的值应该下降，并最终变得稳定。

2.3 TensorFlow 中的优化器

高中数学学过，函数在一阶导数为零的地方达到其最大值和最小值。梯度下降算法基于相同的原理——调整系数（权重和偏置）使损失函数的梯度下降。在回归中，使用梯度下降来优化损失函数并获得系数。本节将介绍如何使用 TensorFlow 的梯度下降优化器及其变体。

准备工作

按照损失函数的负梯度成比例地对系数（W 和 b）进行更新。根据训练样本的大小，有三种梯度下降的变体：

- **Vanilla 梯度下降**：在 Vanilla 梯度下降（也称作**批梯度下降**）中，在每个循环中计算整个训练集的损失函数的梯度。该方法可能很慢并且难以处理非常大的数据集。该方法能保证收敛到凸损失函数的全局最小值，但对于非凸损失函数可能会稳定在局部极小值处。
- **随机梯度下降**：在随机梯度下降中，一次提供一个训练样本用于更新权重和偏置，从而使损失函数的梯度减小，然后再转向下一个训练样本。整个过程重复了若干个循环。由于每次更新一次，所以它比 Vanilla 快，但由于频繁更新，所以损失函数值的方差会比较大。
- **小批量梯度下降**：该方法结合了前两者的优点，利用一批训练样本来更新参数。

具体做法

1. 首先确定想用的优化器。TensorFlow 为你提供了各种各样的优化器。这里从最流行、最简单的梯度下降优化器开始：

```
tf.train.GradientDescentOptimizer(learning_rate)
```

2. `GradientDescentOptimizer` 中的 `learning_rate` 参数可以是一个常数或张量。它的值介于 0 和 1 之间。

3. 必须为优化器给定要优化的函数。使用它的方法实现最小化。该方法计算梯度并将梯度应用于系数的学习。该函数在 TensorFlow 文档中的定义如下：

```
minimize(
    loss,
    global_step=None,
    var_list=None,
    gate_gradients=GATE_OP,
    aggregation_method=None,
    colocate_gradients_with_ops=False,
    name=None,
    grad_loss=None
)
```

4. 综上所述，这里定义计算图：

```
...
optimizer = tf.train.GradientDescentOptimizer(learning_rate=0.01)
train_step = optimizer.minimize(loss)
...

#Execution Graph
with tf.Session() as sess:
    ...
    sess.run(train_step, feed_dict = {X:X_data, Y:Y_data})
    ...
```

5. 馈送给 `feed_dict` 的 X 和 Y 数据可以是 X 和 Y 个点（随机梯度）、整个训练集（Vanilla）或成批次的。

6. 梯度下降中的另一个变化是增加了动量项（详见第3章）。为此，使用优化器 `tf.train.MomentumOptimizer()`。它可以把 `learning_rate` 和 `momentum` 作为初始化参数：

```
optimizer = tf.train.MomentumOtimizer(learning_rate=0.01,
momentum=0.5).minimize(loss)
```

7. 可以使用 `tf.train.AdadeltaOptimizer()` 来实现一个自适应的、单调递减的学习率，它使用两个初始化参数 `learning_rate` 和衰减因子 `rho`：

```
optimizer = tf.train.AdadeltaOptimizer(learning_rate=0.8,
rho=0.95).minimize(loss)
```

8. TensorFlow 也支持 Hinton 的 RMSprop，其工作方式类似于 Adadelta 的 `tf.train.RMSpropOptimizer()`：

```
optimizer = tf.train.RMSpropOptimizer(learning_rate=0.01,
decay=0.8, momentum=0.1).minimize(loss)
```

> Adadelta 和 RMSprop 之间的细微不同可参考 http://www.cs.toronto.edu/~tijmen/csc321/slides/lecture_slides_lec6.pdf 和 https://arxiv.org/pdf/1212.5701.pdf。

9. 另一种 TensorFlow 支持的常用优化器是 Adam 优化器。该方法利用梯度的一阶和二阶矩对不同的系数计算不同的自适应学习率：

```
optimizer = tf.train.AdamOptimizer().minimize(loss)
```

10. 除此之外，TensorFlow 还提供了以下优化器：

```
tf.train.AdagradOptimizer   #Adagrad Optimizer
tf.train.AdagradDAOptimizer #Adagrad Dual Averaging optimizer
tf.train.FtrlOptimizer #Follow the regularized leader optimizer
tf.train.ProximalGradientDescentOptimizer #Proximal GD optimizer
tf.train.ProximalAdagradOptimizer # Proximal Adagrad optimizer
```

更多内容

通常建议你从较大学习率开始，并在学习过程中将其降低。这有助于对训练进行微调。可以使用 TensorFlow 中的 `tf.train.exponential_decay` 方法来实现这一点。根据 TensorFlow 文档：

在训练模型时，通常建议在训练过程中降低学习率。该函数利用指数衰减函数初始化学习率。需要一个 global_step 值来计算衰减的学习率。可以传递一个在每个训练步骤中递增的 TensorFlow 变量。函数返回衰减的学习率。

变量：

learning_rate：标量 float32 或 float64 张量或者 Python 数字。初始学习率。

global_step：标量 int32 或 int64 张量或者 Python 数字。用于衰减计算的全局步数，非负。

decay_steps：标量 int32 或 int64 张量或者 Python 数字。正数，参考之前所述的衰减计算。

decay_rate：标量 float32 或 float64 张量或者 Python 数字。衰减率。

staircase：布尔值。若为真则以离散的间隔衰减学习率。

name：字符串。可选的操作名。默认为 ExponentialDecay。

返回：

与 learning_rate 类型相同的标量张量。衰减的学习率。

实现指数衰减学习率的代码如下：

```
global_step = tf.Variable(0, trainable = false)
initial_learning_rate = 0.2
learning_rate = tf.train.exponential_decay(initial_learning_rate,
global_step, decay_steps=100000, decay_rate=0.95, staircase=True)
# Pass this learning rate to optimizer as before.
```

拓展阅读

下面是不同优化器的链接：

- https://arxiv.org/pdf/1609.04747.pdf：该文章提供了不同优化器算法的综述。
- https://www.tensorflow.org/api_guides/python/train#Optimizers：这是 TensorFlow.org 链接，详述了 TensorFlow 中优化器的使用方法。
- https://arxiv.org/pdf/1412.6980.pdf：关于 Adam 优化器的论文。

2.4 读取 CSV 文件和数据预处理

大多数人了解 Pandas 及其在处理大数据文件方面的实用性。TensorFlow 提供了读取这种文件的方法。在第 1 章中，介绍了如何在 TensorFlow 中读取文件，本节将重点介绍如何从 CSV 文件中读取数据并在训练之前对数据进行预处理。

准备工作

采用哈里森和鲁宾菲尔德于 1978 年收集的波士顿房价数据集（http://lib.stat.cmu.edu/datasets/boston）。该数据集包括 506 个样本场景，每个房屋含 14 个特征：

- CRIM：城镇人均犯罪率
- ZN：占地 25000 平方英尺（1 英尺 =0.3048 米）以上的住宅用地比例

- INDUS：每个城镇的非零售商业用地比例
- CHAS：查尔斯河（Charles River）变量（若土地位于河流边界，则为 1；否则为 0）
- NOX：一氧化氮浓度（每千万）
- RM：每个寓所的平均房间数量
- AGE：1940 年以前建成的自住单元比例
- DIS：到 5 个波士顿就业中心的加权距离
- RAD：径向高速公路可达性指数
- TAX：每万美元的全价值物业税税率
- PTRATIO：镇小学老师比例
- B：$1000(Bk-0.63)^2$，其中 Bk 是城镇黑人的比例
- LSTAT：低地位人口的百分比
- MEDV：1000 美元自有住房的中位值

具体做法

1. 导入所需的模块并声明全局变量：

```
import tensorflow as tf

# Global parameters
DATA_FILE = 'boston_housing.csv'
BATCH_SIZE = 10
NUM_FEATURES = 14
```

2. 定义一个将文件名作为参数的函数，并返回大小等于 `BATCH_SIZE` 的张量：

```
def data_generator(filename):
    """
    Generates Tensors in batches of size Batch_SIZE.
    Args: String Tensor
    Filename from which data is to be read
    Returns: Tensors
    feature_batch and label_batch
    """
```

3. 定义 `f_queue` 和 `reader` 为文件名：

```
f_queue = tf.train.string_input_producer(filename)
reader = tf.TextLineReader(skip_header_lines=1) # Skips the first line
_, value = reader.read(f_queue)
```

4. 这里指定要使用的数据以防数据丢失。对 `.csv` 解码并选择需要的特征。例如，选择 RM、PTRATIO 和 LSTAT 特征：

```
record_defaults = [ [0.0] for _ in range(NUM_FEATURES)]
data = tf.decode_csv(value, record_defaults=record_defaults)
features = tf.stack(tf.gather_nd(data,[[5],[10],[12]]))
label = data[-1]
```

5. 定义参数来生成批并使用 `tf.train.shuffle_batch()` 来随机重新排列张量。该函数返回张量 `feature_batch` 和 `label_batch`：

```
# minimum number elements in the queue after a
dequeuemin_after_dequeue = 10 * BATCH_SIZE

# the maximum number of elements in the queue
capacity = 20 * BATCH_SIZE

# shuffle the data to generate BATCH_SIZE sample pairs
feature_batch, label_batch = tf.train.shuffle_batch([features,
label], batch_size=BATCH_SIZE,
                                                    capacity=capacity,
min_after_dequeue=min_after_dequeue)

return feature_batch, label_batch
```

6. 这里定义了另一个函数在会话中生成批：

```
def generate_data(feature_batch, label_batch):
    with tf.Session() as sess:
        # intialize the queue threads
        coord = tf.train.Coordinator()
        threads = tf.train.start_queue_runners(coord=coord)
        for _ in range(5): # Generate 5 batches
            features, labels = sess.run([feature_batch,
label_batch])
            print (features, "HI")
        coord.request_stop()
        coord.join(threads)
```

7. 使用这两个函数得到批中的数据。这里，仅打印数据；在学习训练时，将在这里执行优化步骤：

```
if __name__ =='__main__':
    feature_batch, label_batch = data_generator([DATA_FILE])
    generate_data(feature_batch, label_batch)
```

更多内容

用第 1 章提到的 TensorFlow 控制操作和张量来对数据进行预处理。例如，对于波士顿房价的情况，大约有 16 个数据行的 MEDV 是 50.0。在大多数情况下，这些数据点包含缺失或删减的值，因此建议不要考虑用这些数据训练。可以使用下面的代码在训练数据集中删除它们：

```
condition = tf.equal(data[13], tf.constant(50.0))
data = tf.where(condition, tf.zeros(NUM_FEATURES), data[:])
```

这里定义了一个张量布尔条件，若 MEDV 等于 50.0 则为真。如果条件为真则可使用 TensorFlow `tf.where()` 操作赋为零值。

2.5 房价估计——简单线性回归

本节将针对波士顿房价数据集的房间数量（RM）采用简单线性回归。

准备工作

目标是预测在最后一列（MEDV）给出的房价。本小节直接从 TensorFlow contrib 数据集加载数据。使用随机梯度下降优化器优化单个训练样本的系数。

具体做法

1. 导入需要的所有软件包：

```
import tensorflow as tf
import numpy as np
import matplotlib.pyplot as plt
```

2. 在神经网络中，所有的输入都线性增加。为了使训练有效，输入应该被归一化，所以这里定义一个函数来归一化输入数据：

```
def normalize(X):
    """ Normalizes the array X"""
    mean = np.mean(X)
    std = np.std(X)
    X = (X - mean)/std
    return X
```

3. 现在使用 TensorFlow contrib 数据集加载波士顿房价数据集，并将其分解为 X_train 和 Y_train。可以对数据进行归一化处理：

```
# Data
boston = tf.contrib.learn.datasets.load_dataset('boston')
X_train, Y_train = boston.data[:,5], boston.target
#X_train = normalize(X_train)   # This step is optional here
n_samples = len(X_train)
```

4. 为训练数据声明 TensorFlow 占位符：

```
# Placeholder for the Training Data
X = tf.placeholder(tf.float32, name='X')
Y = tf.placeholder(tf.float32, name='Y')
```

5. 创建 TensorFlow 的权重和偏置变量且初始值为零：

```
# Variables for coefficients initialized to 0
b = tf.Variable(0.0)
w = tf.Variable(0.0)
```

6. 定义用于预测的线性回归模型：

```
# The Linear Regression Model
Y_hat = X * w + b
```

7. 定义损失函数：

```
# Loss function
loss = tf.square(Y - Y_hat, name='loss')
```

8. 选择梯度下降优化器：

```
# Gradient Descent with learning rate of 0.01 to minimize loss
optimizer = tf.train.GradientDescentOptimizer(learning_rate=0.01).minimize(loss)
```

9. 声明初始化操作符：

```
# Initializing Variables
init_op = tf.global_variables_initializer()
total = []
```

10. 现在，开始计算图，训练 100 次：

```
# Computation Graph
with tf.Session() as sess:
    # Initialize variables
    sess.run(init_op)
    writer = tf.summary.FileWriter('graphs', sess.graph)
    # train the model for 100 epochs
    for i in range(100):
        total_loss = 0
        for x,y in zip(X_train,Y_train):
            _, l = sess.run ([optimizer, loss], feed_dict={X:x, Y:y})
            total_loss += l
        total.append(total_loss / n_samples)
        print('Epoch {0}: Loss {1}'.format(i, total_loss/n_samples))
    writer.close()
    b_value, w_value = sess.run([b,w])
```

11. 查看结果：

```
Y_pred = X_train * w_value + b_value
print('Done')
# Plot the result
plt.plot(X_train, Y_train, 'bo', label='Real Data')
plt.plot(X_train,Y_pred,  'r', label='Predicted Data')
plt.legend()
plt.show()
plt.plot(total)
plt.show()
```

解读分析

从下图中可以看到，简单线性回归器试图拟合给定数据集的线性线：

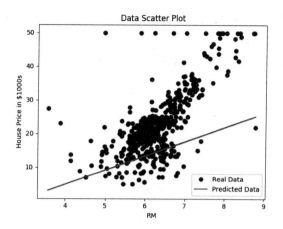

在下图中可以看到，随着模型不断学习数据，损失函数不断下降：

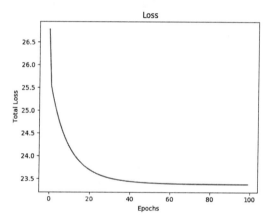

下图是简单线性回归器的 TensorBoard 图：

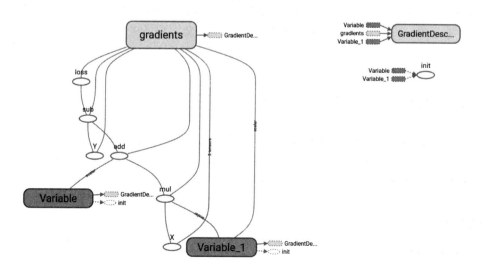

该图有两个名称范围节点 Variable 和 Variable_1，它们分别是表示偏置和权重的高级节点。以梯度命名的节点也是一个高级节点，展开节点，可以看到它需要 7 个输入并使用 `GradientDescentOptimizer` 计算**梯度**，对权重和偏置进行更新：

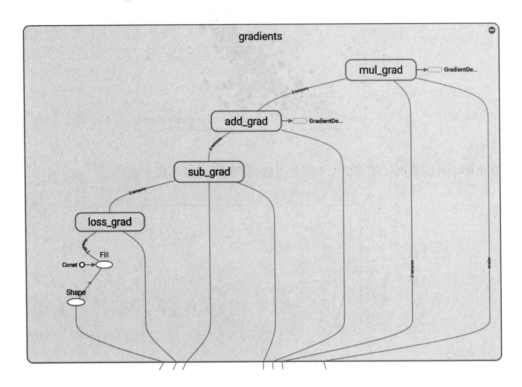

更多内容

这里进行了简单的线性回归，但是如何定义模型的性能呢？有多种方法可以做到这一点。统计上来说，可以计算 R^2 或将数据分为训练集和交叉验证集，并检查验证集的准确性（损失项）。

2.6 房价估计——多元线性回归

可以通过在权重和占位符的声明中稍作修改来对相同的数据进行多元线性回归。在多元线性回归的情况下，由于每个特征具有不同的值范围，归一化变得至关重要。这里是波士顿房价数据集的多重线性回归的代码，使用 13 个输入特征。

具体做法

1. 导入需要的所有软件包：

```
import tensorflow as tf
import numpy as np
import matplotlib.pyplot as plt
```

2. 因为各特征的数据范围不同,需要归一化特征数据。为此定义一个归一化函数。另外,这里添加一个额外的固定输入值将权重和偏置结合起来。为此定义函数 `append_bias_reshape()`。该技巧有时可有效简化编程。

```
def normalize(X)
    """ Normalizes the array X """
    mean = np.mean(X)
    std = np.std(X)
    X = (X - mean)/std
    return X

def append_bias_reshape(features,labels):
    m = features.shape[0]
    n = features.shape[1]
    x = np.reshape(np.c_[np.ones(m),features],[m,n + 1])
    y = np.reshape(labels,[m,1])
    return x, y
```

3. 现在使用 TensorFlow contrib 数据集加载波士顿房价数据集,并将其划分为 X_train 和 Y_train。注意到 X_train 包含所需要的特征。可以选择在这里对数据进行归一化处理,也可以添加偏置并对网络数据重构:

```
# Data
boston = tf.contrib.learn.datasets.load_dataset('boston')
X_train, Y_train = boston.data, boston.target
X_train = normalize(X_train)
X_train, Y_train = append_bias_reshape(X_train, Y_train)
m = len(X_train)    #Number of training examples
n = 13 + 1    # Number of features + bias
```

4. 为训练数据声明 TensorFlow 占位符。观测占位符 X 的形状变化。

```
# Placeholder for the Training Data
X = tf.placeholder(tf.float32, name='X', shape=[m,n])
Y = tf.placeholder(tf.float32, name='Y')
```

5. 为权重和偏置创建 TensorFlow 变量。通过随机数初始化权重:

```
# Variables for coefficients
w = tf.Variable(tf.random_normal([n,1]))
```

6. 定义要用于预测的线性回归模型。现在需要矩阵乘法来完成这个任务:

```
# The Linear Regression Model
Y_hat = tf.matmul(X, w)
```

7. 为了更好地求微分,定义损失函数:

```
# Loss function
loss = tf.reduce_mean(tf.square(Y - Y_hat, name='loss'))
```

8. 选择正确的优化器:

```
# Gradient Descent with learning rate of 0.01 to minimize loss
optimizer = 
tf.train.GradientDescentOptimizer(learning_rate=0.01).minimize(loss
)
```

9. 定义初始化操作符：

```
# Initializing Variables
init_op = tf.global_variables_initializer()
total = []
```

10. 开始计算图：

```
with tf.Session() as sess:
    # Initialize variables
    sess.run(init_op)
    writer = tf.summary.FileWriter('graphs', sess.graph)
     # train the model for 100 epcohs
    for i in range(100):
        _, l = sess.run([optimizer, loss], feed_dict={X: X_train, Y: Y_train})
        total.append(l)
        print('Epoch {0}: Loss {1}'.format(i, l))
    writer.close()
    w_value, b_value = sess.run([w, b])
```

11. 绘制损失函数：

```
plt.plot(total)
plt.show()
```

在这里，我们发现损失随着训练过程的进行而减少：

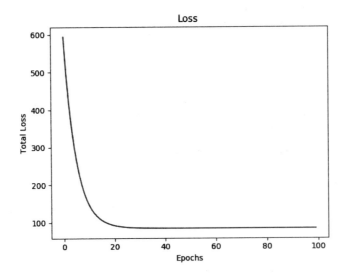

解读分析

本节使用了13个特征来训练模型。简单线性回归和多元线性回归的主要不同在于权重，且系数的数量始终等于输入特征的数量。下图为所构建的多元线性回归模型的 TensorBoard 图：

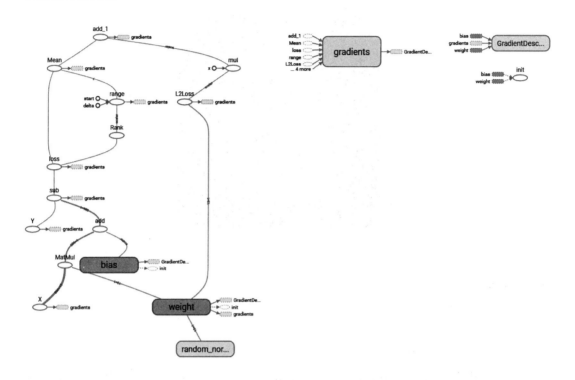

更多内容

现在可以使用从模型中学到的系数来预测房价：

```
N= 500
X_new = X_train [N,:]
Y_pred =  (np.matmul(X_new, w_value) + b_value).round(1)
print('Predicted value: ${0}  Actual value: / ${1}'.format(Y_pred[0]*1000,
Y_train[N]*1000) , '\nDone')
```

2.7 MNIST 数据集的逻辑回归

本节基于回归学习对 https://www.tensorflow.org/get_started/mnist/beginners 提供的 MNIST 数据集进行处理，但将添加一些 TensorBoard 总结以便更好地理解 MNIST 数据集。大部分人已经对 MNIST 数据集很熟悉了，它是机器学习的基础，包含手写数字的图像及其标签来说明它是哪个数字。

对于逻辑回归，对输出 y 使用独热（one-hot）编码。因此，有 10 位表示输出，每位的值为 1 或 0，独热意味着对于每个图片的标签 y，10 位中仅有一位的值为 1，其余的为 0。

因此，对于手写数字 8 的图像，其编码值为 [0 0 0 0 0 0 0 0 1 0]：

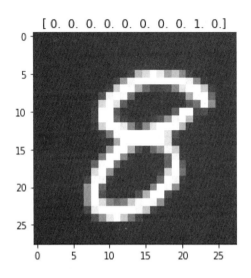

具体做法

1. 导入所需的模块：

```
import tensorflow as tf
import matplotlib.pyplot as plt, matplotlib.image as mpimg
```

2. 可以从模块 `input_data` 给出的 TensorFlow 示例中获取 MNIST 的输入数据。该 `one_hot` 标志设置为真，以使用标签的 `one_hot` 编码。这产生了两个张量，大小为 [55000,784] 的 `mnist.train.images` 和大小为 [55000,10] 的 `mnist.train.labels`。`mnist.train.images` 的每项都是一个范围介于 0 到 1 的像素强度：

```
from tensorflow.examples.tutorials.mnist import input_data
mnist = input_data.read_data_sets("MNIST_data/", one_hot=True)
```

3. 在 TensorFlow 图中为训练数据集的输入 x 和标签 y 创建占位符：

```
x = tf.placeholder(tf.float32, [None, 784], name='X')
y = tf.placeholder(tf.float32, [None, 10],name='Y')
```

4. 创建学习变量、权重和偏置：

```
W = tf.Variable(tf.zeros([784, 10]), name='W')
b = tf.Variable(tf.zeros([10]), name='b')
```

5. 创建逻辑回归模型。TensorFlow OP 给出了 `name_scope ("wx_b")`：

```
name_scope("wx_b"):

    with tf.name_scope("wx_b") as scope:
        y_hat = tf.nn.softmax(tf.matmul(x,W) + b)
```

6. 训练时添加 summary 操作来收集数据。使用直方图以便看到权重和偏置随时间相对于彼此值的变化关系。可以通过 TensorBoard Histogtam 选项卡看到：

```
w_h = tf.summary.histogram("weights", W)
b_h = tf.summary.histogram("biases", b)
```

7. 定义交叉熵（cross-entropy）和损失（loss）函数，并添加 name scope 和 summary 以实现更好的可视化。使用 scalar summary 来获得随时间变化的损失函数。scalar summary 在 Events 选项卡下可见：

```
with tf.name_scope('cross-entropy') as scope:
    loss = tf.reduce_mean(tf.nn.softmax_cross_entropy_with_logits(labels=y, logits=y_hat))
    tf.summary.scalar('cross-entropy', loss)
```

8. 采用 TensorFlow GradientDescentOptimizer，学习率为 0.01。为了更好地可视化，定义一个 name_scope：

```
with tf.name_scope('Train') as scope:
    optimizer = tf.train.GradientDescentOptimizer(0.01).minimize(loss)
```

9. 为变量进行初始化：

```
# Initializing the variables
init = tf.global_variables_initializer()
```

10. 组合所有的 summary 操作：

```
merged_summary_op = tf.summary.merge_all()
```

11. 现在，可以定义会话并将所有的 summary 存储在定义的文件夹中：

```
with tf.Session() as sess:
    sess.run(init)  # initialize all variables
    summary_writer = tf.summary.FileWriter('graphs', sess.graph)  # Create an event file
    # Training
    for epoch in range(max_epochs):
        loss_avg = 0
        num_of_batch = int(mnist.train.num_examples/batch_size)
        for i in range(num_of_batch):
            batch_xs, batch_ys = mnist.train.next_batch(100)  # get the next batch of data
            _, l, summary_str = sess.run([optimizer,loss, merged_summary_op], feed_dict={x: batch_xs, y: batch_ys})  # Run the optimizer
            loss_avg += l
            summary_writer.add_summary(summary_str, epoch*num_of_batch + i)  # Add all summaries per batch
        loss_avg = loss_avg/num_of_batch
        print('Epoch {0}: Loss {1}'.format(epoch, loss_avg))
    print('Done')
    print(sess.run(accuracy, feed_dict={x: mnist.test.images,y: mnist.test.labels}))
```

12. 经过 30 个周期，准确率达到了 86.5%；经过 50 个周期，准确率达到了 89.36%；经过 100 个周期，准确率提高到了 90.91%。

解读分析

这里使用张量 `tensorboard --logdir=garphs` 运行 TensorBoard。在浏览器中，导航到网址 `localhost:6006` 查看 TensorBoard。该模型图如下：

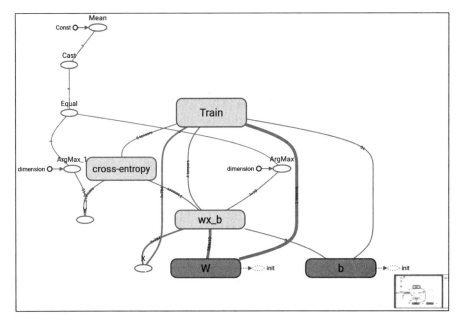

在 Histogram 选项卡下，可以看到**权重**（weights）和**偏置**（biases）的直方图：

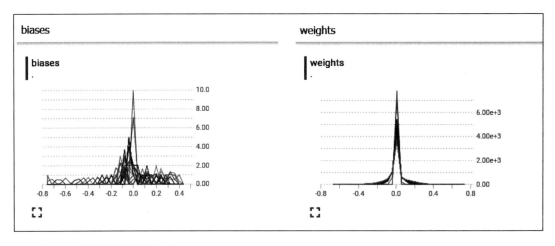

权重和**偏置**的分布如下：

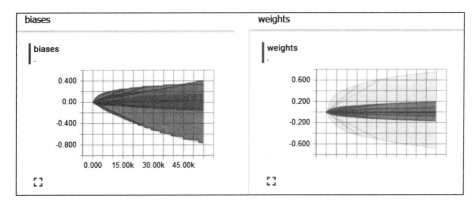

可以看到，随着时间的推移，偏置和权重都发生了变化。在该示例中，根据 TensorBoard 中的分布可知偏置变化的范围更大。在 Events 选项卡下，可以看到 scalar summary，即本示例中的交叉熵。下图显示交叉熵损失随时间不断减少：

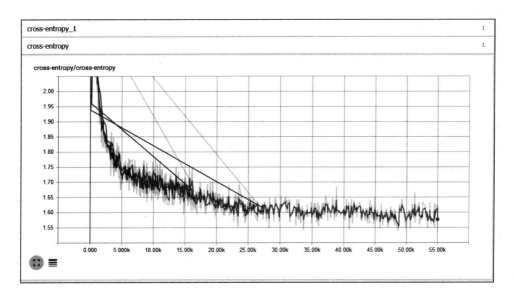

拓展阅读

如果你有兴趣了解更多相关知识，可以查看下面这些资源：

- 关于 TensorBoard 与可视化：https://www.tensorflow.org/get_started/summaries_and_tensorboard
- 关于统计与概率的课程：https://www.khanacademy.org/math/statistics-probability/describing-relationships-quantitative-data
- 更多关于回归的细节：https://onlinecourses.science.psu.edu/ stat501/node/250

CHAPTER 3

第 3 章

神经网络——感知机

最近十年以来，神经网络一直处于机器学习研究和应用的前沿。**深度神经网络**（DNN）、迁移学习以及计算高效的图形处理器（GPU）的普及使得图像识别、语音识别甚至文本生成领域取得了重大进展。在本章中，将集中讨论基本的神经网络感知机。它具有全连接人造神经元的分层结构。

3.1 引言

神经网络受人类大脑的启发，也被称为**连接模型**。像人脑一样，神经网络是大量被称为**权重**的突触相互连接的人造神经元的集合。就像我们通过年长者提供的例子来学习一样，人造神经网络通过向它们提供的例子来学习，这些例子被称为训练数据集。有了足够数量的训练数据集，人造神经网络可以提取信息，并用于它们没有见过的数据。

神经网络并不是最近才出现的。第一个神经网络模型 McCulloch Pitts（MCP）（`http://vordenker.de/ggphilosophy/mcculloch_a-logical-calculus.pdf`）早在 1943 年就被提出来了，该模型可以执行类似与、或、非的逻辑操作。MCP 模型的权重和偏置是固定的，因此不具备学习的可能。这个问题在若干年后的 1958 年由 Frank Rosenblatt 解决（`https://blogs.umass.edu/brain-wars/files/2016/03/rosenblatt-1957.pdf`）。他提出了第一个具有学习能力的神经网络，称之为**感知机**（perceptron）。

从那时起，人们就知道添加多层神经元并建立一个深的、稠密的网络将有助于神经网络解决复杂的任务。就像母亲为孩子的成就感到自豪一样，科学家和工程师对使用**神经网络**（`https://www.youtube.com/watch?v=jPHUlQiwD9Y`）所能实现的功能做出了高度的评价。这些评价并不是虚假的，但是由于硬件计算的限制和网络结构的复杂，当时根本无法实现。这导致了在 20 世纪 70 年代和 80 年代出现了被称为 **AI 寒冬**的时期。在这段时期，由于人工智能项目得不到资助，导致这一领域的进展放缓。

随着 DNN 和 GPU 的出现，情况发生了变化。今天，可以利用一些技术通过微调参数来获得表现更好的网络，比如 dropout 和迁移学习等技术，这缩短了训练时间。最后，硬件公司提出了使用专门的硬件芯片快速地执行基于神经网络的计算。

人造神经元是所有神经网络的核心。它由两个主要部分构成：一个加法器，将所有输入加权求和到神经元上；一个处理单元，根据预定义函数产生一个输出，这个函数被称为**激活函数**。每个神经元都有自己的一组权重和阈值（偏置），它通过不同的学习算法学习这些权重和阈值：

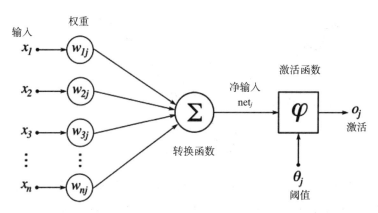

来源：https://commons.wikimedia.org/wiki/File:Rosenblattperceptron.png

当只有一层这样的神经元存在时，它被称为感知机。输入层被称为**第零层**，因为它只是缓冲输入。存在的唯一一层神经元形成输出层。输出层的每个神经元都有自己的权重和阈值。当存在许多这样的层时，网络被称为**多层感知机**（MLP）。MLP 有一个或多个隐藏层。这些隐藏层具有不同数量的隐藏神经元。每个隐藏层的神经元具有相同的激活函数：

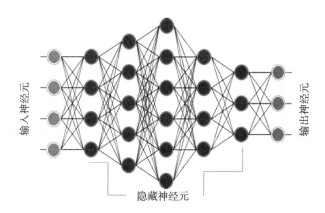

上图的 MLP 具有一个有 4 个输入的输入层，5 个分别有 4、5、6、4 和 3 个神经元的隐藏层，以及一个有 3 个神经元的输出层。在该 MLP 中，下层的所有神经元都连接到其相邻的上层的所有神经元。因此，MLP 也被称为**全连接层**。MLP 中的信息流通常是从输入到输出，目前没有反馈或跳转，因此这些网络也被称为**前馈网络**。

感知机使用**梯度下降算法**进行训练。第 2 章已经介绍了梯度下降，在这里再深入一点。感知机通过监督学习算法进行学习，也就是给网络提供训练数据集的理想输出。在输出端，定义了一个误差函数或目标函数 $J(W)$，这样当网络完全学习了所有的训练数据后，目标函数将是最小的。

输出层和隐藏层的权重被更新，使得目标函数的梯度减小：

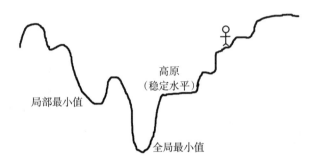

为了更好地理解它，想象一个充满山丘、高原和凹坑的地形。目标是走到地面（目标函数的全局最小值）。如果你站在最上面，必须往下走，那么很明显你将会选择下山，即向负坡度（或负梯度）方向移动。相同的道理，感知机的权重与目标函数梯度的负值成比例地变化。

梯度的值越大，权值的变化越大，反之亦然。现在，这一切都很好，但是当到达高原时，可能会遇到问题，因为梯度是零，所以权重没有变化。当进入一个小坑（局部最小值）时，也会遇到问题，因为尝试移动到任何一边，梯度都会增加，迫使网络停留在坑中。

正如第 2 章所述，针对增加网络的收敛性提出了梯度下降的各种变种使得网络避免陷入局部最小值或高原的问题，比如添加动量、可变学习率。

TensorFlow 会在不同的优化器的帮助下自动计算这些梯度。然而，需要注意的重要一点是，由于 TensorFlow 将计算梯度，这也将涉及激活函数的导数，所以你选择的激活函数必须是可微分的，并且在整个训练场景中具有非零梯度。

感知机中的梯度下降与第 2 章中的梯度下降的一个主要不同是，输出层的目标函数已经被定义好了，但它也用于隐藏层神经元的权值更新。这是使用**反向传播**（BPN）算法完成的，输出中的误差向后传播到隐藏层并用于确定权重变化。

3.2 激活函数

每个神经元都必须有激活函数。它们为神经元提供了模拟复杂非线性数据集所必需的非线性特性。该函数取所有输入的加权和，进而生成一个输出信号。你可以把它看作输入和输出之间的转换。使用适当的激活函数，可以将输出值限定在一个定义的范围内。

如果 x_i 是第 j 个输入，W_j 是连接第 j 个输入到神经元的权重，b 是神经元的偏置，神经元的输出（在生物学术语中，神经元的激活）由激活函数决定，并且在数学上表示如下：

$$Y_{\text{hat}} = g\left(\sum_{j=1}^{N} W_j x_j + b\right)$$

这里，g 表示激活函数。激活函数的参数 $\sum W_j x_j + b$ 被称为**神经元的活动**。

准备工作

这里对给定输入刺激的反应是由神经元的激活函数决定的。有时回答是二元的（是或不是）。例如，当有人开玩笑的时候，要么笑，要么不笑。在其他时候，反应似乎是线性的，例如，由于疼痛而哭泣。有时，答复似乎是在一个范围内。

模仿类似的行为，人造神经元使用许多不同的激活函数。在这里，你将学习如何定义和使用 TensorFlow 中的一些常用激活函数。

具体做法

下面继续讨论激活函数：

1. 阈值激活函数：这是最简单的激活函数。在这里，如果神经元的激活值大于零，那么神经元就会被激活；否则，它还是处于抑制状态。下面绘制阈值激活函数的图，随着神经元的激活值的改变在 TensorFlow 中实现阈值激活函数：

```
import tensorflow as tf
import numpy as np
import matplotlib.pyplot as plt

# Threshold Activation function
def threshold (x):
    cond = tf.less(x, tf.zeros(tf.shape(x), dtype = x.dtype))
    out = tf.where(cond, tf.zeros(tf.shape(x)), tf.ones(tf.shape(x)))
    return out
# Plotting Threshold Activation Function
h = np.linspace(-1,1,50)
out = threshold(h)
init = tf.global_variables_initializer()
with tf.Session() as sess:
    sess.run(init)
    y = sess.run(out)
    plt.xlabel('Activity of Neuron')
    plt.ylabel('Output of Neuron')
    plt.title('Threshold Activation Function')
    plt.plot(h, y)
```

上述代码的输出如下图所示：

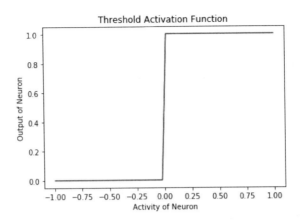

2. **Sigmoid 激活函数**：在这种情况下，神经元的输出由函数 g(x)=1/(1+exp(-x)) 确定。在 TensorFlow 中，方法是 `tf.sigmoid`，它提供了 Sigmoid 激活函数。这个函数的范围在 0 到 1 之间。在形状上，它看起来像字母 S，因此名字叫 Sigmoid：

```
# Plotting Sigmoidal Activation function
h = np.linspace(-10,10,50)
out = tf.sigmoid(h)
init = tf.global_variables_initializer()
with tf.Session() as sess:
    sess.run(init)
    y = sess.run(out)
    plt.xlabel('Activity of Neuron')
    plt.ylabel('Output of Neuron')
    plt.title('Sigmoidal Activation Function')
    plt.plot(h, y)
```

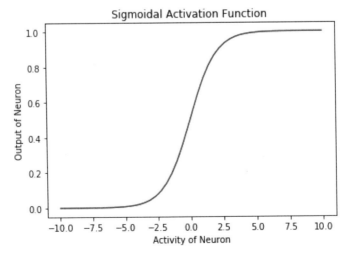

3. **双曲正切激活函数**：在数学上，它表示为 (1−exp(−2x)/(1+exp(−2x)))。在形状上，它类似于 Sigmoid 函数，但是它的中心位置是 0，其范围是从 −1 到 1。TensorFlow 有一个

内置函数 `tf.tanh`，用来实现双曲正切激活函数：

```
# Plotting Hyperbolic Tangent Activation function
h = np.linspace(-10,10,50)
out = tf.tanh(h)
init = tf.global_variables_initializer()
with tf.Session() as sess:
    sess.run(init)
    y = sess.run(out)
    plt.xlabel('Activity of Neuron')
    plt.ylabel('Output of Neuron')
    plt.title('Hyperbolic Tangent Activation Function')
    plt.plot(h, y)
```

以下是上述代码的输出：

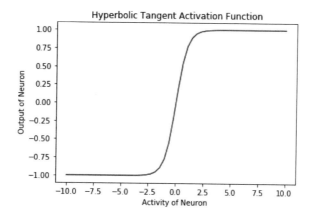

4. 线性激活函数：在这种情况下，神经元的输出与神经元的输入值相同。这个函数的任何一边都不受限制：

```
# Linear Activation Function
b = tf.Variable(tf.random_normal([1,1], stddev=2))
w = tf.Variable(tf.random_normal([3,1], stddev=2))
linear_out = tf.matmul(X_in, w) + b
init = tf.global_variables_initializer()
with tf.Session() as sess:
    sess.run(init)
    out = sess.run(linear_out)
print(out)
```

5. 整流线性单元（ReLU）激活函数也被内置在 TensorFlow 库中。这个激活函数类似于线性激活函数，但有一个大的改变：对于负的输入值，神经元不会激活（输出为零），对于正的输入值，神经元的输出与输入值相同：

```
# Plotting ReLU Activation function
h = np.linspace(-10,10,50)
out = tf.nn.relu(h)
init = tf.global_variables_initializer()
with tf.Session() as sess:
```

```
        sess.run(init)
        y = sess.run(out)
plt.xlabel('Activity of Neuron')
plt.ylabel('Output of Neuron')
plt.title('ReLU Activation Function')
plt.plot(h, y)
```

以下是 ReLU 激活函数的输出:

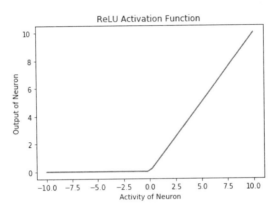

6. **Softmax 激活函数**是一个归一化的指数函数。一个神经元的输出不仅取决于其自身的输入值,还取决于该层中存在的所有其他神经元的输入的总和。这样做的一个优点是使得神经元的输出小,因此梯度不会过大。数学表达式为 $y_i = \exp(x_i) / \sum_j \exp(x_j)$:

```
# Plotting Softmax Activation function
h = np.linspace(-5,5,50)
out = tf.nn.softmax(h)
init = tf.global_variables_initializer()
with tf.Session() as sess:
    sess.run(init)
    y = sess.run(out)
    plt.xlabel('Activity of Neuron')
    plt.ylabel('Output of Neuron')
    plt.title('Softmax Activation Function')
    plt.plot(h, y)
```

以下是上述代码的输出:

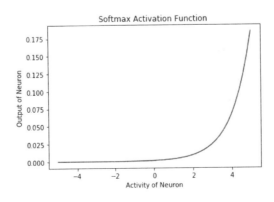

解读分析

以下是上述函数的解释：

- **阈值激活函数**用于 McCulloch Pitts 神经元和原始的感知机。这是不可微的，在 $x=0$ 时是不连续的。因此，使用这个激活函数来进行基于梯度下降或其变体的训练是不可能的。
- **Sigmoid 激活函数**一度很受欢迎，从曲线来看，它像一个连续版的阈值激活函数。它受到梯度消失问题的困扰，即函数的梯度在两个边缘附近变为零。这使得训练和优化变得困难。
- **双曲正切激活函数**在形状上也是 S 形并具有非线性特性。该函数以 0 为中心，与 Sigmoid 函数相比具有更陡峭的导数。与 Sigmoid 函数一样，它也受到梯度消失问题的影响。
- **线性激活函数**是线性的。该函数是双边都趋于无穷的 [-inf, inf]。它的线性是主要问题。线性函数之和是线性函数，线性函数的线性函数也是线性函数。因此，使用这个函数，不能表示复杂数据集中存在的非线性。
- **ReLU 激活函数**是线性激活功能的整流版本，这种整流功能允许其用于多层时捕获非线性。使用 ReLU 的主要优点之一是导致稀疏激活。在任何时刻，所有神经元的负的输入值都不会激活神经元。就计算量来说，这使得网络在计算方面更轻便。ReLU 神经元存在死亡 ReLU 的问题，也就是说，那些没有激活的神经元的梯度为零，因此将无法进行任何训练，并停留在死亡状态。尽管存在这个问题，但 ReLU 仍是隐藏层最常用的激活函数之一。
- **Softmax 激活函数**被广泛用作输出层的激活函数，该函数的范围是 [0,1]。在多类分类问题中，它被用来表示一个类的概率。所有单位输出和总是 1。

更多内容

神经网络已被用于各种任务。这些任务可以大致分为两类：函数逼近（回归）和分类。根据手头的任务，一个激活函数可能比另一个更好。一般来说，隐藏层最好使用 ReLU 神经元。对于分类任务，Softmax 通常是更好的选择；对于回归问题，最好使用 Sigmoid 函数或双曲正切函数。

拓展阅读

链接 https://www.tensorflow.org/versions/r0.12/api_docs/python/nn/activation_functions_ 提供了在 TensorFlow 中定义的激活函数的详细信息以及如何使用它们。

链接 https://en.wikipedia.org/wiki/Activation_function 是对激活函数的一个很好的总结。

3.3 单层感知机

简单感知机是一个单层神经网络。它使用阈值激活函数，正如 Marvin Minsky 在论文中所证明的，它只能解决线性可分的问题。虽然这限制了单层感知机只能应用于线性可分问题，但它具有学习能力已经很好了。

准备工作

当感知机使用阈值激活函数时，不能使用 TensorFlow 优化器来更新权重。我们将不得不使用权重更新规则：

$$\Delta W = \eta X^T (Y - Y_{hat})$$

η 是学习率。为了简化编程，当输入固定为 +1 时，偏置可以作为一个额外的权重。那么，上面的公式可以用来同时更新权重和偏置。

具体做法

下面讨论如何实现单层感知机：

1. 导入所需的模块：

```
import tensorflow as tf
import numpy as np
```

2. 定义要使用的超参数：

```
# Hyper parameters
eta = 0.4  # learning rate parameter
epsilon = 1e-03 # minimum accepted error
max_epochs = 100 # Maximum Epochs
```

3. 定义 `threshold`（阈值）函数：

```
# Threshold Activation function
def threshold (x):
    cond = tf.less(x, tf.zeros(tf.shape(x), dtype = x.dtype))
    out = tf.where(cond, tf.zeros(tf.shape(x)), tf.ones(tf.shape(x)))
    return out
```

4. 指定训练数据。在这个例子中，取三个输入神经元（A，B，C）并训练它学习逻辑 AB+BC：

```
# Training Data  Y = AB + BC, sum of two linear functions.
T, F = 1., 0.
X_in = [
    [T, T, T, T],
    [T, T, F, T],
    [T, F, T, T],
    [T, F, F, T],
    [F, T, T, T],
    [F, T, F, T],
```

```
    [F, F, T, T],
    [F, F, F, T],
    ]
Y = [
    [T],
    [T],
    [F],
    [F],
    [T],
    [F],
    [F],
    [F]
]
```

5. 定义要用到的变量和用于计算更新的计算图，最后执行计算图：

```
W = tf.Variable(tf.random_normal([4,1], stddev=2, seed = 0))
h = tf.matmul(X_in, W)
Y_hat = threshold(h)
error = Y - Y_hat
mean_error = tf.reduce_mean(tf.square(error))
dW =  eta * tf.matmul(X_in, error, transpose_a=True)
train = tf.assign(W, W+dW)
init = tf.global_variables_initializer()
err = 1
epoch = 0
with tf.Session() as sess:
    sess.run(init)
    while err > epsilon and epoch < max_epochs:
        epoch += 1
        err, _ = sess.run([mean_error, train])
        print('epoch: {0}  mean error: {1}'.format(epoch, err))
    print('Training complete')
```

以下是上述代码的输出：

```
epoch: 1  mean error: 0.625
epoch: 2  mean error: 0.125
epoch: 3  mean error: 0.125
epoch: 4  mean error: 0.375
epoch: 5  mean error: 0.125
epoch: 6  mean error: 0.125
epoch: 7  mean error: 0.375
epoch: 8  mean error: 0.125
epoch: 9  mean error: 0.125
epoch: 10  mean error: 0.125
epoch: 11  mean error: 0.125
epoch: 12  mean error: 0.0
Training complete
```

更多内容

如果使用 Sigmoid 激活函数，而不是阈值激活函数，会发生什么？

你猜对了，首先，可以使用 TensorFlow 优化器来更新权重。其次，网络将表现得像逻辑回归。

3.4 计算反向传播算法的梯度

BPN 算法是神经网络中研究最多、使用最多的算法之一。它用于将输出层中的误差传播到隐藏层的神经元，然后用于更新权重。整个学习可以分成两个过程：正向传播和反向传播。

正向传播：输入被馈送到网络，信号从输入层通过隐藏层传播到输出层。在输出层，计算误差和损失函数。

反向传播：在反向传播中，首先计算输出层神经元损失函数的梯度，然后计算隐藏层神经元损失函数的梯度。接下来用梯度更新权重。

这两个过程重复迭代直到收敛。

准备工作

首先给网络提供 M 个训练对 (X, Y)，X 为输入，Y 为期望的输出。输入通过激活函数 $g(h)$ 和隐藏层传播到输出层。输出 Y_{hat} 是网络的输出，得到 error=$Y-Y_{hat}$。
其损失函数 $J(W)$ 如下：

$$J(W) = \frac{1}{2M} \sum_{i=1}^{N} \left(Y_i - Y_{hat_i}\right)^2$$

其中，i 取遍所有输出层的神经元（1 到 N）。然后可以使用 $J(W)$ 的梯度并使用链式法则求导，来计算连接第 i 个输出层神经元到第 j 个隐藏层神经元的权重 W_{ij} 的变化：

$$\Delta W_{ij} = -\eta \frac{\partial J}{\partial W_{ij}} = \eta \frac{1}{M} \frac{\partial \left(Y_i - Y_{hat_i}\right)^2}{\partial Y_{hat_i}} \frac{\partial Y_{hat_i}}{\partial h_i} \frac{\partial h_i}{\partial W_{ij}} = \eta \frac{1}{M} \left(Y_i - Y_{hat_i}\right) g'(h_i) O_j$$

这里，O_j 是隐藏层神经元的输出，h 表示隐藏层的输入值。这很容易理解，但现在怎么更新连接第 n 个隐藏层的神经元 k 到第 $n+1$ 个隐藏层的神经元 j 的权值 W_{jk}？过程是相同的：将使用损失函数的梯度和链式法则求导，但这次计算 W_{jk}：

$$\Delta W_{jk} = -\eta \frac{\partial J}{\partial W_{jk}} = \eta \frac{1}{M} \sum_i \left\{ \frac{\partial \left(Y_i - Y_{hat_i}\right)^2}{\partial Y_{hat_i}} \frac{\partial Y_{hat_i}}{\partial h_i} \frac{\partial h_i}{\partial O_j} \right\} \frac{\partial O_j}{\partial W_{jk}}$$

$$= \eta \frac{1}{M} \sum_i \left\{ \left(Y_i - Y_{hat_i}\right) g'(h_i) W_{ij} \right\} g'(h_k) O_k$$

现在已经有方程了，看看如何在 TensorFlow 中做到这一点。在这里，还是使用 MNIST 数据集（http://yann.lecun.com/exdb/MNIST/）。

具体做法

现在开始使用反向传播算法：

1. 导入模块:

```
import tensorflow as tf
from tensorflow.examples.tutorials.mnist import input_data
```

2. 加载数据集，通过设置 one_hot = True 来使用独热编码标签:

```
mnist = input_data.read_data_sets("MNIST_data/", one_hot=True)
```

3. 定义超参数和其他常量。这里，每个手写数字的尺寸是 $28 \times 28 = 784$ 像素。数据集被分为 10 类，以 0 到 9 之间的数字表示。这两点是固定的。学习率、最大迭代周期数、每次批量训练的批量大小以及隐藏层中的神经元数量都是超参数。可以通过调整这些超参数，看看它们是如何影响网络表现的:

```
# Data specific constants
n_input = 784 # MNIST data input (img shape: 28*28)
n_classes = 10 # MNIST total classes (0-9 digits)

# Hyperparameters
max_epochs = 10000
learning_rate = 0.5
batch_size = 10
seed = 0
n_hidden = 30  # Number of neurons in the hidden layer
```

4. 需要 Sigmoid 函数的导数来进行权重更新，所以定义它:

```
def sigmaprime(x):
    return tf.multiply(tf.sigmoid(x), tf.subtract(tf.constant(1.0), tf.sigmoid(x)))
```

5. 为训练数据创建占位符:

```
x_in = tf.placeholder(tf.float32, [None, n_input])
y = tf.placeholder(tf.float32, [None, n_classes])
```

6. 创建模型:

```
def multilayer_perceptron(x, weights, biases):
    # Hidden layer with RELU activation
    h_layer_1 = tf.add(tf.matmul(x, weights['h1']), biases['h1'])
    out_layer_1 = tf.sigmoid(h_layer_1)
    # Output layer with linear activation
    h_out = tf.matmul(out_layer_1, weights['out']) + biases['out']
    return tf.sigmoid(h_out), h_out, out_layer_1, h_layer_1
```

7. 定义权重和偏置变量:

```
weights = {
    'h1': tf.Variable(tf.random_normal([n_input, n_hidden], seed = seed)),
    'out': tf.Variable(tf.random_normal([n_hidden, n_classes], seed = seed)) }

biases = {
    'h1': tf.Variable(tf.random_normal([1, n_hidden], seed =
```

```
seed)),
    'out': tf.Variable(tf.random_normal([1, n_classes], seed =
seed))}
```

8. 为正向传播、误差、梯度和更新计算创建计算图：

```
# Forward Pass
 y_hat, h_2, o_1, h_1 = multilayer_perceptron(x_in, weights,
biases)

 # Error
 err = y_hat - y

 # Backward Pass
 delta_2 = tf.multiply(err, sigmaprime(h_2))
 delta_w_2 = tf.matmul(tf.transpose(o_1), delta_2)

 wtd_error = tf.matmul(delta_2, tf.transpose(weights['out']))
 delta_1 = tf.multiply(wtd_error, sigmaprime(h_1))
 delta_w_1 = tf.matmul(tf.transpose(x_in), delta_1)

 eta = tf.constant(learning_rate)

 # Update weights
 step = [
     tf.assign(weights['h1'],tf.subtract(weights['h1'],
tf.multiply(eta, delta_w_1)))
    , tf.assign(biases['h1'],tf.subtract(biases['h1'],
tf.multiply(eta, tf.reduce_mean(delta_1, axis=[0]))))
    , tf.assign(weights['out'], tf.subtract(weights['out'],
tf.multiply(eta, delta_w_2)))
    , tf.assign(biases['out'], tf.subtract(biases['out'],
tf.multiply(eta,tf.reduce_mean(delta_2, axis=[0]))))
 ]
```

9. 定义计算精度 accuracy 的操作：

```
acct_mat = tf.equal(tf.argmax(y_hat, 1), tf.argmax(y, 1))
accuracy = tf.reduce_sum(tf.cast(acct_mat, tf.float32))
```

10. 初始化变量：

```
init = tf.global_variables_initializer()
```

11. 执行图：

```
with tf.Session() as sess:
    sess.run(init)
    for epoch in range(max_epochs):
        batch_xs, batch_ys = mnist.train.next_batch(batch_size)
        sess.run(step, feed_dict = {x_in: batch_xs, y : batch_ys})
        if epoch % 1000 == 0:
            acc_test = sess.run(accuracy, feed_dict =
                    {x_in: mnist.test.images,
                     y : mnist.test.labels})
            acc_train = sess.run(accuracy, feed_dict=
            {x_in: mnist.train.images,
```

```
                  y: mnist.train.labels}))
            print('Epoch: {0}   Accuracy Train%: {1}   Accuracy
Test%: {2}'
                  .format(epoch,acc_train/600,(acc_test/100)))
```

结果如下：

```
Epoch: 0    Accuracy Train%: 10.13  Accuracy Test%: 11.04
Epoch: 1000 Accuracy Train%: 77.14166666666667  Accuracy Test%: 84.6
Epoch: 2000 Accuracy Train%: 80.05333333333333  Accuracy Test%: 87.27
Epoch: 3000 Accuracy Train%: 82.18666666666667  Accuracy Test%: 90.16
Epoch: 4000 Accuracy Train%: 82.99833333333333  Accuracy Test%: 90.71
Epoch: 5000 Accuracy Train%: 82.82666666666667  Accuracy Test%: 90.58
Epoch: 6000 Accuracy Train%: 83.99166666666666  Accuracy Test%: 91.85
Epoch: 7000 Accuracy Train%: 84.29333333333334  Accuracy Test%: 92.07
Epoch: 8000 Accuracy Train%: 84.62833333333333  Accuracy Test%: 92.1
Epoch: 9000 Accuracy Train%: 84.45  Accuracy Test%: 92.16

Process finished with exit code 0
```

解读分析

在这里，训练网络时的批量大小为 10，如果增加批量的值，网络性能就会下降。另外，需要在测试数据集上检测训练好的网络的精度，这里测试数据集的大小是 1000。

更多内容

单隐藏层多层感知机在训练数据集上的准确率为 84.45，在测试数据集上的准确率为 92.1。这是好的，但不够好。MNIST 数据集被用作机器学习中分类问题的基准。接下来，看一下如何使用 TensorFlow 的内置优化器影响网络性能。

拓展阅读

- MNIST 数据集：http://yann.lecun.com/exdb/mnist/
- 反向传播算法的简化讲解：http://neuralnetworksanddeeplearning.com/chap2.html
- 反向传播算法的另一个直观解释：http://cs231n.github.io/optimization-2/
- 反向传播算法的详细信息，以及如何将它应用于不同的网络：https://page.mi.fu-berlin.de/rojas/neural/chapter/K7.pdf

3.5 使用 MLP 实现 MNIST 分类器

TensorFlow 支持自动求导，可以使用 TensorFlow 优化器来计算和使用梯度。它使用梯度自动更新用变量定义的张量。本节将使用 TensorFlow 优化器来训练网络。

准备工作

3.4 节定义了层、权重、损失、梯度以及通过梯度更新权重。用公式实现可以帮助我们更好地理解，但随着网络层数的增加，这可能非常麻烦。

本节将使用 TensorFlow 的一些强大功能，如 Contrib（层）来定义神经网络层及使用 TensorFlow 自带的优化器来计算和使用梯度。第 2 章介绍了如何使用 TensorFlow 的优化器。Contrib 可以用来添加各种层到神经网络模型，如添加构建块。这里使用的一个方法是 `tf.contrib.layers.fully_connected`，在 TensorFlow 文档中定义如下：

```
fully_connected(
    inputs,
    num_outputs,
    activation_fn=tf.nn.relu,
    normalizer_fn=None,
    normalizer_params=None,
    weights_initializer=initializers.xavier_initializer(),
    weights_regularizer=None,
    biases_initializer=tf.zeros_initializer(),
    biases_regularizer=None,
    reuse=None,
    variables_collections=None,
    outputs_collections=None,
    trainable=True,
    scope=None
)
```

这样就添加了一个全连接层。

> 上面那段代码创建了一个称为权重的变量，表示全连接的权重矩阵，该矩阵与输入相乘产生隐藏层单元的张量。如果提供了 `normalizer_fn`（比如 `batch_norm`），那么就会归一化。否则，如果 `normalizer_fn` 是 `None`，并且设置了 `biases_initializer`，则会创建一个偏置变量并将其添加到隐藏层单元中。最后，如果 `activation_fn` 不是 `None`，它也会被应用到隐藏层单元。

具体做法

1. 第一步是改变损失函数，尽管对于分类任务，最好使用交叉熵损失函数。这里继续使用**均方误差（MSE）**：

```
loss = tf.reduce_mean(tf.square(y - y_hat, name='loss'))
```

2. 接下来，使用 `GradientDescentOptimizer`：

```
optimizer = tf.train.GradientDescentOptimizer(learning_rate=learning_rate)
 train = optimizer.minimize(loss)
```

3. 对于同一组超参数，只有这两处改变，在测试数据集上的准确率只有 61.3%。增加

max_epoch，可以提高准确性，但不能有效地发挥 TensorFlow 的能力。

4. 这是一个分类问题，所以最好使用交叉熵损失，隐藏层使用 ReLU 激活函数，输出层使用 softmax 函数。做些必要的修改，完整代码如下所示：

```python
import tensorflow as tf
import tensorflow.contrib.layers as layers

from tensorflow.python import debug as tf_debug

# Network Parameters
n_hidden = 30
n_classes = 10
n_input = 784

# Hyperparameters
batch_size = 200
eta = 0.001
max_epoch = 10

# MNIST input data
from tensorflow.examples.tutorials.mnist import input_data
mnist = input_data.read_data_sets("/tmp/data/", one_hot=True)

def multilayer_perceptron(x):
    fc1 = layers.fully_connected(x, n_hidden, activation_fn=tf.nn.relu, scope='fc1')
    #fc2 = layers.fully_connected(fc1, 256, activation_fn=tf.nn.relu, scope='fc2')
    out = layers.fully_connected(fc1, n_classes, activation_fn=None, scope='out')
    return out

# build model, loss, and train op
x = tf.placeholder(tf.float32, [None, n_input], name='placeholder_x')
y = tf.placeholder(tf.float32, [None, n_classes], name='placeholder_y')
y_hat = multilayer_perceptron(x)

loss = tf.reduce_mean(tf.nn.softmax_cross_entropy_with_logits(logits=y_hat, labels=y))
train = tf.train.AdamOptimizer(learning_rate= eta).minimize(loss)
init = tf.global_variables_initializer()




with tf.Session() as sess:
    sess.run(init)
    for epoch in range(10):
        epoch_loss = 0.0
        batch_steps = int(mnist.train.num_examples / batch_size)
        for i in range(batch_steps):
            batch_x, batch_y = mnist.train.next_batch(batch_size)
            _, c = sess.run([train, loss],
```

```
                        feed_dict={x: batch_x, y: batch_y})
                epoch_loss += c / batch_steps
            print ('Epoch %02d, Loss = %.6f' % (epoch, epoch_loss))

        # Test model
        correct_prediction = tf.equal(tf.argmax(y_hat, 1),
tf.argmax(y, 1))
        accuracy = tf.reduce_mean(tf.cast(correct_prediction,
tf.float32))
        print ("Accuracy%:", accuracy.eval({x: mnist.test.images, y:
mnist.test.labels}))
```

解读分析

修改后的 MNIST MLP 分类器在测试数据集上只用了一个隐藏层，并且在 10 个 epoch 内，只需要几行代码，就可以得到 96% 的精度。这就是 TensorFlow 的强大之处。

```
Epoch 00, Loss = 0.319608
Epoch 01, Loss = 0.178234
Epoch 02, Loss = 0.139025
Epoch 03, Loss = 0.116732
Epoch 04, Loss = 0.103431
Epoch 05, Loss = 0.093936
Epoch 06, Loss = 0.085003
Epoch 07, Loss = 0.076824
Epoch 08, Loss = 0.071777
Epoch 09, Loss = 0.065691
Accuracy%: 96.5300142765

Process finished with exit code 0
```

3.6 使用 MLP 逼近函数来预测波士顿房价

Hornik 等人的工作（http://www.cs.cmu.edu/~bhiksha/courses/deeplearning/Fall.2016 /notes/Sonia_Hornik.pdf）证明了下面这句话：

"只有一个隐藏层的多层前馈网络足以逼近任何函数，同时还可以保证很高的精度和令人满意的效果。"

本节将展示如何使用 MLP 进行函数逼近，具体来说，是预测波士顿的房价。第 2 章使用回归技术对房价进行预测，现在使用 MLP 完成相同的任务。

准备工作

对于函数逼近，这里的损失函数是 MSE。输入应该归一化，隐藏层是 ReLU，输出层最好是 Sigmoid。

具体做法

下面是如何使用 MLP 进行函数逼近的示例。

1. 导入需要用到的模块：`sklearn`，该模块可以用来获取数据集，预处理数据，并将其分成训练集和测试集；`pandas`，可以用来分析数据集；`matplotlib` 和 `seaborn` 可以用来可视化。

```
import tensorflow as tf
import tensorflow.contrib.layers as layers
from sklearn import datasets
import matplotlib.pyplot as plt
from sklearn.model_selection import train_test_split
from sklearn.preprocessing import MinMaxScaler
import pandas as pd
import seaborn as sns
%matplotlib inline
```

2. 加载数据集并创建 Pandas 数据帧来分析数据：

```
# Data
boston = datasets.load_boston()
df = pd.DataFrame(boston.data, columns=boston.feature_names)
df['target'] = boston.target
```

3. 了解一些关于数据的细节：

```
#Understanding Data
df.describe()
```

下表很好地描述了数据：

	CRIM	ZN	INDUS	CHAS	NOX	RM	AGE	DIS	RAD	TAX	PTRATIO	B	LSTAT	target
count	506	506	506	506	506	506	506	506	506	506	506	506	506	506
mean	3.59376	11.3636	11.1368	0.06917	0.5547	6.28463	68.5749	3.79504	9.54941	408.237	18.4555	356.674	12.6531	22.5328
std	8.59678	23.3225	6.86035	0.25399	0.11588	0.70262	28.1489	2.10571	8.70726	168.537	2.16495	91.2949	7.14106	9.1971
min	0.00632	0	0.46	0	0.385	3.561	2.9	1.1296	1	187	12.6	0.32	1.73	5
25%	0.08205	0	5.19	0	0.449	5.8855	45.025	2.10018	4	279	17.4	375.378	6.95	17.025
50%	0.25651	0	9.69	0	0.538	6.2085	77.5	3.20745	5	330	19.05	391.44	11.36	21.2
75%	3.64742	12.5	18.1	0	0.624	6.6235	94.075	5.18843	24	666	20.2	396.225	16.955	25
max	88.9762	100	27.74	1	0.871	8.78	100	12.1265	24	711	22	396.9	37.97	50

4. 找到输入的不同特征与输出之间的关联：

```
# Plotting correlation
color map _ , ax = plt.subplots( figsize =( 12 , 10 ) )
corr = df.corr(method='pearson')
cmap = sns.diverging_palette( 220 , 10 , as_cmap = True )
_ = sns.heatmap( corr, cmap = cmap, square=True, cbar_kws={
'shrink' : .9 }, ax=ax, annot = True, annot_kws = { 'fontsize' : 12
})
```

以下是上述代码的输出：

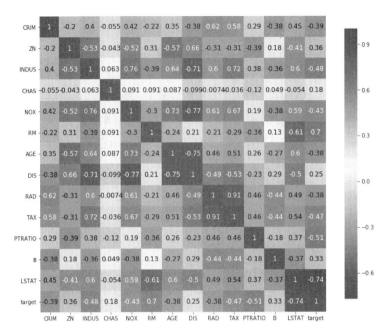

5. 从前面的代码中，可以看到三个参数 `RM`、`PTRATIO` 和 `LSTAT` 在幅度上与输出之间具有大于 0.5 的相关性。选择它们进行训练。将数据集分解为训练数据集和测试数据集。使用 `MinMaxScaler` 来规范数据集。需要注意的一个重要变化是，由于神经网络使用 Sigmoid 激活函数（Sigmoid 的输出只能在 0～1 之间），所以还必须对目标值 Y 进行归一化：

```
# Create Test Train Split
X_train, X_test, y_train, y_test = train_test_split(df [['RM',
'LSTAT', 'PTRATIO']], df[['target']], test_size=0.3,
random_state=0)
# Normalize data
X_train = MinMaxScaler().fit_transform(X_train)
y_train = MinMaxScaler().fit_transform(y_train)
X_test = MinMaxScaler().fit_transform(X_test)
y_test = MinMaxScaler().fit_transform(y_test)
```

6. 定义常量和超参数：

```
#Network Parameters
m = len(X_train)
n = 3 # Number of features
n_hidden = 20 # Number of hidden neurons
# Hyperparameters
batch_size = 200
eta = 0.01
max_epoch = 1000
```

7. 创建一个单隐藏层的多层感知机模型：

```
def multilayer_perceptron(x):
    fc1 = layers.fully_connected(x, n_hidden,
activation_fn=tf.nn.relu, scope='fc1')
```

```
    out = layers.fully_connected(fc1, 1, activation_fn=tf.sigmoid,
scope='out')
    return out
```

8. 声明训练数据的占位符并定义损失和优化器：

```
# build model, loss, and train op
x = tf.placeholder(tf.float32, name='X', shape=[m,n])
y = tf.placeholder(tf.float32, name='Y')
y_hat = multilayer_perceptron(x)
correct_prediction = tf.square(y - y_hat)
mse = tf.reduce_mean(tf.cast(correct_prediction, "float"))
train = tf.train.AdamOptimizer(learning_rate= eta).minimize(mse)
init = tf.global_variables_initializer()
```

9. 执行计算图：

```
# Computation Graph
with tf.Session() as sess: # Initialize variables
    sess.run(init) writer = tf.summary.FileWriter('graphs', sess.graph)
# train the model for 100 epcohs
    for i in range(max_epoch):
        _, l, p = sess.run([train, loss, y_hat], feed_dict={x: X_train, y: y_train})
        if i%100 == 0:
            print('Epoch {0}: Loss {1}'.format(i, l))
    print("Training Done")
print("Optimization Finished!")
# Test model correct_prediction = tf.square(y - y_hat)
# Calculate accuracy
accuracy = tf.reduce_mean(tf.cast(correct_prediction, "float"))
print(" Mean Error:", accuracy.eval({x: X_train, y: y_train}))
plt.scatter(y_train, p)
writer.close()
```

解读分析

在只有一个隐藏层的情况下，该模型在训练数据集上预测房价的平均误差为 0.0071。下图显示了房屋估价与实际价格的关系：

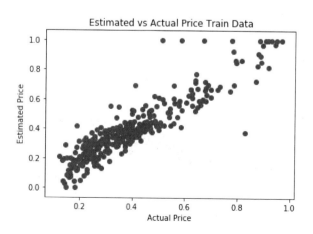

更多内容

在这里,使用 TensorFlow 操作层(Contrib)来构建神经网络层。这使得工作稍微容易一些,因为避免了分别为每层声明权重和偏置。如果使用像 Keras 这样的 API,工作可以进一步简化。下面是 Keras 中以 TensorFlow 作为后端的代码:

```
#Network Parameters
m = len(X_train)
n = 3 # Number of features
n_hidden = 20 # Number of hidden neurons

# Hyperparameters
batch = 20
eta = 0.01
max_epoch = 100
# Build Model
model = Sequential()
model.add(Dense(n_hidden, input_dim=n, activation='relu'))
model.add(Dense(1, activation='sigmoid'))
model.summary()
# Summarize the model
#Compile model
model.compile(loss='mean_squared_error', optimizer='adam')
#Fit the model
model.fit(X_train, y_train, validation_data=(X_test,
y_test),epochs=max_epoch, batch_size=batch, verbose=1)
#Predict the values and calculate RMSE and R2 score
y_test_pred = model.predict(X_test)
y_train_pred = model.predict(X_train)
r2 = r2_score( y_test, y_test_pred )
rmse = mean_squared_error( y_test, y_test_pred )
print( "Performance Metrics R2 : {0:f}, RMSE : {1:f}".format( r2, rmse ) )
```

前面的代码给出了预测值和实际值之间的结果。可以看到,通过去除异常值(一些房屋价格与其他参数无关,比如最右边的点),可以改善结果:

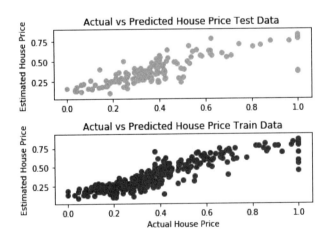

3.7 调整超参数

正如你目前所看到的,神经网络的性能非常依赖超参数。因此,了解这些参数如何影响网络变得至关重要。常见的超参数是学习率、正则化器、正则化系数、隐藏层的维数、初始权重值,甚至选择什么样的优化器优化权重和偏置。

具体做法

1. 调整超参数的第一步是构建模型。与之前一样,在 TensorFlow 中构建模型。
2. 添加一种方法将模型保存在 `model_file` 中。在 TensorFlow 中,可以使用 `Saver` 对象来完成。然后保存在会话中:

```
... saver = tf.train.Saver() ... with tf.Session() as sess: ... #Do
the training steps ... save_path = saver.save(sess,
"/tmp/model.ckpt") print("Model saved in file: %s" % save_path)
```

3. 确定要调整的超参数。
4. 为超参数选择可能的值。在这里,你可以做随机的选择、固定间隔值或手动选择。三者分别称为随机搜索、网格搜索和手动搜索。例如,下面是用来调节学习率的代码:

```
# Random Choice: generate 5 random values of learning rate
# lying between 0 and 1
learning_rate = np.random.rand(5)
#Grid Search: generate 5 values starting from 0, separated by
# 0.2
learning_rate = [i for i in np.arange(0,1,0.2)]
#Manual Search: give any values you seem plausible manually
learning_rate = [0.5, 0.6, 0.32, 0.7, 0.01]
```

5. 选择对损失函数给出最佳响应的参数。所以,可以在开始时将损失函数的最大值定义为 `best_loss`(如果是精度,可以选择将自己期望得到的准确率设为模型的最低精度):

```
best_loss = 2
# It can be any number, but it would be better if you keep it same
as the loss you achieved from your base model defined in steps 1
and 2
```

6. 把你的模型放在 for 循环中,然后保存任何能更好估计损失的模型:

```
... # Load and preprocess data
... # Hyperparameters
Tuning epochs = [50, 60, 70]
batches = [5, 10, 20]
rmse_min = 0.04
for epoch in epochs:
    for batch in batches:
        model = get_model()
        model.compile(loss='mean_squared_error', optimizer='adam')
        model.fit(X_train, y_train, validation_data=(X_test,
y_test),epochs=epoch, batch_size=batch, verbose=1)
        y_test_pred = model.predict(X_test)
```

```
                rmse = mean_squared_error( y_test, y_test_pred )
                if rmse < rmse_min:
                    rmse_min = rmse
                    # serialize model to JSON
                    model_json = model.to_json()
with open("model.json", "w") as json_file:
    json_file.write(model_json)
    # serialize weights to HDF5
    model.save_weights("model.hdf5")
    print("Saved model to disk")
```

更多内容

贝叶斯优化也可以用来调整超参数。其中，用高斯过程定义了一个采集函数。高斯过程使用一组先前评估的参数和得出的精度来假定未观察到的参数。采集函数使用这一信息来推测下一组参数。`https://github.com/lucfra/RFHO` 上有一个包装器用于基于梯度的超参数优化。

拓展阅读

- 对于超参数优化，`https://roamanalytics.com/2016/09/15/optimizing-the-hyperparameter-of-of-which-hyperparameter-optimizer-to-use/` 是对两个优秀的开源软件包 Hyperopt 和 scikit 很好的介绍。
- 关于超参数优化的另一个资源：`http://fastml.com/optimizing-hyperparams-with-hyperopt/`。
- Bengio 和其他人关于超参数优化的各种算法的详细论文：`https://papers.nips.cc/paper/4443-algorithms-for-hyper-parameter-optimization.pdf`

3.8 高级 API——Keras

Keras 是与 TensorFlow 一起使用的更高级别的作为后端的 API。添加层就像添加一行代码一样简单。在模型架构之后，使用一行代码，你可以编译和拟合模型。之后，它可以用于预测。变量声明、占位符甚至会话都由 API 管理。

具体做法

1. 定义模型的类型。Keras 提供了两种类型的模型：序列和模型类 API。Keras 提供各种类型的神经网络层：

```
# Import the model and layers needed
from keras.model import Sequential
```

```
from keras.layers import Dense
model = Sequential()
```

2. 在 `model.add()` 的帮助下将层添加到模型中。依照 Keras 文档描述，Keras 提供全连接层的选项（针对密集连接的神经网络）：

```
layer Dense(units, activation=None, use_bias=True,
kernel_initializer='glorot_uniform', bias_initializer='zeros',
kernel_regularizer=None, bias_regularizer=None,
activity_regularizer=None, kernel_constraint=None,
bias_constraint=None)
```

> 密集层实现的操作：`output=activation(dot(input, kernel) + bias)`，其中 `activation` 是元素激活函数，是作为激活参数传递的，`kernel` 是由该层创建的权重矩阵，`bias` 是由该层创建的偏置向量（仅在 `use_bias` 为 `True` 时适用）。

3. 可以使用它来添加尽可能多的层，每个隐藏层都由前一层提供输入。只需要为第一层指定输入维度：

```
#This will add a fully connected neural network layer with 32
neurons, each taking 13 inputs, and with activation function ReLU
mode.add(Dense(32, input_dim=13, activation='relu')) ))
model.add(10, activation='sigmoid')
```

4. 一旦模型被定义，需要选择一个损失函数和优化器。Keras 提供了多种损失函数（`mean_squared_error`、`mean_absolute_error`、`mean_absolute_percentage_error`、`categorical_crossentropy` 和 优 化 器（`sgd`、`RMSprop`、`Adagrad`、`Adadelta`、`Adam` 等）。损失函数和优化器确定后，可以使用 `compile(self, optimizer, loss, metrics = None, sample_weight_mode = None)` 来配置学习过程：

```
model.compile(optimizer='rmsprop',
        loss='categorical_crossentropy',
        metrics=['accuracy'])
```

5. 使用 `fit` 方法训练模型：

```
model.fit(data, labels, epochs=10, batch_size=32)
```

6. 可以在 `predict` 方法 `predict(self, x, batch_size=32, verbose=0)` 的帮助下进行预测：

```
model.predict(test_data, batch_size=10)
```

更多内容

Keras 提供选项来添加卷积层、池化层、循环层，甚至是局部连接层。每种方法的详细描述在 Keras 的官方文档中可以找到：`https://keras.io/models/sequential/`。

拓展阅读

- McCulloch, Warren S., and Walter Pitts. *A logical calculus of the ideas immanent in nervous activity* The bulletin of mathematical biophysics 5.4 (1943): 115-133. `http://vordenker.de/ggphilosophy/mcculloch_a-logical-calculus.pdf`
- Rosenblatt, Frank (1957), The Perceptron--a perceiving and recognizing automaton. Report 85-460-1, Cornell Aeronautical Laboratory. `https://blogs.umass.edu/brain-wars/files/2016/03/rosenblatt-1957.pdf`
- *The Thinking Machine*, CBS Broadcast `https://www.youtube.com/watch?v=jPHUlQiwD9Y`
- Hornik, Kurt, Maxwell Stinchcombe, and Halbert White. *Multilayer feedforward networks are universal approximators*. Neural networks 2.5 (1989): 359-366.

CHAPTER 4

第 4 章

卷积神经网络

卷积神经网络（CNN 或有时被称为 ConvNet）是很吸引人的。在短时间内，它们变成了一种颠覆性的技术，打破了从文本、视频到语音等多个领域所有最先进的算法，远远超出了其最初在图像处理的应用范围。

4.1 引言

CNN 由许多神经网络层组成。卷积和池化这两种不同类型的层通常是交替的。网络中每个滤波器的深度从左到右增加。最后通常由一个或多个全连接的层组成：

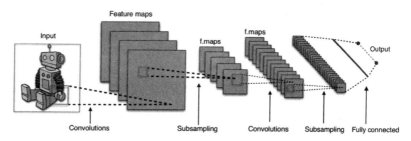

卷积神经网络的一个例子：https://commons.wikimedia.org/wiki/File:Typical_cnn.png

Convnets 背后有三个关键动机：**局部感受野**、**共享权重**和**池化**。让我们一起看一下。

局部感受野

如果想保留图像中的空间信息，那么用像素矩阵表示每个图像是很方便的。然后，编码局部结构的简单方法是将相邻输入神经元的子矩阵连接成属于下一层的单隐藏层神经元。这个单隐藏层神经元代表一个局部感受野。请注意，此操作名为"**卷积**"，此类网络也因此而得名。

当然，可以通过重叠的子矩阵来编码更多的信息。例如，假设每个子矩阵的大小是

5×5，并且将这些子矩阵应用到 28×28 像素的 MNIST 图像。然后，就能够在下一隐藏层中生成 23×23 的局部感受野。事实上，在触及图像的边界之前，只需要滑动子矩阵 23 个位置。

定义从一层到另一层的特征图。当然，可以有多个独立从每个隐藏层学习的特征映射。例如，可以从 28×28 输入神经元开始处理 MNIST 图像，然后（还是以 5×5 的步幅）在下一个隐藏层中得到每个大小为 23×23 的神经元的 k 个特征图。

共享权重和偏置

假设想要从原始像素表示中获得移除与输入图像中位置信息无关的相同特征的能力。一个简单的直觉就是对隐藏层中的所有神经元使用相同的权重和偏置。通过这种方式，每层将从图像中学习到独立于位置信息的潜在特征。

数学例子

理解卷积的一个简单方法是考虑作用于矩阵的滑动窗函数。在下面的例子中，给定输入矩阵 **I** 和核 **K**，得到卷积输出。将 3×3 核 **K**（有时称为**滤波器**或**特征检测器**）与输入矩阵逐元素地相乘以得到输出卷积矩阵中的一个元素。所有其他元素都是通过在 **I** 上滑动窗口获得的：

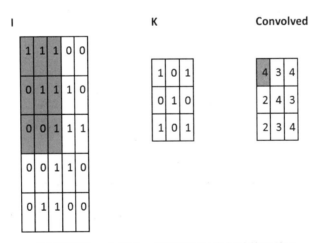

卷积运算的一个例子：用粗体表示参与计算的单元

在这个例子中，一触及 **I** 的边界就停止滑动窗口（所以输出是 3×3）。或者，可以选择用零填充输入（以便输出为 5×5），这是有关**填充**的选择。

另一个选择是关于滑窗所采用的滑动方式的**步幅**。步幅可以是 1 或大于 1。大步幅意味着核的应用更少以及更小的输出尺寸，而小步幅产生更多的输出并保留更多的信息。

滤波器的大小、步幅和填充类型是超参数，可以在训练网络时进行微调。

TensorFlow 中的 ConvNet

在 TensorFlow 中,如果想添加一个卷积层,可以这样写:

```
tf.nn.conv2d(input, filter, strides, padding, use_cudnn_on_gpu=None,
data_format=None, name=None)
```

参数说明如下:
- `input`:张量,必须是 half、`float32`、`float64` 三种类型之一。
- `filter`:张量必须具有与输入相同的类型。
- `strides`:整数列表。长度是 4 的一维向量。输入的每一维度的滑动窗口步幅。必须与指定格式维度的顺序相同。
- `padding`:可选字符串为 SAME、VALID。要使用的填充算法的类型。
- `use_cudnn_on_gpu`:一个可选的布尔值,默认为 `True`。
- `data_format`:可选字符串为 NHWC、NCHW,默认为 NHWC。指定输入和输出数据的数据格式。使用默认格式 NHWC,数据按照以下顺序存储:[batch, in_height, in_width, in_channels]。或者,格式可以是 NCHW,数据存储顺序为:[batch, in_channels, in_height, in_width]。
- `name`:操作的名称(可选)。

下图提供了一个卷积的例子:

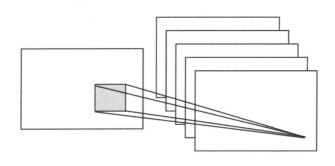

卷积运算的例子

池化层

假设我们要总结一个特征映射的输出。我们可以使用从单个特征映射产生的输出的空间邻接性,并将子矩阵的值聚合成单个输出值,从而合成地描述与该物理区域相关联的含义。

最大池化

一个简单而通用的选择是所谓的**最大池化算子**,它只是输出在区域中观察到的最大输入值。在 TensorFlow 中，如果想要定义一个大小为 2×2 的最大池化层，可以这样写：

```
tf.nn.max_pool(value, ksize, strides, padding, data_format='NHWC',
name=None)
```

参数说明如下：
- `value`：形状为 `[batch, height, width, channels]` 和类型是 `tf.float32` 的四维张量。
- `ksize`：长度 >= 4 的整数列表。输入张量的每个维度的窗口大小。
- `strides`：长度 >= 4 的整数列表。输入张量的每个维度的滑动窗口的步幅。
- `padding`：一个字符串，可以是 `VALID` 或 `SAME`。
- `data_format`：一个字符串，支持 `NHWC` 和 `NCHW`。
- `name`：操作的可选名称。

下图给出了最大池化操作的示例：

池化操作的一个例子

平均池化

另一个选择是平均池化，它简单地将一个区域聚合成在该区域观察到的输入值的平均值。

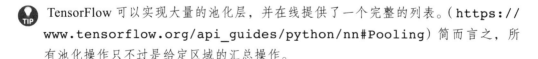

TensorFlow 可以实现大量的池化层，并在线提供了一个完整的列表。（https://www.tensorflow.org/api_guides/python/nn#Pooling）简而言之，所有池化操作只不过是给定区域的汇总操作。

ConvNet 总结

CNN 基本上是几层具有非线性激活函数的卷积，以及将池化层应用于卷积的结果。每层应用不同的滤波器（成百上千个）。理解的关键是滤波器不是预先设定好的，而是在训练阶段学习的，以使得恰当的损失函数被最小化。已经观察到，较低层会学习检测基本特征，而较高层检测更复杂的特征，例如形状或面部。请注意，由于有池化层，靠后的层中的神经元看到的更多的是原始图像，因此它们能够编辑前几层中学习的基本特征。

到目前为止，描述了 ConvNet 的基本概念。CNN 在时间维度上对音频和文本数据进行一维卷积和池化操作，沿（高度 × 宽度）维度对图像进行二维处理，沿（高度 × 宽度 × 时间）维度对视频进行三维处理。对于图像，在输入上滑动滤波器会生成一个特征图，为每个空间位置提供滤波器的响应。

> 换句话说，一个 ConvNet 由多个滤波器堆叠在一起，学习识别在图像中独立于位置信息的具体视觉特征。这些视觉特征在网络的前面几层很简单，然后随着网络的加深，组合成更加复杂的全局特征。

4.2 创建一个 ConvNet 来分类手写 MNIST 数字

在这一节中，你将学习如何创建一个简单的三层卷积网络来预测 MNIST 数字。这个深层网络由两个带有 ReLU 和 maxpool 的卷积层以及两个全连接层组成。

准备工作

MNIST 由 60000 个手写体数字的图片组成。本节的目标是高精度地识别这些数字。

具体做法

1. 导入 `tensorflow`、`matplotlib`、`random` 和 `numpy`。然后，导入 `mnist` 数据集并进行独热编码。请注意，TensorFlow 有一些内置的库来处理 `MNIST`，我们也会用到它们：

```
from __future__ import division, print_function
import tensorflow as tf
import matplotlib.pyplot as plt
import numpy as np
# Import MNIST data
from tensorflow.examples.tutorials.mnist import input_data
mnist = input_data.read_data_sets("MNIST_data/", one_hot=True)
```

2. 仔细观察一些数据有助于理解 MNIST 数据集。了解训练数据集中有多少张图片，测试数据集中有多少张图片。可视化一些数字，以便了解它们是如何表示的。这种输出可以对于识别手写体数字的难度有一种视觉感知，即使是对于人类来说也是如此。

```
def train_size(num):
    print ('Total Training Images in Dataset = ' + str(mnist.train.images.shape))
    print ('----------------------------------------')
    x_train = mnist.train.images[:num,:]
    print ('x_train Examples Loaded = ' + str(x_train.shape))
```

```
    y_train = mnist.train.labels[:num,:]
    print ('y_train Examples Loaded = ' + str(y_train.shape))
    print('')
    return x_train, y_train
def test_size(num):
    print ('Total Test Examples in Dataset = ' + str(mnist.test.images.shape))
    print ('----------------------------------------------')
    x_test = mnist.test.images[:num,:]
    print ('x_test Examples Loaded = ' + str(x_test.shape))
    y_test = mnist.test.labels[:num,:]
    print ('y_test Examples Loaded = ' + str(y_test.shape))
    return x_test, y_test
def display_digit(num):
    print(y_train[num])
    label = y_train[num].argmax(axis=0)
    image = x_train[num].reshape([28,28])
    plt.title('Example: %d  Label: %d' % (num, label))
    plt.imshow(image, cmap=plt.get_cmap('gray_r'))
    plt.show()
def display_mult_flat(start, stop):
    images = x_train[start].reshape([1,784])
    for i in range(start+1,stop):
        images = np.concatenate((images, x_train[i].reshape([1,784])))
    plt.imshow(images, cmap=plt.get_cmap('gray_r'))
    plt.show()
x_train, y_train = train_size(55000)
display_digit(np.random.randint(0, x_train.shape[0]))
display_mult_flat(0,400)
```

上述代码的输出：

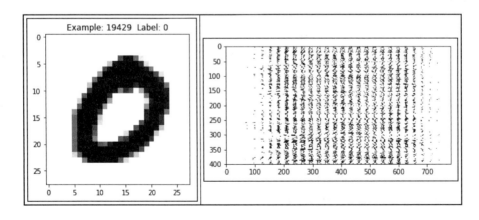

MNIST 手写数字的一个例子

3. 设置学习参数 `batch_size` 和 `display_step`。另外，MNIST 图片都是 28×28 像素，因此设置 `n_input = 784`，`n_classes = 10` 代表输出数字 [0-9]，并且 dropout 概率是 0.85，则：

```
# Parameters
learning_rate = 0.001
training_iters = 500
batch_size = 128
display_step = 10
# Network Parameters
n_input = 784
# MNIST data input (img shape: 28*28)
n_classes = 10
# MNIST total classes (0-9 digits)
dropout = 0.85
# Dropout, probability to keep units
```

4. 设置 TensorFlow 计算图的输入。定义两个占位符来存储预测值和真实标签:

```
x = tf.placeholder(tf.float32, [None, n_input])
y = tf.placeholder(tf.float32, [None, n_classes])
keep_prob = tf.placeholder(tf.float32)
```

5. 定义一个输入为 x, 权值为 W, 偏置为 b, 给定步幅的卷积层。激活函数是 ReLU, padding 设定为 SAME 模式:

```
def conv2d(x, W, b, strides=1):
    x = tf.nn.conv2d(x, W, strides=[1, strides, strides, 1], padding='SAME')
    x = tf.nn.bias_add(x, b)
    return tf.nn.relu(x)
```

6. 定义一个输入是 x 的 maxpool 层, 卷积核为 ksize 并且 padding 为 SAME:

```
def maxpool2d(x, k=2):
    return tf.nn.max_pool(x, ksize=[1, k, k, 1], strides=[1, k, k, 1], padding='SAME')
```

7. 定义 convnet, 其构成是两个卷积层, 然后是全连接层, 一个 dropout 层, 最后是输出层:

```
def conv_net(x, weights, biases, dropout):
    # reshape the input picture
    x = tf.reshape(x, shape=[-1, 28, 28, 1])
    # First convolution layer
    conv1 = conv2d(x, weights['wc1'], biases['bc1'])
    # Max Pooling used for downsampling
    conv1 = maxpool2d(conv1, k=2)
    # Second convolution layer
    conv2 = conv2d(conv1, weights['wc2'], biases['bc2'])
    # Max Pooling used for downsampling
    conv2 = maxpool2d(conv2, k=2)
    # Reshape conv2 output to match the input of fully connected layer
    fc1 = tf.reshape(conv2, [-1, weights['wd1'].get_shape().as_list()[0]])
    # Fully connected layer
fc1 = tf.add(tf.matmul(fc1, weights['wd1']), biases['bd1'])
fc1 = tf.nn.relu(fc1)
# Dropout
```

```
fc1 = tf.nn.dropout(fc1, dropout)
# Output the class prediction
out = tf.add(tf.matmul(fc1, weights['out']), biases['out'])
return out
```

8. 定义网络层的权重和偏置。第一个 conv 层有一个 5×5 的卷积核，1 个输入和 32 个输出。第二个 conv 层有一个 5×5 的卷积核，32 个输入和 64 个输出。全连接层有 7×7×64 个输入和 1024 个输出，而第二层有 1024 个输入和 10 个输出对应于最后的数字数目。所有的权重和偏置用 `randon_normal` 分布完成初始化：

```
weights = {
    # 5x5 conv, 1 input, and 32 outputs
    'wc1': tf.Variable(tf.random_normal([5, 5, 1, 32])),
    # 5x5 conv, 32 inputs, and 64 outputs
    'wc2': tf.Variable(tf.random_normal([5, 5, 32, 64])),
    # fully connected, 7*7*64 inputs, and 1024 outputs
    'wd1': tf.Variable(tf.random_normal([7*7*64, 1024])),
    # 1024 inputs, 10 outputs for class digits
    'out': tf.Variable(tf.random_normal([1024, n_classes]))
}
biases = {
    'bc1': tf.Variable(tf.random_normal([32])),
    'bc2': tf.Variable(tf.random_normal([64])),
    'bd1': tf.Variable(tf.random_normal([1024])),
    'out': tf.Variable(tf.random_normal([n_classes]))
}
```

9. 建立一个给定权重和偏置的 convnet。定义基于 `cross_entropy_with_logits` 的损失函数，并使用 Adam 优化器进行损失最小化。优化后，计算精度：

```
pred = conv_net(x, weights, biases, keep_prob)
cost =
tf.reduce_mean(tf.nn.softmax_cross_entropy_with_logits(logits=pred,
labels=y))
optimizer =
tf.train.AdamOptimizer(learning_rate=learning_rate).minimize(cost)
correct_prediction = tf.equal(tf.argmax(pred, 1), tf.argmax(y, 1))
accuracy = tf.reduce_mean(tf.cast(correct_prediction, tf.float32))
init = tf.global_variables_initializer()
```

10. 启动计算图并迭代 `training_iterats` 次，其中每次输入 `batch_size` 个数据进行优化。请注意，用从 mnist 数据集分离出的 mnist.train 数据进行训练。每进行 `display_step` 次迭代，会计算当前的精度。最后，在 2048 个测试图片上计算精度，此时无 dropout。

```
train_loss = []
train_acc = []
test_acc = []
with tf.Session() as sess:
    sess.run(init)
    step = 1
    while step <= training_iters:
        batch_x, batch_y = mnist.train.next_batch(batch_size)
        sess.run(optimizer, feed_dict={x: batch_x, y: batch_y,
```

```
                                  keep_prob: dropout})
            if step % display_step == 0:
                loss_train, acc_train = sess.run([cost, accuracy],
                                           feed_dict={x: batch_x,
                                                      y: batch_y,
                                                      keep_prob:
1.})
                print "Iter " + str(step) + ", Minibatch Loss= " + \
                    "{:.2f}".format(loss_train) + ", Training
Accuracy= " + \
                    "{:.2f}".format(acc_train)
                # Calculate accuracy for 2048 mnist test images.
                # Note that in this case no dropout
                acc_test = sess.run(accuracy,
                                feed_dict={x: mnist.test.images,
                                           y: mnist.test.labels,
                                           keep_prob: 1.})
                print "Testing Accuracy:" + \
                    "{:.2f}".format(acc_train)
                train_loss.append(loss_train)
                train_acc.append(acc_train)
                test_acc.append(acc_test)
            step += 1
```

11. 画出每次迭代的Softmax损失以及训练和测试的精度:

```
eval_indices = range(0, training_iters, display_step)
# Plot loss over time
plt.plot(eval_indices, train_loss, 'k-')
plt.title('Softmax Loss per iteration')
plt.xlabel('Iteration')
plt.ylabel('Softmax Loss')
plt.show()
# Plot train and test accuracy
plt.plot(eval_indices, train_acc, 'k-', label='Train Set Accuracy')
plt.plot(eval_indices, test_acc, 'r--', label='Test Set Accuracy')
plt.title('Train and Test Accuracy')
plt.xlabel('Generation')
plt.ylabel('Accuracy')
plt.legend(loc='lower right')
plt.show()
```

以下是上述代码的输出。首先看一下每次迭代的Softmax损失:

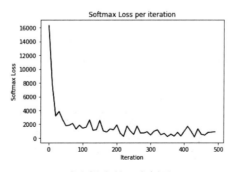

减少损失的一个例子

再来看一下训练和测试的精度：

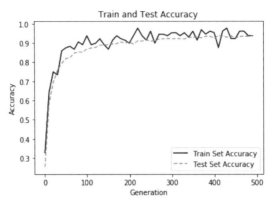

训练和测试精度上升的一个例子

解读分析

使用 ConvNet，在 MNIST 数据集上的表现提高到了近 95% 的精度。ConvNet 的前两层网络由卷积、ReLU 激活函数和最大池化部分组成，然后是两层全连接层（含 dropout）。训练的 batch 大小为 128，使用 Adam 优化器，学习率为 0.001，最大迭代次数为 500 次。

4.3 创建一个 ConvNet 来分类 CIFAR-10 数据集

在这一节中，你将学习如何对 CIFAR-10 中的图片进行分类。CIFAR-10 数据集由 10 类 60000 张 32×32 像素的彩色图片组成，每类有 6000 张图片。有 50000 张训练图片和 10000 张测试图片。下面的图片取自 `https://www.cs.toronto.edu/~kriz/cifar.html`：

CIFAR 图像的例子

准备工作

在这一节,将使用 TFLearn(一个更高层次的框架),它抽象了一些 TensorFlow 的内部细节,能够专注于深度网络的定义。可以在 http://tflearn.org/ 上了解 TFLearn 的信息,这里的代码是标准发布的一部分,网址为 https://github.com/tflearn/tflearn/tree/master/examples。

具体做法

1. 导入几个 `utils` 和核心层用于实现 ConvNet、`dropout`、`fully_connected` 和 `max_pool`。另外,导入一些对图像处理和图像增强有用的模块。请注意,TFLearn 为 ConvNet 提供了一些已定义的更高级别的层,这能够专注于代码的定义:

```
from __future__ import division, print_function, absolute_import
import tflearn
from tflearn.data_utils import shuffle, to_categorical
from tflearn.layers.core import input_data, dropout, fully_connected
from tflearn.layers.conv import conv_2d, max_pool_2d
from tflearn.layers.estimator import regression
from tflearn.data_preprocessing import ImagePreprocessing
from tflearn.data_augmentation import ImageAugmentation
```

2. 加载 CIFAR-10 数据,并将其分为 `X_train` 和 `Y_train`,`X_test` 用于测试,`Y_test` 是测试集的标签。对 `X` 和 `Y` 进行混洗可能是有用的,因为这样能避免训练依赖于特定的数据配置。最后一步是对 `X` 和 `Y` 进行独热编码:

```
# Data loading and preprocessing
from tflearn.datasets import cifar10
(X, Y), (X_test, Y_test) = cifar10.load_data()
X, Y = shuffle(X, Y)
Y = to_categorical(Y, 10)
Y_test = to_categorical(Y_test, 10)
```

3. 使用 `ImagePreprocessing()` 对数据集进行零中心化(即对整个数据集计算平均值),同时进行 STD 标准化(即对整个数据集计算标准差)。TFLearn 数据流旨在通过 CPU 先对数据进行预处理,然后在 GPU 上加速模型训练。

```
# Real-time data preprocessing
img_prep = ImagePreprocessing()
img_prep.add_featurewise_zero_center()
img_prep.add_featurewise_stdnorm()
```

4. 通过随机左右翻转和随机旋转来增强数据集。这一步是一个简单的技巧,用于增加可用于训练的数据:

```
# Real-time data augmentation
img_aug = ImageAugmentation()
img_aug.add_random_flip_leftright()
img_aug.add_random_rotation(max_angle=25.)
```

5. 使用之前定义的图片预处理和图片增强操作创建卷积网络。网络由三个卷积层组成。第一层有 32 个卷积核，尺寸是 3×3，激活函数用 ReLU，这一层后使用 `max_pool` 层用于缩小尺寸。然后是两个卷积核级联，卷积核的个数是 64，尺寸是 3×3，激活函数是 ReLU。之后依次是 `max_pool` 层，具有 512 个神经元、激活函数为 ReLU 的全连接的网络，设置 dropout 概率为 50%。最后一层是全连接层，利用 10 个神经元和激活函数 `softmax` 对 10 个手写数字进行分类。请注意，这种特殊类型的 ConvNet 在 CIFAR-10 中非常有效。其中，使用 Adam 优化器（`categorical_crossentropy`）学习率是 0.001：

```
# Convolutional network building
network = input_data(shape=[None, 32, 32, 3],
                     data_preprocessing=img_prep,
                     data_augmentation=img_aug)
network = conv_2d(network, 32, 3, activation='relu')
network = max_pool_2d(network, 2)
network = conv_2d(network, 64, 3, activation='relu')
network = conv_2d(network, 64, 3, activation='relu')
network = max_pool_2d(network, 2)
network = fully_connected(network, 512, activation='relu')
network = dropout(network, 0.5)
network = fully_connected(network, 10, activation='softmax')
network = regression(network, optimizer='adam',
                     loss='categorical_crossentropy',
                     learning_rate=0.001)
```

6. 实例化 ConvNet 并以 `batch_size = 96` 训练 50 个 epoch：

```
# Train using classifier
model = tflearn.DNN(network, tensorboard_verbose=0)
model.fit(X, Y, n_epoch=50, shuffle=True, validation_set=(X_test, Y_test),
          show_metric=True, batch_size=96, run_id='cifar10_cnn')
```

解读分析

TFLearn 隐藏了许多 TensorFlow 的实现细节，在许多情况下，它可专注于具有更高抽象级别的 ConvNet 的定义。我们的设计在 50 次迭代中达到了 88% 的准确度。以下图片是在 Jupyter notebook 中执行的快照：

```
In [4]:  # Convolutional network building
         network = input_data(shape=[None, 32, 32, 3],
                              data_preprocessing=img_prep,
                              data_augmentation=img_aug)
         network = conv_2d(network, 32, 3, activation='relu')
         network = max_pool_2d(network, 2)
         network = conv_2d(network, 64, 3, activation='relu')
         network = conv_2d(network, 64, 3, activation='relu')
         network = max_pool_2d(network, 2)
         network = fully_connected(network, 512, activation='relu')
         network = dropout(network, 0.5)
         network = fully_connected(network, 10, activation='softmax')
         network = regression(network, optimizer='adam',
                              loss='categorical_crossentropy',
                              learning_rate=0.001)
```

```
In [5]: # Train using classifier
        model = tflearn.DNN(network, tensorboard_verbose=0)
        model.fit(X, Y, n_epoch=50, shuffle=True, validation_set=(X_test, Y_test),
                show_metric=True, batch_size=96, run_id='cifar10_cnn')

Training Step: 26049 | total loss: 0.32852 | time: 119.623s
| Adam | epoch: 050 | loss: 0.32852 - acc: 0.8853 -- iter: 49920/50000
Training Step: 26050 | total loss: 0.32454 | time: 127.685s
| Adam | epoch: 050 | loss: 0.32454 - acc: 0.8853 | val_loss: 0.64020 - v
al_acc: 0.8192 -- iter: 50000/50000
--
```

Jupyter 执行 CIFAR10 分类的一个例子

更多内容

要安装 TFLearn，请参阅"安装指南"（`http://tflearn.org/installation`），如果你想查看更多示例，可以在网上找到已经很成熟的解决方案列表（`http://tflearn.org/examples/`）。

4.4 用 VGG19 做风格迁移的图像重绘

在这一节中，你会教计算机如何绘画。关键的想法是有一个神经网络从模型图像推断其绘画风格。然后将这种风格转移到另一张重新绘制的图片上。本节是对用 log0 开发的代码的改进，可在网址 `https://github.com/log0/neural-style-painting/blob/master/TensorFlow%20Implementation%20of%20A%20Neural%20Algorithm%20of%20Artistic%20Sty le.ipynb` 上获得相关代码。

准备工作

我们要实现 Leon A. Gatys、Alexander S. Ecker 和 Matthias Bethge 在论文 *A Neural Algorithm of Artistic Style*（`https://arxiv.org/abs/1508.06576`）中描述的算法。因此，最好先读一下这篇论文。这次还会使用预训练好的模型 VGG19（`http://www.vlfeat.org/matconvnet/models/beta16/imagenet-vgg-verydeep-19.mat`），应将其下载到本地。风格图片是梵高的一幅名画（`https://commons.wikimedia.org/wiki/File:VanGogh-starry_night.jpg`），内容图片是从维基百科（`https://commons.wikimedia.org/wiki/File:Marilyn_Monroe_in_1952.jpg`）下载的玛丽莲·梦露的图片。内容图片将根据梵高风格重新绘制。

具体做法

1. 导入一些模块，如 `numpy`、`scipy`、`tensorflow` 和 `matplotlib`。然后导入 `PIL` 来处理图片。注意，由于此代码在 Jupyter notebook 上运行，你可以在线下载，片段 `%matplotlib inline` 已经被添加进去：

```
import os
import sys
import numpy as np
import scipy.io
import scipy.misc
import tensorflow as tf
import matplotlib.pyplot as plt
from matplotlib.pyplot
import imshow
from PIL
import Image %matplotlib inline from __future__
import division
```

2. 设置用于学习的风格图像和要重绘的内容图像的输入路径：

```
OUTPUT_DIR = 'output/'
# Style image
STYLE_IMAGE = 'data/StarryNight.jpg'
# Content image to be repainted
CONTENT_IMAGE = 'data/Marilyn_Monroe_in_1952.jpg'
```

3. 设置生成图像的噪声比，将重点放在重绘内容图像时的内容损失和风格损失上。除此之外，存储了预训练的 VGG 模型的路径以及在 VGG 预训练期间计算的平均值，这个平均值是已知的，并从输入图像减去该均值之后再输入到 VGG 模型：

```
# how much noise is in the image
NOISE_RATIO = 0.6
# How much emphasis on content loss.
BETA = 5
# How much emphasis on style loss.
ALPHA = 100
# the VGG 19-layer pre-trained model
VGG_MODEL = 'data/imagenet-vgg-verydeep-19.mat'
# The mean used when the VGG was trained
# It is subtracted from the input to the VGG model. MEAN_VALUES =
np.array([123.68, 116.779, 103.939]).reshape((1,1,1,3))
```

4. 显示内容图片只是为了了解它是什么样子：

```
content_image = scipy.misc.imread(CONTENT_IMAGE)
imshow(content_image)
```

这里是前面代码的输出（这个图片的地址是 https://commons.wikimedia.org/wiki/File:Marilyn_Monroe_in_1952.jpg）：

5. 调整风格图像的大小，显示它也是为了了解它是什么样子。请注意，内容图像和风格图像现在具有相同的尺寸和相同数量的颜色通道：

```
style_image = scipy.misc.imread(STYLE_IMAGE)
# Get shape of target and make the style image the same
target_shape = content_image.shape
print "target_shape=", target_shape
print "style_shape=", style_image.shape
#ratio = target_shape[1] / style_image.shape[1]
#print "resize ratio=", ratio
style_image = scipy.misc.imresize(style_image, target_shape)
scipy.misc.imsave(STYLE_IMAGE, style_image)
imshow(style_image)
```

下图是前面代码的输出：

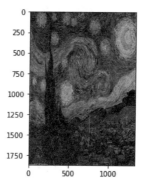

梵高绘画的一个例子，见 https://commons.wikimedia.org/wiki/File:VanGogh-starry_night_ballance1.jpg

6. 根据论文中的描述定义 VGG 模型。注意深度学习网络是相当复杂的，因为它将多个 ConvNet 层与 ReLU 激活函数和最大池化结合在一起。另外值得注意的是，在风格迁移的论文中的许多实验表明，平均池化的表现要好于最大池化。因此，这里使用平均池化：

```
def load_vgg_model(path, image_height, image_width,
color_channels):
    """
    Returns the VGG model as defined in the paper
        0 is conv1_1 (3, 3, 3, 64)
        1 is relu
        2 is conv1_2 (3, 3, 64, 64)
        3 is relu
        4 is maxpool
        5 is conv2_1 (3, 3, 64, 128)
        6 is relu
        7 is conv2_2 (3, 3, 128, 128)
        8 is relu
        9 is maxpool
        10 is conv3_1 (3, 3, 128, 256)
        11 is relu
        12 is conv3_2 (3, 3, 256, 256)
        13 is relu
```

```
            14 is conv3_3  (3, 3, 256, 256)
            15 is relu
            16 is conv3_4  (3, 3, 256, 256)
            17 is relu
            18 is maxpool
            19 is conv4_1  (3, 3, 256, 512)
            20 is relu
            21 is conv4_2  (3, 3, 512, 512)
            22 is relu
            23 is conv4_3  (3, 3, 512, 512)
            24 is relu
            25 is conv4_4  (3, 3, 512, 512)
            26 is relu
            27 is maxpool
            28 is conv5_1  (3, 3, 512, 512)
            29 is relu
            30 is conv5_2  (3, 3, 512, 512)
            31 is relu
            32 is conv5_3  (3, 3, 512, 512)
            33 is relu
            34 is conv5_4  (3, 3, 512, 512)
            35 is relu
            36 is maxpool
            37 is fullyconnected (7, 7, 512, 4096)          38 is relu
            39 is fullyconnected (1, 1, 4096, 4096)
            40 is relu
            41 is fullyconnected (1, 1, 4096, 1000)
            42 is softmax
    """
    vgg = scipy.io.loadmat(path)
    vgg_layers = vgg['layers']

    def _weights(layer, expected_layer_name):
        """        Return the weights and bias from the VGG model for a given layer.
"""
        W = vgg_layers[0][layer][0][0][0][0][0]
        b = vgg_layers[0][layer][0][0][0][0][1]
        layer_name = vgg_layers[0][layer][0][0][-2]
        assert layer_name == expected_layer_name
        return W, b
    def _relu(conv2d_layer):
        """
        Return the RELU function wrapped over a TensorFlow layer. Expects a
        Conv2d layer input.
        """
        return tf.nn.relu(conv2d_layer)

    def _conv2d(prev_layer, layer, layer_name):
        """
        Return the Conv2D layer using the weights, biases from the VGG
        model at 'layer'.
        """
        W, b = _weights(layer, layer_name)
        W = tf.constant(W)
```

```python
            b = tf.constant(np.reshape(b, (b.size))))
        return tf.nn.conv2d(
            prev_layer, filter=W, strides=[1, 1, 1, 1],
padding='SAME') + b

    def _conv2d_relu(prev_layer, layer, layer_name):
        """
        Return the Conv2D + RELU layer using the weights, biases
from the VGG
        model at 'layer'.
        """
        return _relu(_conv2d(prev_layer, layer, layer_name))

    def _avgpool(prev_layer):
        """
        Return the AveragePooling layer.
        """
        return tf.nn.avg_pool(prev_layer, ksize=[1, 2, 2, 1],
strides=[1, 2, 2, 1], padding='SAME')

    # Constructs the graph model.
    graph = {}
    graph['input']    = tf.Variable(np.zeros((1,
                                    image_height,
image_width, color_channels)),
                                    dtype = 'float32')
    graph['conv1_1']  = _conv2d_relu(graph['input'], 0, 'conv1_1')
    graph['conv1_2']  = _conv2d_relu(graph['conv1_1'], 2, 'conv1_2')
    graph['avgpool1'] = _avgpool(graph['conv1_2'])
    graph['conv2_1']  = _conv2d_relu(graph['avgpool1'], 5,
'conv2_1')
    graph['conv2_2']  = _conv2d_relu(graph['conv2_1'], 7, 'conv2_2')
    graph['avgpool2'] = _avgpool(graph['conv2_2'])
    graph['conv3_1']  = _conv2d_relu(graph['avgpool2'], 10,
'conv3_1')
    graph['conv3_2']  = _conv2d_relu(graph['conv3_1'], 12,
'conv3_2')
    graph['conv3_3']  = _conv2d_relu(graph['conv3_2'], 14,
'conv3_3')
    graph['conv3_4']  = _conv2d_relu(graph['conv3_3'], 16,
'conv3_4')
    graph['avgpool3'] = _avgpool(graph['conv3_4'])
    graph['conv4_1']  = _conv2d_relu(graph['avgpool3'], 19,
'conv4_1')
    graph['conv4_2']  = _conv2d_relu(graph['conv4_1'], 21,
'conv4_2')
    graph['conv4_3']  = _conv2d_relu(graph['conv4_2'], 23,
'conv4_3')
    graph['conv4_4']  = _conv2d_relu(graph['conv4_3'], 25,
'conv4_4')
    graph['avgpool4'] = _avgpool(graph['conv4_4'])
    graph['conv5_1']  = _conv2d_relu(graph['avgpool4'], 28,
'conv5_1')
    graph['conv5_2']  = _conv2d_relu(graph['conv5_1'], 30,
'conv5_2')
    graph['conv5_3']  = _conv2d_relu(graph['conv5_2'], 32,
'conv5_3')
```

```
    graph['conv5_4']  = _conv2d_relu(graph['conv5_3'], 34,
'conv5_4')
    graph['avgpool5'] = _avgpool(graph['conv5_4'])
    return graph
```

7. 根据论文描述定义内容损失函数：

```
def content_loss_func(sess, model):
""" Content loss function as defined in the paper. """

def _content_loss(p, x):
# N is the number of filters (at layer l).
N = p.shape[3]
# M is the height times the width of the feature map (at layer l).
M = p.shape[1] * p.shape[2] return (1 / (4 * N * M)) *
tf.reduce_sum(tf.pow(x - p, 2))
return _content_loss(sess.run(model['conv4_2']), model['conv4_2'])
```

8. 定义那些将要重新使用的VGG层。如果想要更平滑的特征，需要增加更高层的权重（`conv5_1`），并减少低层（`conv1_1`）的权重。如果想要提取更尖锐的特征，需要反向操作：

```
STYLE_LAYERS = [
('conv1_1', 0.5),
('conv2_1', 1.0),
('conv3_1', 1.5),
('conv4_1', 3.0),
('conv5_1', 4.0),
]
```

9. 按照论文原文定义风格损失函数：

```
def style_loss_func(sess, model):
    """
    Style loss function as defined in the paper.
    """

    def _gram_matrix(F, N, M):
        """
        The gram matrix G.
        """
        Ft = tf.reshape(F, (M, N))
        return tf.matmul(tf.transpose(Ft), Ft)

    def _style_loss(a, x):
        """
        The style loss calculation.
        """
        # N is the number of filters (at layer l).
        N = a.shape[3]
        # M is the height times the width of the feature map (at layer l).
        M = a.shape[1] * a.shape[2]
        # A is the style representation of the original image (at layer l).
        A = _gram_matrix(a, N, M)
        # G is the style representation of the generated image (at layer l).
        G = _gram_matrix(x, N, M)
        result = (1 / (4 * N**2 * M**2)) * tf.reduce_sum(tf.pow(G - A, 2))
```

```
        return result
    E = [_style_loss(sess.run(model[layer_name]), model[layer_name])
        for layer_name, _ in STYLE_LAYERS]
    W = [w for _, w in STYLE_LAYERS]
    loss = sum([W[l] * E[l] for l in range(len(STYLE_LAYERS))])
    return loss
```

10.定义一个函数来生成噪声图像,并将其按给定的比例与内容图像混合。定义两个预处理和保存图像的辅助方法:

```
def generate_noise_image(content_image, noise_ratio = NOISE_RATIO):
    """ Returns a noise image intermixed with the content image at a certain
    ratio.
    """
    noise_image = np.random.uniform(
            -20, 20,
            (1,
             content_image[0].shape[0],
             content_image[0].shape[1],
             content_image[0].shape[2])).astype('float32')
    # White noise image from the content representation. Take a weighted average
    # of the values
    input_image = noise_image * noise_ratio + content_image * (1 - noise_ratio)
    return input_image

def process_image(image):
    # Resize the image for convnet input, there is no change but just
    # add an extra dimension.
    image = np.reshape(image, ((1,) + image.shape))
    # Input to the VGG model expects the mean to be subtracted.
    image = image - MEAN_VALUES
    return image

def save_image(path, image):
    # Output should add back the mean.
    image = image + MEAN_VALUES
    # Get rid of the first useless dimension, what remains is the image.
    image = image[0]
    image = np.clip(image, 0, 255).astype('uint8')
    scipy.misc.imsave(path, image)
```

11.开始一个TensorFlow交互式会话:

```
sess = tf.InteractiveSession()
```

12.加载已处理的内容图像并显示它:

```
content_image = load_image(CONTENT_IMAGE) imshow(content_image[0])
```

得到前面代码的输出如下(注意使用了 https://commons.wikimedia.org/wiki/File:Marilyn_Monroe_ in_1952.jpg 中的图片):

13. 加载已处理的风格图像并显示它:

```
style_image = load_image(STYLE_IMAGE) imshow(style_image[0])
```

输出如下:

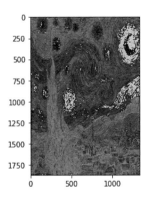

14. 加载模型并显示它:

```
model = load_vgg_model(VGG_MODEL, style_image[0].shape[0],
style_image[0].shape[1], style_image[0].shape[2]) print(model)
```

15. 生成用于引导重绘的随机噪声图像:

```
input_image = generate_noise_image(content_image)
imshow(input_image[0])
```

16. 运行 TensorFlow 会话:

```
sess.run(tf.initialize_all_variables())
```

17. 用相应的图像构建 content_loss 和 sytle_loss:

```
# Construct content_loss using content_image.
sess.run(model['input'].assign(content_image))
content_loss = content_loss_func(sess, model)
# Construct style_loss using style_image.
sess.run(model['input'].assign(style_image))
style_loss = style_loss_func(sess, model)
```

18. 内容损失和风格损失的加权组合作为总的损失：

```
# Construct total_loss as weighted combination of content_loss and
sytle_loss
total_loss = BETA * content_loss + ALPHA * style_loss
```

19. 建立一个优化器来减少总损失。在这里，采用 Adam 优化器：

```
# The content is built from one layer, while the style is from five
# layers. Then we minimize the total_loss
optimizer = tf.train.AdamOptimizer(2.0)
train_step = optimizer.minimize(total_loss)
```

20. 用输入图像引导网络：

```
sess.run(tf.initialize_all_variables())
sess.run(model['input'].assign(input_image))
```

21. 运行模型进行固定次数的迭代，并生成中间重绘的图像：

```
sess.run(tf.initialize_all_variables())
sess.run(model['input'].assign(input_image))
print "started iteration"
for it in range(ITERATIONS):
   sess.run(train_step)
   print it , " "
   if it%100 == 0:
       # Print every 100 iteration.
       mixed_image = sess.run(model['input'])
       print('Iteration %d' % (it))
       print('sum : ',
sess.run(tf.reduce_sum(mixed_image)))
       print('cost: ', sess.run(total_loss))
       if not os.path.exists(OUTPUT_DIR):
           os.mkdir(OUTPUT_DIR)
       filename = 'output/%d.png' % (it)
       save_image(filename, mixed_image)
```

22. 下图分别展示了在 200 次、400 次和 600 次迭代之后重绘出来的内容图片：

风格迁移示例

解读分析

在本节中,我们已经看到了如何使用风格迁移重新绘制内容图像。提供一张风格图像作为神经网络的输入,然后网络学习画家采用的关键风格信息,并将这些风格信息迁移至当前的内容图像。

更多内容

自 2015 年这一方法被提出以来,风格迁移一直是一个活跃的研究领域。已经提出了许多新的想法来加速计算,并将风格迁移延伸到视频分析。其中有两个工作值得一提,Logan Engstrom 提出的快速风格迁移 `https://github.com/lengstrom/fast-style-transfer/`,提出了一种快速的实现方法,同时也适用于视频。

网站 `https://deepart.io` 允许你用喜欢的艺术家的风格重绘自己的图片。在 Android App、iPhone App 和 Web 上都可以使用。

4.5 使用预训练的 VGG16 网络进行迁移学习

本节讨论迁移学习,它是一个非常强大的深度学习技术,在不同领域有很多应用。动机很简单,可以打个比方来解释。假设你想学习一种新的语言,比如西班牙语,那么从你已经掌握的另一种语言(比如英语)学起,可能是有用的。

按照这种思路,计算机视觉研究人员通常使用预训练 CNN 来生成新任务的表示,其中数据集可能不够大,无法从头开始训练整个 CNN。另一个常见的策略是采用在 ImageNet 上预训练好的网络,然后通过微调整个网络来适应新任务。这里提出的例子受启于 Francois Chollet 写的关于 Keras 的一个非常有名的博客(`https://blog.keras.io/building-powerful-image-classification-models-using-very-little-data.html`)。

准备工作

这个想法是使用在像 ImageNet 这样的大型数据集上预先训练的 VGG16 网络。注意,训练的计算量可能相当大,因此使用已经预训练的网络是有意义的:

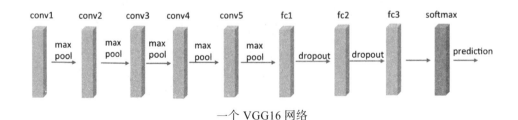

一个 VGG16 网络

那么，如何使用VGG16呢？Keras使其变得容易，因为有一个标准的VGG16模型可以作为一个库来使用，预先计算好的权重会自动下载。请注意，这里省略了最后一层，并将其替换为自定义层，该层将在预定义的VGG16的顶部进行微调。在这个例子中，你将学习如何分类Kaggle提供的狗和猫的图片。

具体做法

1. 从Kaggle（https://www.kaggle.com/c/dogs-vs-cats/data）下载狗和猫的数据，并创建一个包含两个子目录（train和validation）的数据目录，每个子目录有两个额外的子目录，分别是dogs和cats。

2. 导入稍后将用于计算的Keras模块，并保存一些有用的常量：

```
from keras import applications
from keras.preprocessing.image import ImageDataGenerator
from keras import optimizers
from keras.models import Sequential, Model
from keras.layers import Dropout, Flatten, Dense
from keras import optimizers
img_width, img_height = 256, 256
batch_size = 16
epochs = 50
train_data_dir = 'data/dogs_and_cats/train'
validation_data_dir = 'data/dogs_and_cats/validation'
#OUT CATEGORIES
OUT_CATEGORIES=1
#number of train, validation samples
nb_train_samples = 2000
nb_validation_samples =
```

3. 加载ImageNet上预训练的VGG16网络，省略最后一层，因为这里将在预建的VGG16网络的顶部添加自定义分类网络，并替换原来VGG16的分类层：

```
# load the VGG16 model pretrained on imagenet
base_model = applications.VGG16(weights = "imagenet",
include_top=False, input_shape = (img_width, img_height, 3))
base_model.summary()
```

上述代码的输出如下：

Layer (type)	Output Shape	Param #
input_1 (InputLayer)	(None, 256, 256, 3)	0
block1_conv1 (Conv2D)	(None, 256, 256, 64)	1792
block1_conv2 (Conv2D)	(None, 256, 256, 64)	36928
block1_pool (MaxPooling2D)	(None, 128, 128, 64)	0
block2_conv1 (Conv2D)	(None, 128, 128, 128)	73856
block2_conv2 (Conv2D)	(None, 128, 128, 128)	147584

```
block2_pool (MaxPooling2D)    (None, 64, 64, 128)    0
block3_conv1 (Conv2D)         (None, 64, 64, 256)    295168
block3_conv2 (Conv2D)         (None, 64, 64, 256)    590080
block3_conv3 (Conv2D)         (None, 64, 64, 256)    590080
block3_pool (MaxPooling2D)    (None, 32, 32, 256)    0
block4_conv1 (Conv2D)         (None, 32, 32, 512)    1180160
block4_conv2 (Conv2D)         (None, 32, 32, 512)    2359808
block4_conv3 (Conv2D)         (None, 32, 32, 512)    2359808
block4_pool (MaxPooling2D)    (None, 16, 16, 512)    0
block5_conv1 (Conv2D)         (None, 16, 16, 512)    2359808
block5_conv2 (Conv2D)         (None, 16, 16, 512)    2359808
block5_conv3 (Conv2D)         (None, 16, 16, 512)    2359808
block5_pool (MaxPooling2D)    (None, 8, 8, 512)      0
=================================================================
Total params: 14,714,688
Trainable params: 14,714,688
Non-trainable params: 0
```

4. 冻结预训练的 VGG16 网络的一定数量的较低层。在这里决定冻结最前面的 15 层：

```
# Freeze the 15 lower layers for layer in base_model.layers[:15]:
layer.trainable = False
```

5. 为了分类，添加一组自定义的顶层：

```
# Add custom to layers # build a classifier model to put on top of
the convolutional model top_model = Sequential()
top_model.add(Flatten(input_shape=base_model.output_shape[1:]))
top_model.add(Dense(256, activation='relu'))
top_model.add(Dropout(0.5)) top_model.add(Dense(OUT_CATEGORIES,
activation='sigmoid'))
```

6. 自定义网络应该单独进行预训练，为了简单起见，这里省略了这部分，将此任务交给读者：

```
#top_model.load_weights(top_model_weights_path)
```

7. 创建一个新的网络，这是预训练的 VGG16 网络和预训练的定制网络的组合体：

```
# creating the final model, a composition of
# pre-trained and
model = Model(inputs=base_model.input,
outputs=top_model(base_model.output))
# compile the model
model.compile(loss = "binary_crossentropy", optimizer =
optimizers.SGD(lr=0.0001, momentum=0.9), metrics=["accuracy"])
```

8. 重新训练组合的新模型，仍然保持 VGG16 的 15 个最低层处于冻结状态。在这个特定的例子中，也使用 Image Augumentator 来增强训练集：

```
# Initiate the train and test generators with data Augumentation
train_datagen = ImageDataGenerator(
rescale = 1./255,
horizontal_flip = True)
test_datagen = ImageDataGenerator(rescale=1. / 255)
train_generator = train_datagen.flow_from_directory(
    train_data_dir,
    target_size=(img_height, img_width),
    batch_size=batch_size,
    class_mode='binary')
validation_generator = test_datagen.flow_from_directory(
    validation_data_dir,
    target_size=(img_height, img_width),
    batch_size=batch_size,
    class_mode='binary', shuffle=False)
model.fit_generator(
    train_generator,
    steps_per_epoch=nb_train_samples // batch_size,
    epochs=epochs,
    validation_data=validation_generator,
    validation_steps=nb_validation_samples // batch_size,
    verbose=2, workers=12)
```

9. 在组合网络上评估结果：

```
score = model.evaluate_generator(validation_generator,
nb_validation_samples/batch_size)
scores = model.predict_generator(validation_generator,
nb_validation_samples/batch_size)
```

解读分析

一个标准的 VGG16 网络已经在整个 ImageNet 上进行了预训练，并且使用了预先计算好的从网上下载的权值。这个网络和一个已经被单独训练的定制网络并置在一起。然后，并置的网络作为一个整体被重新训练，同时保持 VGG16 的 15 个低层的参数不变。

这个组合非常有效。它可以节省大量的计算能力，重新利用已经工作的 VGG16 网络进行迁移学习，该网络已经在 ImageNet 上完成了学习，可以将此学习应用到新的特定领域，通过微调去完成分类任务。

更多内容

根据具体的分类任务，有几条经验法则需要考虑：

- 如果新的数据集很小，并且与 ImageNet 数据集相似，那么可以冻结所有的 VGG16 网络并仅重新训练定制网络。这样，也可以最小化组合网络过度拟合的风险。
可运行代码 `base_model.layers: layer.trainable = False` 冻结所有低层参数。
- 如果新数据集很大并且与 ImageNet 数据集相似，那么可以重新训练整个并置网络。仍然保持预先计算的权重作为训练起点，并通过几次迭代进行微调：

可运行代码model.layers: layer.trainable = True取消冻结所有低层的参数。
- 如果新数据集与 ImageNet 数据集有很大的不同，实际上仍然可以使用预训练模型的权值进行初始化。在这种情况下，将有足够的数据和信心通过整个网络进行微调。更多信息请访问 http://cs231n.github.io/transfer-learning/。

4.6 创建 DeepDream 网络

Google 于 2014 年在 ImageNet 大型视觉识别竞赛（ILSVRC）训练了一个神经网络，并于 2015 年 7 月开放源代码。其算法可参考论文 *Going Deeper with Convolutions*（由 Christian Szegedy、Wei Liu、Yangqing Jia、Pierre Sermanet、Scott Reed、Dragomir Anguelov、Dumitru Erhan、Vincent Vanhoucke e Andrew Rabinovich 撰写，网址为 https://arxiv.org/abs/1409.4842）。该网络学习了每张图片的表示。低层学习低级特征，比如线条和边缘，而高层学习更复杂的模式，比如眼睛、鼻子、嘴巴等。因此，如果试图在网络中表示更高层次的特征，我们会看到从原始 ImageNet 中提取的不同特征的组合，例如鸟的眼睛和狗的嘴巴。考虑到这一点，如果拍摄一张新的图片，并尝试最大化与网络高层的相似性，那么结果会得到一张新的视觉体验的图片。在这张新视觉体验的图片中，由高层学习的一些模式如同是原始图像的梦境一般。下图是一张想象图片的例子：

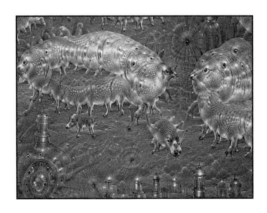

Google Deep Dreams 的示例（https://commons.wikimedia.org/wiki/File:Aurelia-aurita-3-0009.jpg）

准备工作

从网上下载预训练的 Inception 模型（https://github.com/martinwicke/tensorflow-tutorial/blob/master/tensorflow_inception_graph.pb）。

具体做法

1. 导入 `numpy` 进行数值计算，`functools` 定义一个或多个参数已经填充的偏函数，Pillow 用于图像处理，`matplotlib` 用于产生图像：

```
import numpy as np from functools
import partial import PIL.Image
import tensorflow as tf
import matplotlib.pyplot as plt
```

2. 设置内容图像和预训练模型的路径。从随机噪声的种子图像开始：

```
content_image = 'data/gulli.jpg'
# start with a gray image with a little noise
img_noise = np.random.uniform(size=(224,224,3)) + 100.0
model_fn = 'data/tensorflow_inception_graph.pb'
```

3. 以 graph 的形式加载从网上下载的 Inception 网络。初始化一个 TensorFlow 会话，用 `FastGFile(..)` 加载这个 graph，并用 `ParseFromstring(..)` 解析该 graph。之后，使用 `placeholder(..)` 方法创建一个占位符作为输入。`imagenet_mean` 是预先计算的常数，这里的内容图像减去该值以实现数据标准化。事实上，这是训练得到的平均值，规范化使得收敛更快。该值将从输入中减去并存储在 `t_preprocessed` 变量中，然后用于加载 graph 定义：

```
# load the graph
graph = tf.Graph()
sess = tf.InteractiveSession(graph=graph)
with tf.gfile.FastGFile(model_fn, 'rb') as f:
        graph_def = tf.GraphDef()
        graph_def.ParseFromString(f.read())
t_input = tf.placeholder(np.float32, name='input') # define the input tensor
imagenet_mean = 117.0
t_preprocessed = tf.expand_dims(t_input-imagenet_mean, 0)
tf.import_graph_def(graph_def, {'input':t_preprocessed})
```

4. 定义一些 util 函数来可视化图像，并将 TF-graph 生成函数转换为常规 Python 函数（请参阅下面的示例）：

```
# helper
#pylint: disable=unused-variable
def showarray(a):
    a = np.uint8(np.clip(a, 0, 1)*255)
    plt.imshow(a)
    plt.show()
def visstd(a, s=0.1):
    '''Normalize the image range for visualization'''
    return (a-a.mean())/max(a.std(), 1e-4)*s + 0.5

def T(layer):
    '''Helper for getting layer output tensor'''
    return graph.get_tensor_by_name("import/%s:0"%layer)
```

```
def tffunc(*argtypes):
    '''Helper that transforms TF-graph generating function into a regular
one.
    See "resize" function below.
    '''
    placeholders = list(map(tf.placeholder, argtypes))
    def wrap(f):
        out = f(*placeholders)
        def wrapper(*args, **kw):
            return out.eval(dict(zip(placeholders, args)),
session=kw.get('session'))
        return wrapper
    return wrap

def resize(img, size):
    img = tf.expand_dims(img, 0)
    return tf.image.resize_bilinear(img, size)[0,:,:,:]
resize = tffunc(np.float32, np.int32)(resize)
```

5. 计算图像的梯度上升值。为了提高效率，应用平铺计算，其中在不同的图块上计算单独的梯度上升。通过多次迭代对图像应用随机偏移以模糊图块的边界：

```
def calc_grad_tiled(img, t_grad, tile_size=512):
    '''Compute the value of tensor t_grad over the image in a tiled way.
    Random shifts are applied to the image to blur tile boundaries over
    multiple iterations.'''
    sz = tile_size
    h, w = img.shape[:2]
    sx, sy = np.random.randint(sz, size=2)
    img_shift = np.roll(np.roll(img, sx, 1), sy, 0)
    grad = np.zeros_like(img)
    for y in range(0, max(h-sz//2, sz),sz):
        for x in range(0, max(w-sz//2, sz),sz):
            sub = img_shift[y:y+sz,x:x+sz]
            g = sess.run(t_grad, {t_input:sub})
            grad[y:y+sz,x:x+sz] = g

    return np.roll(np.roll(grad, -sx, 1), -sy, 0)
```

6. 定义用来减少输入层均值的优化对象。通过考虑输入张量，该梯度函数可以计算优化张量的符号梯度。为了提高效率，图像被分割成几块，然后调整大小并添加到块数组中。对于每个块，使用 `calc_grad_tiled` 函数：

```
def render_deepdream(t_obj, img0=img_noise,
                     iter_n=10, step=1.5, octave_n=4, octave_scale=1.4):
    t_score = tf.reduce_mean(t_obj) # defining the optimization objective
    t_grad = tf.gradients(t_score, t_input)[0] # behold the power of automatic differentiation!
    # split the image into a number of octaves
    img = img0
    octaves = []
    for _ in range(octave_n-1):
        hw = img.shape[:2]
        lo = resize(img, np.int32(np.float32(hw)/octave_scale))
        hi = img-resize(lo, hw)
```

```
            img = lo
            octaves.append(hi)
    # generate details octave by octave
        for octave in range(octave_n):
            if octave>0:
                hi = octaves[-octave]
                img = resize(img, hi.shape[:2])+hi
            for _ in range(iter_n):
                g = calc_grad_tiled(img, t_grad)
                img += g*(step / (np.abs(g).mean()+1e-7))
                #this will usually be like 3 or 4 octaves
                #Step 5 output deep dream image via matplotlib
            showarray(img/255.0)
```

7. 加载特定的内容图像，并开始想象。在这个例子中，作者的脸被转化成类似于狼的模样：

深度转换的例子，其中一个作者变成了狼

解读分析

神经网络存储训练图像的抽象描述：较低层存储线条和边缘等特征，较高层存储较复杂的图像特征，如眼睛、脸部和鼻子。通过应用梯度上升过程，这里使损失函数最大化并促使发现类似于由较高层记忆的图案的内容图片模式。这样网络就生成了令人致幻的图片。

更多内容

许多网站允许直接尝试 DeepDreaming。其中，我非常喜欢 DeepArt.io（https://deepar.tio/），它允许你上传内容图像和风格图像，并在云上进行学习。

拓展阅读

在 2015 年提出初步结果之后，又发表了许多关于 DeepDreaming 的论文和博客文章：

- *DeepDream: A code example to visualize Neural Networksh*（`https://research.googleblog.com/2015/07/deepdream-code-example-for-visualizing.html`）。
- 机器人何时产生幻觉（When Robots Hallucinate），LaFrance 和 Adrienne（`https://www.theatlantic.com/technology/archive/2015/09/robots-hallucinate-dream/403498/`）。

另外，理解如何将预训练网络的每一层可视化，并更直观地理解网络如何记忆下层中的基本特征和更高层中的复杂特征。这是有关此主题一篇有趣的博客：

- *How convolutional neural networks see the world*（`https://blog.keras.io/category/demo.html`）。

CHAPTER 5

第 5 章

高级卷积神经网络

本章将讨论如何使用**卷积神经网络**（CNN）在图像以及其他领域进行深度学习。这里将主要关注文本分析和**自然语言处理**（NLP）领域。

5.1 引言

前一章介绍了如何将卷积网络应用于图像。本章将把相似的想法应用于文本。

文本和图像有什么共同之处？乍一看很少。但是，如果将句子或文档表示为矩阵，则该矩阵与其中每个单元是像素的图像矩阵没有什么区别。接下来的问题是，如何能够将文本表示为矩阵？好吧，这很简单：矩阵的每一行都是一个表示文本的向量。当然，现在需要定义一个基本单位。一个简单方法是将基本单位表示为字符。另一种做法是将一个单词看作基本单位，将相似的单词聚合在一起，然后用表示符号表示每个聚合（有时称为聚类或嵌入）。

> **TIP** 请注意，无论如何选择基本单位，都需要完成一个从基本单位到整数值地址的一一映射，以便可以将文本视为矩阵。例如，有 10 行文字，每行都是一个 100 维的嵌入，那么将其表示为 10×100 的矩阵。在这个特别的文本图像中，一个像素表示该句子 x 在位置 y 处有相应的嵌入。你也许会注意到，文本并不是一个真正的矩阵，而是一个矢量，因为位于相邻行中的两个单词几乎没有什么关联。实际上，位于相邻列中的两个单词最有可能具有某种相关性，这是文本矩阵与图像的主要差异。

现在你可能想问：我明白你是想把文本当成一个向量，但是这样做就失去了这个词的位置信息，这个位置信息应该是很重要的，不是吗？

其实，事实证明，在很多真实的应用程序中，知道一个句子是否包含一个特定的基本单位（一个字符、一个单词或一个聚合体）是非常准确的信息，即使不去记住其在句子中的确切位置。

5.2 为情感分析创建一个 ConvNet

本节将使用 TFLearn 创建一个基于 CNN 的情感分析深度学习网络。正如前一节所讨论的，这里的 CNN 是一维的。这里将使用 IMDb 数据集，收集 45000 个高度受欢迎的电影评论样本进行训练，并用 5000 个样本进行测试。

准备工作

TFLearn 有从网络自动下载数据集的库，便于创建卷积网络，所以可以直接编写代码。

具体做法

1. 导入 TensorFlow、`tflearn` 以及构建网络所需要的模块。然后导入 IMDb 库并执行独热编码和填充：

```
import tensorflow as tf
import tflearn
from tflearn.layers.core import input_data, dropout, fully_connected
from tflearn.layers.conv import conv_1d, global_max_pool
from tflearn.layers.merge_ops import merge
from tflearn.layers.estimator import regression
from tflearn.data_utils import to_categorical, pad_sequences
from tflearn.datasets import imdb
```

2. 加载数据集，用 0 填充整个句子至句子的最大长度，然后在标签上进行独热编码，其中两个数值分别对应 true 和 false 值。请注意，参数 n_words 是词汇表中单词的个数。表外的单词均设为未知。此外，请注意 trainX 和 trainY 是稀疏向量，因为每个评论可能仅包含整个单词集的一个子集。

```
# IMDb Dataset loading
train, test, _ = imdb.load_data(path='imdb.pkl', n_words=10000,
valid_portion=0.1)
trainX, trainY = train
testX, testY = test
#pad the sequence
trainX = pad_sequences(trainX, maxlen=100, value=0.)
testX = pad_sequences(testX, maxlen=100, value=0.)
#one-hot encoding
trainY = to_categorical(trainY, nb_classes=2)
testY = to_categorical(testY, nb_classes=2)
```

3. 显示几个维度来检查刚刚处理的数据，并理解数据维度的含义：

```
print ("size trainX", trainX.size)
print ("size testX", testX.size)
print ("size testY:", testY.size)
print ("size trainY", trainY.size)
size trainX 2250000
 size testX 250000
 size testY: 5000
 site trainY 45000
```

4. 为数据集中包含的文本构建一个嵌入。就目前而言，考虑这个步骤是一个黑盒子，它把这些词汇映射聚类，以便类似的词汇可能出现在同一个聚类中。请注意，在之前的步骤中，词汇是离散和稀疏的。通过嵌入操作，这里将创建一个将每个单词嵌入连续密集向量空间的映射。使用这个向量空间表示将给出一个连续的、分布式的词汇表示。如何构建嵌入，将在下节讨论 RNN 时详细讲解：

```
# Build an embedding
network = input_data(shape=[None, 100], name='input')
network = tflearn.embedding(network, input_dim=10000, output_dim=128)
```

5. 创建合适的卷积网络。这里有三个卷积层。由于正在处理文本，这里将使用一维卷积网络，这些图层将并行执行。每一层需要一个 128 维的张量（即嵌入输出），并应用多个具有有效填充的滤波器（分别为 3、4、5）、激活函数 ReLU 和 L2 regularizer。然后将每个图层的输出通过合并操作连接起来。接下来添加最大池层，以 50% 的概率丢弃参数的 dropout 层。最后一层是使用 softmax 激活的全连接层：

```
#Build the convnet
branch1 = conv_1d(network, 128, 3, padding='valid', activation='relu', regularizer="L2")
branch2 = conv_1d(network, 128, 4, padding='valid', activation='relu', regularizer="L2")
branch3 = conv_1d(network, 128, 5, padding='valid', activation='relu', regularizer="L2")
network = merge([branch1, branch2, branch3], mode='concat', axis=1)
network = tf.expand_dims(network, 2)
network = global_max_pool(network)
network = dropout(network, 0.5)
network = fully_connected(network, 2, activation='softmax')
```

6. 学习阶段使用 Adam 优化器以及 `categorical_crossentropy` 作为损失函数：

```
network = regression(network, optimizer='adam', learning_rate=0.001,
loss='categorical_crossentropy', name='target')
```

7. 在训练中，采用 `batch_size=32`，观察在训练和验证集上达到的准确度。正如你所看到的，在通过电影评论预测情感表达时能够获得 79% 的准确性：

```
# Training
model = tflearn.DNN(network, tensorboard_verbose=0)
model.fit(trainX, trainY, n_epoch = 5, shuffle=True, validation_set=(testX,
testY), show_metric=True, batch_size=32)
Training Step: 3519  | total loss: 0.09738 | time: 85.043s
 | Adam | epoch: 005 | loss: 0.09738 - acc: 0.9747 -- iter: 22496/22500
 Training Step: 3520  | total loss: 0.09733 | time: 86.652s
 | Adam | epoch: 005 | loss: 0.09733 - acc: 0.9741 | val_loss: 0.58740 -
val_acc: 0.7944 -- iter: 22500/22500
--
```

解读分析

论文 "*Convolutional Neural Networks for Sentence Classification*"，Yoon Kim, EMNLP 2014(https://arxiv.org/abs/1408.5882) 详细阐述了用于情感分析的一维卷积网

络。请注意，得益于滤波器窗口在连续单词上的操作，文章提出的模型保留了一些位置信息。文中配图给出了网络中的关键点。在开始时，文本被表示为基于标准嵌入的向量，在一维密集空间中提供了紧凑的表示，然后用多个标准的一维卷积层处理这些矩阵。

> 请注意该模型使用了多个具有不同窗口大小的滤波器来获取多个特征。之后，用一个最大池化操作来保留最重要的特征，即每个特征图中具有最高值的特征。为防止过度拟合，文章提出在倒数第二层采用一个 dropout 和用权向量的 L2 范数进行约束。最后一层输出情感为正面或者负面。

为了更好地理解模型，有几个观察结果展示如下：
- 滤波器通常在连续的空间上进行卷积。对于图像来说，这个空间是指高度和宽度上连续的像素矩阵表示。对于文本来说，连续的空间不过是连续词汇自然产生的连续维度。如果只使用独热编码来表示单词，那么空间是稀疏的，如果使用嵌入，则结果空间是密集的，因为相似的单词被聚合。
- 图像通常有三个颜色通道（RGB），而文本自然只有一个通道，因为不需要表示颜色。

更多内容

论文"Convolutional Neural Networks for Sentence Classification"针对句子分类开展了一系列的实验。除了对超参数的微调，具有一层卷积的简单 CNN 在句子分类中表现出色。文章还表明采用一套静态嵌入。(这将在讨论 RNN 时讨论)，并在其上构建一个非常简单的 CNN，可以显著提升情感分析的性能：

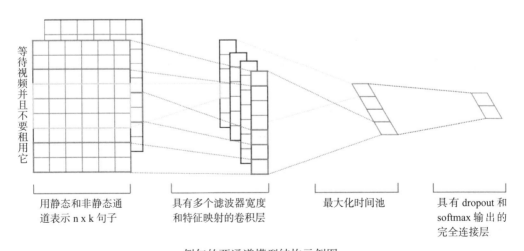

例句的两通道模型结构示例图

一个模型结构的示例的链接：https://arxiv.org/pdf/1408.5882.pdf

使用CNN进行文本分析是一个活跃的研究领域，我建议看看下面的文章：
- *Text Understanding from Scratch*，Xiang Zhang，Yann LeCun(https://arxiv.org/abs/1502.01710)。这篇文章证明可以使用CNN将深度学习应用到从字符级输入一直到抽象文本概念的文本理解。作者将CNN应用到包括本体分类、情感分析和文本分类在内的各种大规模数据集中，并表明它们不需要人类语言中关于词语、短语、句子或任何其他句法或语义结构的先验知识就可以达到让人惊艳的效果，模型适用于英文和中文。

5.3 检验VGG预建网络学到的滤波器

本节将使用keras-vis (https://raghakot.github.io/keras-vis/)，一个用于可视化VGG预建网络学到的不同滤波器的Keras软件包。基本思路是选择一个特定的ImageNet类别并理解VGG16网络如何来学习表示它。

准备工作

第一步是选择ImageNet上的一个特定类别来训练VGG16网络。比如，将下图中的美国北斗鸟类别设定为20。

美国北斗鸟案例，来自 https://commons.wikimedia.org/wiki/File:American_Dipper.jpg

ImageNet类别在网站 https://gist.github.com/yrevar/6135f1bd8dcf2e0cc683 中可以找到，作为一个Python字典包，ImageNet的1000个类别ID被处理为人类可读的标签。

具体做法

1. 导入matplotlib和keras-vis使用的模块。另外还需要载入预建的VGG16模块。Keras可以轻松处理这个预建网络：

```
from matplotlib import pyplot as plt
from vis.utils import utils
from vis.utils.vggnet import VGG16
from vis.visualization import visualize_class_activation
```

2. 通过使用 Keras 中包含的预构建图层获取 VGG16 网络,并使用 ImageNet 权重进行训练:

```
# Build the VGG16 network with ImageNet weights
model = VGG16(weights='imagenet', include_top=True)
model.summary()
print('Model loaded.')
```

3. 这是 VGG16 网络的内部结构。许多卷积层与最大池化层交替。一个平坦(flatten)层连接着三个密集层。其中最后一层被称为**预测层**,这个图层应该能够检测高级特征,比如面部特征,在此例中,是鸟的形状。请注意,顶层显式的包含在网络中,因为希望可视化它学到的东西:

```
Layer (type) Output Shape Param #
=================================================================
input_2 (InputLayer) (None, 224, 224, 3) 0
_____
block1_conv1 (Conv2D) (None, 224, 224, 64) 1792
_____
block1_conv2 (Conv2D) (None, 224, 224, 64) 36928
_____
block1_pool (MaxPooling2D) (None, 112, 112, 64) 0
_____
block2_conv1 (Conv2D) (None, 112, 112, 128) 73856
_____
block2_conv2 (Conv2D) (None, 112, 112, 128) 147584
_____
block2_pool (MaxPooling2D) (None, 56, 56, 128) 0
_____
block3_conv1 (Conv2D) (None, 56, 56, 256) 295168
_____
block3_conv2 (Conv2D) (None, 56, 56, 256) 590080
_____
block3_conv3 (Conv2D) (None, 56, 56, 256) 590080
_____
block3_pool (MaxPooling2D) (None, 28, 28, 256) 0
_____
block4_conv1 (Conv2D) (None, 28, 28, 512) 1180160
_____
block4_conv2 (Conv2D) (None, 28, 28, 512) 2359808
_____
block4_conv3 (Conv2D) (None, 28, 28, 512) 2359808
_____
block4_pool (MaxPooling2D) (None, 14, 14, 512) 0
_____
block5_conv1 (Conv2D) (None, 14, 14, 512) 2359808
_____
block5_conv2 (Conv2D) (None, 14, 14, 512) 2359808
_____
```

```
block5_conv3 (Conv2D)         (None, 14, 14, 512)    2359808
_____
block5_pool (MaxPooling2D)    (None, 7, 7, 512)      0
_____
flatten (Flatten)             (None, 25088)          0
_____
fc1 (Dense)                   (None, 4096)           102764544
_____
fc2 (Dense)                   (None, 4096)           16781312
_____
predictions (Dense)           (None, 1000)           4097000
=================================================================
Total params: 138,357,544
Trainable params: 138,357,544
Non-trainable params: 0
_____
Model loaded.
```

网络可以进一步抽象，如下图所示：

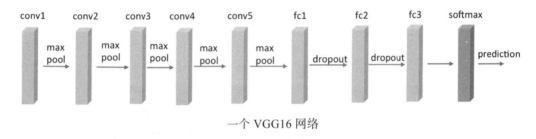

一个 VGG16 网络

4. 现在重点看一下最后的预测层是如何预测出 ID 类别序列为 20 的美国北斗鸟的：

```
layer_name = 'predictions'
layer_idx = [idx for idx, layer in enumerate(model.layers) if layer.name ==
layer_name][0]
# Generate three different images of the same output index.
vis_images = []
for idx in [20, 20, 20]:
    img = visualize_class_activation(model, layer_idx, filter_indices=idx,
max_iter=500)
    img = utils.draw_text(img, str(idx))
    vis_images.append(img)
```

5. 显示给定特征的特定图层的生成图像，并观察网络内部中的美国北斗鸟的概念：

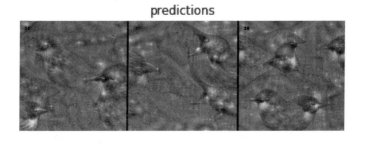

神经网络内部就是这样表示一只鸟的。这是一种虚幻的形象,但我发誓,这正是在没有人为干预的情况下该神经网络自然学到的东西!

6. 如果你还好奇还想了解更多,那么,选择一个网络中更浅的层将其可视化,显示美国北斗鸟的前期训练过程:

```
layer_name = 'block3_conv1'
layer_idx = [idx for idx, layer in enumerate(model.layers) if layer.name ==
layer_name][0]
vis_images = []
for idx in [20, 20, 20]:
    img = visualize_class_activation(model, layer_idx, filter_indices=idx,
max_iter=500)
    img = utils.draw_text(img, str(idx))
    vis_images.append(img)
stitched = utils.stitch_images(vis_images)
plt.axis('off')
plt.imshow(stitched)
plt.title(layer_name)
plt.show()
```

运行代码的输出如下:

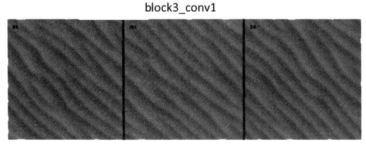

正如预期的那样,这个特定的层学习低层的特征,如曲线。然而,卷积网络的真正威力在于,模型中的网络越深入,越能推断出更复杂的特征。

解读分析

keras-vis 可视化密集层的关键思想是生成一个输入图像以最大化对应于鸟类的最终密集层输出。所以实际上这个模块做的是反转这个过程。给定一个特定的训练密集层与它的权重,生成一个新的最适合该层本身的合成图像。

每个卷积滤波器都使用类似的思路。在这种情况下,请注意,第一个卷积层是可以通过简单地将其权重可视化来解释的,因为它在原始像素上进行操作。

随后的卷积滤波器都对先前的卷积核的输出进行操作,因此直接对它们进行可视化并不一定非常容易理解。但是,如果独立考虑每一层,可以专注于生成最大化滤波器输出的合成输入图像。

更多内容

GitHub (https://github.com/raghakot/keras-vis) 中的 keras-vis 存储库提供了一系列关于如何检查内部网络的可视化示例,包括注意力显著图,其目标是在图像中包含各种类别(例如,草)时检测图像的哪个部分对特定类别(例如,老虎)的训练贡献最大。典型文章有"*Deep Inside Convolutional Networks: Visualising Image Classification Models and Saliency Maps*",由 Karen Simonyan、Andrea Vedaldi 和 Andrew Zisserman 编写(https://arxiv.org/abs/1312.6034),其中一个 Git 库中的图片显示如下,这个案例说明了在网络中一个老虎的显著图样本。

显著性图例 https://gb.com/raghakot/keras-vis

5.4 使用 VGGNet、ResNet、Inception 和 Xception 分类图像

图像分类任务是一个典型的深度学习应用。人们对这个任务的兴趣得益于 ImageNet (http://image-net.org/) 图像数据集根据 WordNet (http://wordnet.princeton.edu/) 层次结构(目前仅有名词)组织,其中检索层次的每个节点包含了成千上万张图片。更确切地说,ImageNet 旨在将图像分类并标注为近 22000 个独立的对象类别。在深度学习的背景下,ImageNet 一般是指论文"ImageNet Large Scale Visual Recognition Challenge"(http://www.image-net.org/challenges/LSVRC/)中的工作,即 ImageNet 大型视觉识别竞赛,简称 ILSVRC。在这种背景下,目标是训练一个模型,可以将输入图像分类为 1000 个独立的对象类别。本节将使用由超过 120 万幅训练图像、50000 幅验证图像和 100000 幅测试图像预训练出的模型。

VGG16 和 VGG19

VGG16 和 VGG19 网络已经被引用到"*Very Deep Convolutional Networks for Large Scale Image Recognition*",由 Karen Simonyan 和 Andrew Zisserman 于 2014 年编写,网址为 https://arxiv.org/abs/1409.1556。该网络使用 3×3 卷积核的卷积层堆叠并交替最大池化层,有两个 4096 维的全连接层,然后是 softmax 分类器。16 和 19 分别代表网

络中权重层的数量（即列 D 和 E）：

ConvNet Configuration					
A	A-LRN	B	C	D	E
11 weight layers	11 weight layers	13 weight layers	16 weight layers	16 weight layers	19 weight layers
input (224 × 224 RGB image)					
conv3-64	conv3-64 LRN	conv3-64 conv3-64	conv3-64 conv3-64	conv3-64 conv3-64	conv3-64 conv3-64
maxpool					
conv3-128	conv3-128	conv3-128 conv3-128	conv3-128 conv3-128	conv3-128 conv3-128	conv3-128 conv3-128
maxpool					
conv3-256 conv3-256	conv3-256 conv3-256	conv3-256 conv3-256	conv3-256 conv3-256 **conv1-256**	conv3-256 conv3-256 conv3-256	conv3-256 conv3-256 conv3-256 **conv3-256**
maxpool					
conv3-512 conv3-512	conv3-512 conv3-512	conv3-512 conv3-512	conv3-512 conv3-512 **conv1-512**	conv3-512 conv3-512 conv3-512	conv3-512 conv3-512 conv3-512 **conv3-512**
maxpool					
conv3-512 conv3-512	conv3-512 conv3-512	conv3-512 conv3-512	conv3-512 conv3-512 **conv1-512**	conv3-512 conv3-512 conv3-512	conv3-512 conv3-512 conv3-512 **conv3-512**
maxpool					
FC-4096					
FC-4096					
FC-1000					
soft-max					

`https://arxiv.org/pdf/1409.1556.pdf` 中所示的深层网络配置示例

在 2015 年，16 层或 19 层网络就可以认为是深度网络，但到了 2017 年，深度网络可达数百层。请注意，VGG 网络训练非常缓慢，并且由于深度和末端的全连接层，使得它们需要较大的权重存储空间。

ResNet

ResNet（残差网络）的提出源自论文"*Deep Residual Learning for Image Recognition*"，由 Kaiming He、XiangyuZhang、ShaoqingRen 和 JianSun 于 2015 年编写，网址为 `https://arxiv.org/abs/1512.03385`。这个网络是非常深的，可以使用一个称为残差模块的标准的网络组件来组成更复杂的网络（可称为网络中的网络），使用标准的随机梯度下降法进行训练。

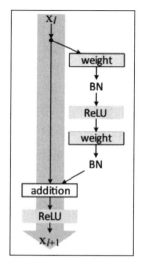

与 VGG 相比，ResNet 更深，但是由于使用全局平均池操作而不是全连接密集层，所以模型的尺寸更小。

Inception

Inception 网络源自文章"*Rethinking the Inception Architecture for Computer Vision*"，由 Christian Szegedy、Vincent Vanhoucke、Sergey Ioffe、Jonathon Shlens 和 Zbigniew Wojna 于 2015 年编写，网址为 https://arxiv.org/abs/1512.00567。其主要思想是使用多个尺度的卷积核提取特征，并在同一模块中同时计算 1×1、3×3 和 5×5 卷积。然后将这些滤波器的输出沿通道维度堆叠并传递到网络中的下一层，如下图所示：

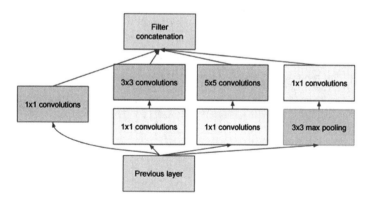

Inception-v3 见论文"*Rethinking the Inception Architecture for Computer Vision*"；Inception-v4 见论文"*Inception-v4*""*Inception-ResNet and the Impact of Residual Connections on Learning*"，由 Christian Szegedy、Sergey Ioffe、Vincent Vanhoucke 和 Alex Alemi 于 2016 年编写，网址为 https://arxiv.org/abs/1602.07261。

Xception

Xception 网络是 Inception 网络的扩展,详见论文"*Xception: Deep Learning with Depthwise Separable Convolutions*",由 François Chollet 于 2016 年编写,网址为 `https://arxiv.org/abs/1610.02357`。Xception 使用了一种叫作深度可分离卷积运算的新概念,它可以在包含 3.5 亿个图像和 17000 个类的大型图像分类数据集上胜过 Inception-v3。由于 Xception 架构具有与 Inception-v3 相同的参数数量,因此性能提升不是由于容量的增加,而是由于更高效地使用了模型参数。

准备工作

本节使用 Keras 因为这个框架有上述模块的预处理模块。Keras 在第一次使用时会自动下载每个网络的权重,并将这些权重存储在本地磁盘上。换句话说,你不需要重新训练网络,而是使用互联网上已有的训练参数。假设你想在 1000 个预定义类别中分类网络,这样做是没问题的。下一节将介绍如何从这 1000 个类别开始,将其扩展到一个定制的集合,这个过程称为迁移学习。

具体做法

1. 导入处理和显示图像所需的预建模型和附加模块:

```
from keras.applications import ResNet50
from keras.applications import InceptionV3
from keras.applications import Xception # TensorFlow ONLY
from keras.applications import VGG16
from keras.applications import VGG19
from keras.applications import imagenet_utils
from keras.applications.inception_v3 import preprocess_input
from keras.preprocessing.image import img_to_array
from keras.preprocessing.image import load_img
import numpy as np
import matplotlib.pyplot as plt
from matplotlib.pyplot import imshow
from PIL import Image
%matplotlib inline
```

2. 定义一个用于记忆训练中图像尺寸的映射,这些是每个模型的一些常量参数:

```
MODELS = {
"vgg16": (VGG16, (224, 224)),
"vgg19": (VGG19, (224, 224)),
"inception": (InceptionV3, (299, 299)),
"xception": (Xception, (299, 299)), # TensorFlow ONLY
"resnet": (ResNet50, (224, 224))
}
```

3. 定义用于加载和转换图像的辅助函数。请注意,预先训练的网络已经在张量上进行了训练,其形状还包括 `batch_size` 的附加维度。所以为了图像兼容性需要补充这个维度:

```
def image_load_and_convert(image_path, model):
pil_im = Image.open(image_path, 'r')
```

```
imshow(np.asarray(pil_im))
# initialize the input image shape
# and the pre-processing function (this might need to be changed
inputShape = MODELS[model][1]
preprocess = imagenet_utils.preprocess_input
image = load_img(image_path, target_size=inputShape)
image = img_to_array(image)
# the original networks have been trained on an additional
# dimension taking into account the batch size
# we need to add this dimension for consistency
# even if we have one image only
image = np.expand_dims(image, axis=0)
image = preprocess(image)
return image
```

4. 定义一个辅助函数,用于对图像进行分类并对预测结果进行循环,显示前5名的预测概率:

```
def classify_image(image_path, model):
    img = image_load_and_convert(image_path, model)
    Network = MODELS[model][0]
    model = Network(weights="imagenet")
    preds = model.predict(img)
    P = imagenet_utils.decode_predictions(preds)
    # loop over the predictions and display the rank-5 predictions
    # along with probabilities
    for (i, (imagenetID, label, prob)) in enumerate(P[0]):
    print("{}. {}: {:.2f}%".format(i + 1, label, prob * 100))
```

5. 测试不同类型的预训练网络:

```
classify_image("images/parrot.jpg", "vgg16")
```

接下来你将看到一个带有各自类别预测概率的预测列表:

1. macaw: 99.92%

2. jacamar: 0.03%

3. lorikeet: 0.02%

4. bee_eater: 0.02%

5. toucan: 0.00%

https://commons.wikimedia.org/wiki/File:Blue-and-Yellow-Macaw.jpg 中的一个示例

```
classify_image("images/parrot.jpg", "vgg19")
```

1. macaw: 99.77%
2. lorikeet: 0.07%
3. toucan: 0.06%
4. hornbill: 0.05%
5. jacamar: 0.01%

```
classify_image("images/parrot.jpg", "resnet")
```

1. macaw: 97.93%
2. peacock: 0.86%
3. lorikeet: 0.23%
4. jacamar: 0.12%
5. jay: 0.12%

```
classify_image("images/parrot_cropped1.jpg", "resnet")
```

1. macaw: 99.98%
2. lorikeet: 0.00%

3. peacock: 0.00%

4. sulphur-crested_cockatoo: 0.00%

5. toucan: 0.00%

classify_image("images/incredible-hulk-180.jpg", "resnet")

1. comic_book: 99.76%

2. book_jacket: 0.19%

3. jigsaw_puzzle: 0.05%

4. menu: 0.00%

5. packet: 0.00%

https://comicvine.gamespot.com/the-incredible-hulk-180-and-the-wind-howls-wendigo/4000-14667/ 中的一个示例

classify_image("images/cropped_panda.jpg", "resnet")

1. giant_panda: 99.04%
2. indri: 0.59%
3. lesser_panda: 0.17%
4. gibbon: 0.07%
5. titi: 0.05%

classify_image("images/space-shuttle1.jpg", "resnet")

1.space_shuttle: 92.38%
2.triceratops: 7.15%
3.warplane: 0.11%
4.cowboy_hat: 0.10%
5.sombrero: 0.04%

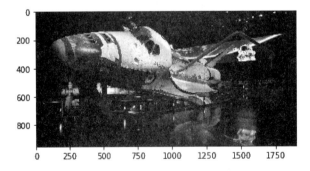

classify_image("images/space-shuttle2.jpg", "resnet")

1.space_shuttle: 99.96%
2.missile: 0.03%
3.projectile: 0.00%

4.steam_locomotive: 0.00%

5.warplane: 0.00%

classify_image("images/space-shuttle3.jpg", "resnet")

1.space_shuttle: 93.21%

2.missile: 5.53%

3.projectile: 1.26%

4.mosque: 0.00%

5.beacon: 0.00%

classify_image("images/space-shuttle4.jpg", "resnet")

1.space_shuttle: 49.61%

2.castle: 8.17%

3.crane: 6.46%

4.missile: 4.62%

5.aircraft_carrier: 4.24%

注意可能报出一些错误，比如：

classify_image("images/parrot.jpg", "inception")

1.stopwatch: 100.00%

2.mink: 0.00%

3.hammer: 0.00%

4.black_grouse: 0.00%

5.web_site: 0.00%

classify_image("images/parrot.jpg", "xception")

1.backpack: 56.69%

2.military_uniform: 29.79%

3.bib: 8.02%

4.purse: 2.14%

5.ping-pong_ball: 1.52%

6. 定义用于显示每个预建和预训练网络的内部架构的函数:

```
def print_model(model):
print ("Model:",model)
Network = MODELS[model][0]
model = Network(weights="imagenet")
model.summary()
print_model('vgg19')

('Model:', 'vgg19')
```

Layer (type)	Output Shape	Param #
input_14 (InputLayer)	(None, 224, 224, 3)	0
block1_conv1 (Conv2D)	(None, 224, 224, 64)	1792
block1_conv2 (Conv2D)	(None, 224, 224, 64)	36928
block1_pool (MaxPooling2D)	(None, 112, 112, 64)	0
block2_conv1 (Conv2D)	(None, 112, 112, 128)	73856
block2_conv2 (Conv2D)	(None, 112, 112, 128)	147584
block2_pool (MaxPooling2D)	(None, 56, 56, 128)	0
block3_conv1 (Conv2D)	(None, 56, 56, 256)	295168
block3_conv2 (Conv2D)	(None, 56, 56, 256)	590080
block3_conv3 (Conv2D)	(None, 56, 56, 256)	590080
block3_conv4 (Conv2D)	(None, 56, 56, 256)	590080
block3_pool (MaxPooling2D)	(None, 28, 28, 256)	0
block4_conv1 (Conv2D)	(None, 28, 28, 512)	1180160

```
block4_conv2 (Conv2D)        (None, 28, 28, 512)   2359808
_____
block4_conv3 (Conv2D)        (None, 28, 28, 512)   2359808
_____
block4_conv4 (Conv2D)        (None, 28, 28, 512)   2359808
_____
block4_pool (MaxPooling2D)   (None, 14, 14, 512)   0
_____
block5_conv1 (Conv2D)        (None, 14, 14, 512)   2359808
_____
block5_conv2 (Conv2D)        (None, 14, 14, 512)   2359808
_____
block5_conv3 (Conv2D)        (None, 14, 14, 512)   2359808
_____
block5_conv4 (Conv2D)        (None, 14, 14, 512)   2359808
_____
block5_pool (MaxPooling2D)   (None, 7, 7, 512)     0
_____
flatten (Flatten)            (None, 25088)         0
_____
fc1 (Dense)                  (None, 4096)          102764544
_____
fc2 (Dense)                  (None, 4096)          16781312
_____
predictions (Dense)          (None, 1000)          4097000
=================================================================
Total params: 143,667,240
Trainable params: 143,667,240
Non-trainable params: 0
```

解读分析

我们已经使用了 Keras 应用，带有预训练权重的预训练 Keras 学习模型是可以获取的，这些模型可用于预测、特征提取以及参数微调。在本例中，使用的是预测模型。将在下一个例子中看到如何使用该模型进行参数微调，以及如何在数据集上构建自定义的分类器，这些分类器在最初训练模型时是不可用的。

更多内容

Inception-v4 在 2017 年 7 月之前不能在 Keras 中直接使用，但可以在线上单独下载（`https://github.com/kentsommer/keras-inceptionV4`）。安装完成后，模块将在第一次使用时自动下载其权重参数。

AlexNet 是最早的堆叠深度网络之一，它只包含八层，前五层是卷积层，后面是全连接层。该网络于 2012 年提出，当年凭借其优异的性能获得冠军（其误差约为 16%，而亚军误差为 26%）。

最近对深度神经网络的研究主要集中在提高精度上。具有相同精度的前提下，轻量化 DNN 体系结构至少有以下三个优点：

- 轻量化 CNN 在分布式训练期间需要更少的服务器通信。
- 轻量化 CNN 需要较少的带宽将新模型从云端导出到模型所在的位置。
- 轻量化 CNN 更易于部署在 FPGA 和其他有限内存的硬件上。为了提供以上优点，论文 "*SqueezeNet: AlexNet-level accuracy with 50x fewer parameters and <0.5MB model size*"（Forrest N. Iandola, Song Han, Matthew W. Moskewicz, Khalid Ashraf, William J. Dally, Kurt Keutzer, 2016, `https://arxiv.org/abs/1602.07360`）提出的 SqueezeNet 在 ImageNet 上实现了 AlexNet 级别的准确性，参数少了 50 倍。另外，由于使用模型压缩技术，可以将 SqueezeNet 压缩到小于 0.5 MB（比 AlexNet 小 510 倍）。Keras 实现的 SqueezeNet 作为一个单独的模块，已在网上开源 (`https://github.com/DT42/squeezenet_demo`)。

5.5 重新利用预建深度学习模型进行特征提取

本节将介绍如何使用深度学习来提取相关的特征。

准备工作

一个非常简单的想法是使用 VGG16 和一般的 DCNN 模型来进行特征提取。这段代码通过从特定图层中提取特征来实现这个想法。

具体做法

1. 导入处理和显示图像所需的预建模型和附加模块：

```
from keras.applications.vgg16 import VGG16
from keras.models import Model
from keras.preprocessing import image
from keras.applications.vgg16 import preprocess_input
import numpy as np
```

2. 从网络中选择一个特定的图层，并获取输出的特征：

```
# pre-built and pre-trained deep learning VGG16 model
base_model = VGG16(weights='imagenet', include_top=True)
for i, layer in enumerate(base_model.layers):
print (i, layer.name, layer.output_shape)
# extract features from block4_pool block
model =
Model(input=base_model.input,
output=base_model.get_layer('block4_pool').output)
```

3. 提取给定图像的特征，代码如下所示：

```
img_path = 'cat.jpg'
img = image.load_img(img_path, target_size=(224, 224))
x = image.img_to_array(img)
```

```
x = np.expand_dims(x, axis=0)
x = preprocess_input(x)
# get the features from this block
features = model.predict(x)
```

解读分析

现在,你可能想知道为什么要从 CNN 中的中间层提取特征。一个直觉是:随着网络的学习将图像分类成不同类别,每一层将学习到进行最终分类所必需的特征。

较低层识别诸如颜色和边缘等较低阶特征,高层将这些较低阶特征组合成较高阶特征,诸如形状或对象。因此,中间层具有从图像中提取重要特征的能力,这些特征有助于不同种类的分类。

这种结构具有多个优点。首先,可以依靠公开的大规模数据集训练,将学习参数迁移到新的领域。其次,可以节省大量训练时间。再次,即使在自己的数据集中没有大量的训练数据,也可以提供合理的解决方案。我们也可以为手头任务准备一个较好的起始网络形状,而不用去猜测它。

5.6 用于迁移学习的深层 InceptionV3 网络

迁移学习是一种非常强大的深度学习技术,在不同的领域有着各种应用。迁移学习的思想很简单,可以用类比来解释。假设你想学习一种新的语言,比如西班牙语,那么从你已经知道的另一种语言,比如说英语开始学起,可能会有所帮助。

遵循这一思路,计算机视觉研究人员通常使用预先训练的 CNN 为新任务生成表示,其中新任务数据集可能不够大,无法从头开始训练整个 CNN。另一个常见的策略是采用预先训练好的 ImageNet 网络,然后对整个网络进行微调以完成新任务。

InceptionV3 网络是由 Google 开发的一个非常深的卷积网络。Keras 实现了完整的网络,如下图所示,它是在 ImageNet 上预先训练好的。这个模型的默认输入尺寸是 299×299,有三个通道。

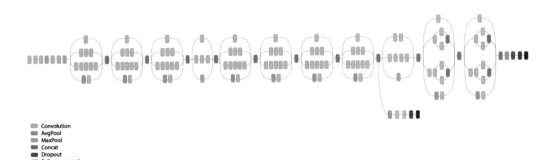

ImageNet v3 网路结构示意图

准备工作

这个框架的例子受 Keras 网站上的在线模型 (https://keras.io/applications/) 启发。假设在一个域中有一个与 ImageNet 不同的训练数据集 D。D 具有 1024 个输入特征和 200 个输出类别。

具体做法

1. 导入预处理模型和处理所需的库：

```
from keras.applications.inception_v3 import InceptionV3
from keras.preprocessing import image
from keras.models import Model
from keras.layers import Dense, GlobalAveragePooling2D
from keras import backend as K
# create the base pre-trained model
base_model = InceptionV3(weights='imagenet', include_top=False)
```

2. 使用一个训练过的 Inception-v3 网络，但是不包括顶层模型，因为想要在 D 上进行微调。顶层是一个密集层，有 1024 个输入，最后一个输出层是一个 softmax 密集层，有 200 个输出类。`x = GlobalAveragePooling2D()(x)` 用于将输入转换为密集层处理的正确形状。实际上，`base_model.output tensor` 的形态有 `dim_ordering="th"`（对应样本，行，列，通道）或者 `dim_ordering="tf"`（对应样本，通道，行，列），但是密集层需要将其调整为（样本，通道）`GlobalAveragePooling2D` 按（行，列）平均。所以如果你看最后四层（`include_top = True`），你会看到这些形状：

```
# layer.name, layer.input_shape, layer.output_shape
('mixed10', [(None, 8, 8, 320), (None, 8, 8, 768), (None, 8, 8, 768),
(None, 8, 8, 192)], (None, 8, 8, 2048))
('avg_pool', (None, 8, 8, 2048), (None, 1, 1, 2048))
('flatten', (None, 1, 1, 2048), (None, 2048))
('predictions', (None, 2048), (None, 1000))
```

3. 如果包含 `_top=False`，则会删除最后三层并显示 mixed_10 层，因此 `GlobalAveragePooling2D` 层将（`None, 8, 8, 2048`）转换为（`None, 2048`），其中（`None, 2048`）张量是（`None, 8, 8, 2048`）张量中每个对应的（8,8）子张量的平均值：

```
# add a global spatial average pooling layer
x = base_model.output
x = GlobalAveragePooling2D()(x)
# let's add a fully-connected layer as first layer
x = Dense(1024, activation='relu')(x)
# and a logistic layer with 200 classes as last layer
predictions = Dense(200, activation='softmax')(x)
# model to train
model = Model(input=base_model.input, output=predictions)
```

4. 所有卷积层都是预先训练好的，所以在整个模型的训练过程中冻结它们。

```
# i.e. freeze all convolutional Inception-v3 layers
for layer in base_model.layers:
    layer.trainable = False
```

5. 对模型进行编译并训练几批次，以便对顶层进行训练：

```
# compile the model (should be done *after* setting layers to non-trainable)
model.compile(optimizer='rmsprop', loss='categorical_crossentropy')
# train the model on the new data for a few epochs
model.fit_generator(...)
```

6. 接下来冻结 Inception 中的顶层并微调 Inception 层。在这个例子中，冻结了前 172 层（一个用来微调的超参数）：

```
# we chose to train the top 2 inception blocks, i.e. we will freeze
# the first 172 layers and unfreeze the rest:
for layer in model.layers[:172]:
   layer.trainable = False
for layer in model.layers[172:]:
   layer.trainable = True
```

7. 重新编译模型进行微调优化。需要重新编译模型以使这些修改生效：

```
# we use SGD with a low learning rate
from keras.optimizers import SGD
model.compile(optimizer=SGD(lr=0.0001, momentum=0.9), loss='categorical_crossentropy')

# we train our model again (this time fine-tuning the top 2 inception blocks
# alongside the top Dense layers
model.fit_generator(...)
```

解读分析

现在我们有了一个新的深度网络，它重新使用了标准的 Inception-v3 网络，但是它通过迁移学习在一个新的领域 D 上进行了训练。当然，有许多参数可以精确调整以达到较好的精度。但是，现在正在通过迁移学习重新使用一个非常大的预训练网络作为起点。这样做可以通过重新使用 Keras 中已有的功能来节省训练成本。

更多内容

截至 2017 年，"计算机视觉"问题（在图像中找到模式的问题）被认为已经解决了，这个问题对生活有很大影响。例如：

- 论文"*Dermatologist-level classification of skin cancer with deep neural networks*"（Andre Esteva, Brett Kuprel, Roberto A. Novoa, Justin Ko, Susan M. Swetter, Helen M. Blau & Sebastian Thrun, 2017, https://www.nature.com/nature/journal/v542/n7639/full/nature21056.html）使用由 2032 种不同疾病组成的 129450 张临床图像的数据集训练 CNN。他们通过 21 位经过认证的皮肤科医师对活检证实的临床图像进行二元分类，分别区分角质形成单元癌与良性脂溢性角化病、恶性黑色素瘤与良性痣。CNN 与人类专家在这两项任务上都达到了同样的水平，证明了人工智能在进行皮肤癌分类中能够与皮肤科医生相媲美。

- 论文 " *High-Resolution Breast Cancer Screening with Multi-View Deep Convolutional Neural Networks* "（Krzysztof J. Geras, Stacey Wolfson, S. Gene Kim, LindaMoy, KyunghyunCho, https://arxiv.org/abs/1703.07047）提出一种有望提高乳腺癌筛查过程效率的新架构，可以处理四种标准的视图或角度。与常用的自然图像 DCN 架构（这种架构适用于 224×224 像素的图像）相比，MV-DCN 还能够使用 2600×2000 像素的分辨率。

5.7 使用扩张 ConvNet、WaveNet 和 NSynth 生成音乐

WaveNet 是生成原始音频波形的深层生成模型。这项突破性的技术已经被 Google DeepMind（https://deepmind.com/）引入（https://deepmind.com/blog/generate-mode-raw-audio/），用于教授如何与计算机对话。结果确实令人惊讶，在网上你可以找到合成声音的例子，电脑学习如何用名人的声音与人们谈话。

所以，你可能想知道为什么学习合成音频是如此困难。听到的每个数字声音都是基于每秒 16000 个样本（有时是 48000 个或更多）建立一个预测模型，在这个模型中学习基于以前所有的样本来重现样本，这是一个非常困难的挑战。尽管如此，有实验表明，WaveNet 已经改进了当前最先进的文本到语音（Text-To-Speech, TTS）系统，降低了英语和普通话之间 50% 的差异。

更酷的是，DeepMind 证明了 WaveNet 可以教会电脑如何产生乐器的声音，比如钢琴音乐。下面给出一些定义。TTS 系统通常分为两个不同的类别：

- 连续 TTS，其中单个语音片段首先被记忆，然后在语音再现时重新组合。这种方法没有大规模应用，因为它只能再现记忆过的语音片段，并且不可能在没有记忆片段的情况下再现新的声音或不同类型的音频。
- 参数 TTS，其中创建模型用于存储要合成的音频的所有特征。在 WaveNet 之前，使用参数 TTS 生成的音频不如连续 TTS 自然。WaveNet 通过直接建模音频声音的生成来改现有技术，而不是使用过去常用的中间信号去处理算法。

原则上，WaveNet 可以看作是一堆卷积层（已经在第 4 章中看到了二维卷积图像），而且步长恒定，没有池化层。请注意，输入和输出的结构具有相同的尺寸，所以 ConvNet 非常适合对音频声音等连续数据进行建模。然而，实验表明，为了达到输出神经元中的感受野大尺寸，有必要使用大量的大型滤波器或者不可避免地增加网络的深度。请记住，一个网络中一层神经元的感受野是前一层神经元对其提供输入的横截面。由于这个原因，纯粹的卷积网络在学习如何合成音频方面效率不高。

WaveNet 的关键在于所谓的扩张因果卷积（有时称为带孔卷积），这就意味着当应用卷积层的滤波器时，一些输入值被跳过。Atrous 来自法语 *àtrous*，意思是 "有孔"，所以 AtrousConvolution 是一个 "带孔" 的卷积。例如，在一个维度上，一个具有扩张 1、大小

为 3 的滤波器 w 将计算如下所示的加权和。

简而言之，在扩张值为 D 的扩张卷积中，通常步长是 1，你也可使用其他的步长。下图给出了一个例子，扩大（孔）尺寸为 0,1,2：

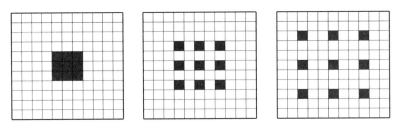

扩张网络的一个例子

由于引入"孔"这个简单的想法，使得堆叠多个扩张的卷积层与指数增加的过滤器、学习长距离输入而不用担心有一个过深的网络成为可能。

因此，WaveNet 属于卷积网络，其卷积层具有各种扩张因子，使得感受野随深度呈指数增长，因此有效地覆盖了数千个音频时间步长。

当训练时，输入是来自人类说话者的录音。这些波形量化为一个固定的整数范围。WaveNet 定义了一个初始卷积层，只访问当前和之前的输入。然后，有一堆扩大的卷积层，仍然只能访问当前和之前的输入。最后，有一系列密集层结合了前面的结果，接下来是分类输出的 softmax 激活函数。

在每个步骤中，从网络预测一个值并将其反馈到输入中。同时，计算下一步的新预测。损失函数是当前步骤的输出和下一步的输入之间的交叉熵。

NSynth (https://magenta.tensorflow.org/nsynth) 是最近由 Google Brain 小组发布的一个 WaveNet 的演变，它不是因果关系，而是旨在看到输入块的整个上下文。如下图所示，神经网络确实是复杂的，但是作为介绍性讨论，知道网络学习如何通过使用基于减少编码/解码期间的误差的方法来再现其输入就足够了：

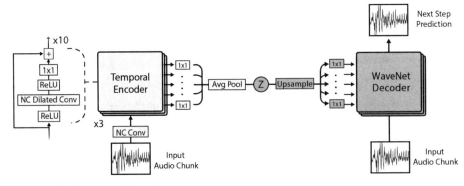

NSynth 架构的一个案例，源自 https://magenta.tensorflow.org/nsynth

准备工作

本节直接使用网上的代码来演示（https://github.com/tensorflow/magenta/tree/master/magenta/models/nsynth），你还可以从 Google Brain（https://aiexperiments.withgoogle.com/sound-maker）中找到一些。感兴趣的读者也可以阅读论文"Neural Audio Synthesis of Musical Notes with WaveNet Autoencoders"（Jesse Engel, Cinjon Resnick, Adam Roberts, Sander Dieleman, Douglas Eck, Karen Simonyan, Mohammad Norouzi, 2017.4，https://arxiv.org/abs/1704.01279）。

具体做法

1. 通过创建单独的 conda 环境来安装 NSynth。使用支持 Jupyter Notebook 的 Python 2.7 创建并激活 Magenta conda 环境：

```
conda create -n magenta python=2.7 jupyter
source activate magenta
```

2. 安装用于读取音频格式的 Magenta pip 软件包和 librosa：

```
pip install magenta
pip install librosa
```

3. 从网上下载安装一个预先建立的模型（http://download.magenta.tensorflow.org/models/nsynth/wavenet-ckpt.tar）并下载示例声音（https://www.freesound.org/people/MustardPlug/sounds/395058/），然后运行 demo 目录中的笔记（在我的例子中是 http://localhost:8888/notebooks/nsynth/Exploring_Neural_Audio_Synthesis_with_NSynth.ipynb）。第一部分包含了稍后将在计算中使用的模块：

```
import os
import numpy as np
import matplotlib.pyplot as plt
from magenta.models.nsynth import utils
from magenta.models.nsynth.wavenet import fastgen
from IPython.display import Audio
%matplotlib inline
%config InlineBackend.figure_format = 'jpg'
```

4. 加载从互联网下载的演示声音，并将其放在与笔记本电脑相同的目录中。这将在约 2.5 秒内将 40000 个样品装入机器：

```
# from https://www.freesound.org/people/MustardPlug/sounds/395058/
fname = '395058__mustardplug__breakbeat-hiphop-a4-4bar-96bpm.wav'
sr = 16000
audio = utils.load_audio(fname, sample_length=40000, sr=sr)
sample_length = audio.shape[0]
print('{} samples, {} seconds'.format(sample_length, sample_length /
float(sr)))
```

5. 使用从互联网上下载的预先训练的 NSynth 模型以非常紧凑的表示方式对音频样本进行编码。每 4 秒给一个 78×16 的尺寸编码，然后可以解码或重新合成。编码是张量 (`#files=1x78x16`)：

```
%time encoding = fastgen.encode(audio, 'model.ckpt-200000', sample_length)
INFO:tensorflow:Restoring parameters from model.ckpt-200000
 CPU times: user 1min 4s, sys: 2.96 s, total: 1min 7s
 Wall time: 25.7 s
print(encoding.shape)

(1, 78, 16)
```

6. 保存稍后用于重新合成的编码。另外，用图形表示快速查看编码形状，并将其与原始音频信号进行比较。如你所见，编码遵循原始音频信号中的节拍：

```
np.save(fname + '.npy', encoding)
fig, axs = plt.subplots(2, 1, figsize=(10, 5))
axs[0].plot(audio);
axs[0].set_title('Audio Signal')
axs[1].plot(encoding[0]);
axs[1].set_title('NSynth Encoding')
```

我们观察如下图所示的音频信号和 NSynth 编码。

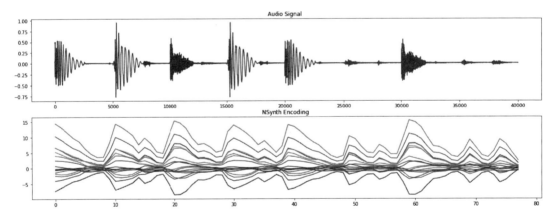

7. 现在对刚刚制作的编码进行解码。换句话说，如果重新合成的声音类似于原来的，这里试图以紧凑表示再现对原始音频的理解。事实上，如果你运行实验并听取原始音频和重新合成的音频，会感觉它们听起来非常相似：

```
%time fastgen.synthesize(encoding, save_paths=['gen_' + fname],
samples_per_save=sample_length)
```

解读分析

WaveNet 是一种卷积网络，卷积层具有各种扩张因子，使得感受野随深度呈指数级增长，因此可以有效地覆盖数千个音频时间步长。NSynth 是 WaveNet 的一种演变，其中原始音频使用类似 WaveNet 的处理来编码,以学习紧凑的表示。然后,这个紧凑的表示被用来再现原始音频。

更多内容

一旦学习如何通过扩张卷积创建一段紧凑的音频,可以发现其中的乐趣。你会发现在互联网上很酷的演示:

1. 例如,可以看到模型如何学习不同乐器的声音:(https://magenta.tensorflow.org/nsynth):

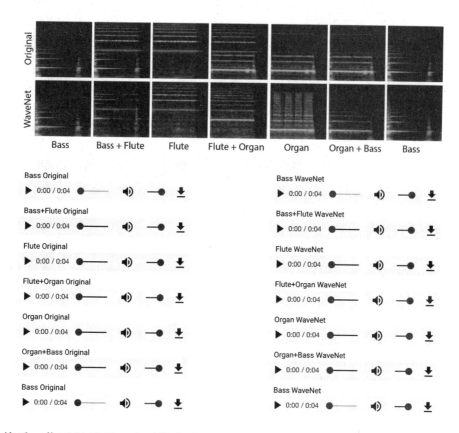

2. 然后,你可以看到一个环境中学习的模型如何在另一个环境中重新混合。例如,通过改变扬声器的身份,可以使用 WaveNet 用不同的声音描述同样的事情(https://deepmind.com/blog/wavenet-generative-model-raw-audio/)。

3. 另一个非常有趣的实验是学习乐器的模型,然后重新混合,这样就可以创造出以前从未听过的新乐器。这真的很酷,它打开了一个新世界。例如,在这个例子中,把西塔琴和电吉他结合起来,形成一种很酷的新乐器。还不够酷?那么把低音贝斯和狗的叫声结合起来怎么样?(https://aiexperiments.withgoogle.com/sound-maker/

view/)玩得开心!

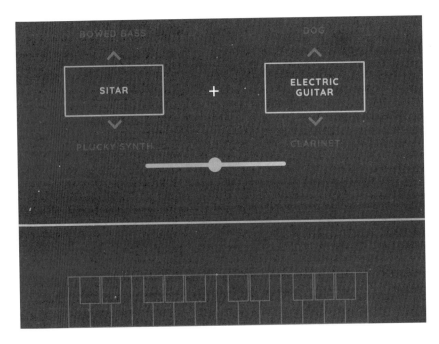

5.8 关于图像的问答

在本节中将学习如何回答关于特定图像内容的问题。这是基于从预先训练的 VGG16 模型提取的视觉特征与词聚类（嵌入）组合的视觉问答的一种强大形式。然后将这两组不同的特征组合成一个网络，其中最后几层由密集层和 dropout 层交替组成。本节基于 Keras 2.0+ 版本运行。

本节将教你：
- 从预训练的 VGG16 网络提取特征。
- 使用预先构建的单词嵌入将单词映射到类似单词聚类的空间中。
- 使用 LSTM 图层来构建语言模型。LSTM 将在第 6 章讨论，现在将把它们用作黑箱。
- 结合不同的异构输入特征来创建组合的特征空间。对于这个任务，使用新的 Keras 2.0 功能 API。
- 连接额外的密集层和 Dropout 层创建一个多层感知器并且增加深度学习网络的泛化能力。

为了简单起见，这里不会重新训练上节中的组合网络，而是使用一组在线预先训练的权重（https://avisingh599.github.io/deeplearning/visual-qa/）。感兴趣的读者可以在由 N 个图像、N 个问题和 N 个答案组成的自己的训练数据集上重新训练网络。这是一个可选的练习。该网络受到论文"*VQA: Visual Question Answering*"（Aishwarya Agrawal, Jiasen Lu, Stanislaw Antol, Margaret Mitchell, C. Lawrence Zitnick, Dhruv Batra, Devi Parikh, 2015，`http://arxiv.org/pdf/1505.00468v4.pdf`）的启发。

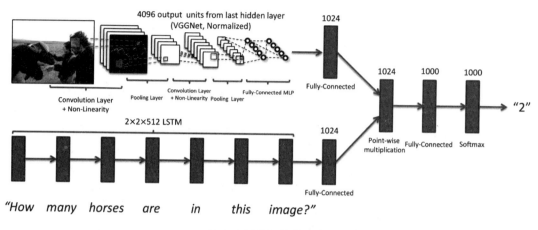

论文中视觉问答的案例

唯一的区别就是这里将图像层产生的特征和语言层产生的特征连接起来。

具体做法

1. 加载需要的 Keras 模块。包括用于词语嵌入的 spaCy，用于图像特征提取的 VGG16 以及用于语言建模的 LSTM。其余几个模块也是标配：

```
%matplotlib inline
import os, argparse
import numpy as np
import cv2 as cv2
import spacy as spacy
import matplotlib.pyplot as plt
from keras.models import Model, Input
from keras.layers.core import Dense, Dropout, Reshape
from keras.layers.recurrent import LSTM
from keras.layers.merge import concatenate
from keras.applications.vgg16 import VGG16
from keras.preprocessing import image
from keras.applications.vgg16 import preprocess_input
from sklearn.externals import joblib
import PIL.Image
```

2. 定义一些常量。请注意，假设问题语料库设定 `max_length_questions=30`，并且将使用 VGG16 来提取描述输入图像的 4096 个特征。另外，知道单词嵌入具有 `length_feature_space=300` 的空间。请注意，将使用从互联网下载的一组预训练权重（`https://github.com/iamaaditya/VQA_Demo`）：

```
# mapping id -> labels for categories
label_encoder_file_name =
'/Users/gulli/Books/TF/code/git/tensorflowBook/Chapter5/FULL_labelencoder_t
rainval.pkl'
# max length across corpus
max_length_questions = 30
# VGG output
length_vgg_features = 4096
# Embedding outout
length_feature_space = 300
# pre-trained weights
VQA_weights_file =
'/Users/gulli/Books/TF/code/git/tensorflowBook/Chapter5/VQA_MODEL_WEIGHTS.h
df5'
```

3. 使用VGG16提取特征。注意从fc2层将特征图像提取出来。给定一幅输入图像，将返回4096个特征。

```
'''image features'''
def get_image_features(img_path, VGG16modelFull):
'''given an image returns a tensor with (1, 4096) VGG16 features'''
# Since VGG was trained as a image of 224x224, every new image
# is required to go through the same transformation
img = image.load_img(img_path, target_size=(224, 224))
x = image.img_to_array(img)
# this is required because of the original training of VGG was batch
# even if we have only one image we need to be consistent
x = np.expand_dims(x, axis=0)
x = preprocess_input(x)
features = VGG16modelFull.predict(x)
model_extractfeatures = Model(inputs=VGG16modelFull.input,
outputs=VGG16modelFull.get_layer('fc2').output)
fc2_features = model_extractfeatures.predict(x)
fc2_features = fc2_features.reshape((1, length_vgg_features))
return fc2_features
```

注意VGG16网络定义如下：

```
Layer (type) Output Shape Param #
=================================================================
input_5 (InputLayer) (None, 224, 224, 3) 0
_____
block1_conv1 (Conv2D) (None, 224, 224, 64) 1792
_____
block1_conv2 (Conv2D) (None, 224, 224, 64) 36928
_____
block1_pool (MaxPooling2D) (None, 112, 112, 64) 0
_____
block2_conv1 (Conv2D) (None, 112, 112, 128) 73856
_____
block2_conv2 (Conv2D) (None, 112, 112, 128) 147584
_____
block2_pool (MaxPooling2D) (None, 56, 56, 128) 0
_____
block3_conv1 (Conv2D) (None, 56, 56, 256) 295168
_____
block3_conv2 (Conv2D) (None, 56, 56, 256) 590080
```

```
block3_conv3 (Conv2D)       (None, 56, 56, 256)    590080
block3_pool (MaxPooling2D)  (None, 28, 28, 256)    0
block4_conv1 (Conv2D)       (None, 28, 28, 512)    1180160
block4_conv2 (Conv2D)       (None, 28, 28, 512)    2359808
block4_conv3 (Conv2D)       (None, 28, 28, 512)    2359808
block4_pool (MaxPooling2D)  (None, 14, 14, 512)    0
block5_conv1 (Conv2D)       (None, 14, 14, 512)    2359808
block5_conv2 (Conv2D)       (None, 14, 14, 512)    2359808
block5_conv3 (Conv2D)       (None, 14, 14, 512)    2359808
block5_pool (MaxPooling2D)  (None, 7, 7, 512)      0
flatten (Flatten)           (None, 25088)          0
fc1 (Dense)                 (None, 4096)           102764544
fc2 (Dense)                 (None, 4096)           16781312
predictions (Dense)         (None, 1000)           4097000
=================================================================
Total params: 138,357,544
Trainable params: 138,357,544
Non-trainable params: 0
```

4. 使用spaCy获取一个单词嵌入并将输入问题映射到空间(`max_length_questions,300`)，其中`max_length_questions`是语料库中问题的最大长度，300是spaCy产生的嵌入的维数。在内部，spaCy使用一个名为gloVe的算法（http://nlp.stanford.edu/projects/glove/）。gloVe算法将给定的标记减少到300维。请注意，关键是从右边由0开始填充`max-lengh-questions`：

```
'''embedding'''
def get_question_features(question):
''' given a question, a unicode string, returns the time series vector
with each word (token) transformed into a 300 dimension representation
calculated using Glove Vector '''
word_embeddings = spacy.load('en', vectors='en_glove_cc_300_1m_vectors')
tokens = word_embeddings(question)
ntokens = len(tokens)
if (ntokens > max_length_questions) :
ntokens = max_length_questions
question_tensor = np.zeros((1, max_length_questions, 300))
for j in xrange(len(tokens)):
question_tensor[0,j,:] = tokens[j].vector
return question_tensor
```

5. 载入一幅图像，利用先前定义的图像特征提取器提取显著特征：

```
image_file_name = 'girl.jpg'
img0 = PIL.Image.open(image_file_name)
img0.show()
#get the salient features
model = VGG16(weights='imagenet', include_top=True)
image_features = get_image_features(image_file_name, model)
print image_features.shape
```

6. 输入一个问题,并通过使用以前定义的句子特征提取方法得到它的显著特征:

```
question = u"Who is in this picture?"
language_features = get_question_features(question)
print language_features.shape
```

7. 将两类不同的特征组合成一个。在这个网络中,有三个 LSTM 层用于语言模型的创建。LSTM 将在第 6 章详细讨论,现在只将它们作为黑箱。最后一个 LSTM 返回 512 个特征,然后用作一系列密集层和 dropout 层的输入。最后一层是一个由 1000 个潜在答案组成的具有 softmax 激活函数的概率空间密集层。

```
'''combine'''
def build_combined_model(
number_of_LSTM = 3,
number_of_hidden_units_LSTM = 512,
number_of_dense_layers = 3,
number_of_hidden_units = 1024,
activation_function = 'tanh',
dropout_pct = 0.5
):
#input image
input_image = Input(shape=(length_vgg_features,),
name="input_image")
model_image = Reshape((length_vgg_features,),
input_shape=(length_vgg_features,))(input_image)
#input language
input_language = Input(shape=(max_length_questions,length_feature_space,),
name="input_language")
#build a sequence of LSTM
model_language = LSTM(number_of_hidden_units_LSTM,
return_sequences=True,
name = "lstm_1")(input_language)
model_language = LSTM(number_of_hidden_units_LSTM,
return_sequences=True,
name = "lstm_2")(model_language)
model_language = LSTM(number_of_hidden_units_LSTM,
return_sequences=False,
name = "lstm_3")(model_language)
#concatenate 4096+512
model = concatenate([model_image, model_language])
#Dense, Dropout
for _ in xrange(number_of_dense_layers):
model = Dense(number_of_hidden_units,
kernel_initializer='uniform')(model)
model = Dropout(dropout_pct)(model)
model = Dense(1000,
activation='softmax')(model)
#create model from tensors
```

```
model = Model(inputs=[input_image, input_language], outputs = model)
return model
```

8. 建立组合的网络,并显示其摘要只是为了了解它的内部。加载预先训练好的权重并使用 rmsprop 优化器,使用 categorical_crossentropy 损失函数编译模型:

```
combined_model = build_combined_model()
combined_model.summary()
combined_model.load_weights(VQA_weights_file)
combined_model.compile(loss='categorical_crossentropy',
optimizer='rmsprop')
```

```
_____
Layer (type)          Output Shape      Param #   Connected to
====================================================================
input_language (InputLayer)  (None, 30, 300)   0
_____
lstm_1 (LSTM)         (None, 30, 512)   1665024   input_language[0][0]
_____
input_image (InputLayer)     (None, 4096)      0
_____
lstm_2 (LSTM)         (None, 30, 512)   2099200   lstm_1[0][0]
_____
reshape_3 (Reshape)   (None, 4096)      0         input_image[0][0]
_____
lstm_3 (LSTM)         (None, 512)       2099200   lstm_2[0][0]
_____
concatenate_3 (Concatenate)  (None, 4608)  0     reshape_3[0][0]
                                                  lstm_3[0][0]
_____
dense_8 (Dense)       (None, 1024)      4719616   concatenate_3[0][0]
_____
dropout_7 (Dropout)   (None, 1024)      0         dense_8[0][0]
_____
dense_9 (Dense)       (None, 1024)      1049600   dropout_7[0][0]
_____
dropout_8 (Dropout)   (None, 1024)      0         dense_9[0][0]
_____
dense_10 (Dense)      (None, 1024)      1049600   dropout_8[0][0]
_____
dropout_9 (Dropout)   (None, 1024)      0         dense_10[0][0]
_____
dense_11 (Dense)      (None, 1000)      1025000   dropout_9[0][0]
====================================================================
```

```
=========================
Total params: 13,707,240
Trainable params: 13,707,240
Non-trainable params: 0
```

9.使用预先训练的组合网络进行预测。请注意,在这种情况下,使用网上已有的权重,但感兴趣的读者可以在自己的训练集上重新训练联合网络:

```
y_output = combined_model.predict([image_features, language_features])
# This task here is represented as a classification into a 1000 top answers
# this means some of the answers were not part of training and thus would
# not show up in the result.
# These 1000 answers are stored in the sklearn Encoder class
labelencoder = joblib.load(label_encoder_file_name)
for label in reversed(np.argsort(y_output)[0,-5:]):
    print str(round(y_output[0,label]*100,2)).zfill(5), "% ",
    labelencoder.inverse_transform(label)
```

解读分析

视觉问答的任务是通过使用不同的深度神经网络的组合来解决的。使用预先训练的 VGG16 来从图像中提取特征,并且使用一系列 LSTM 网络从先前映射到嵌入空间的问题中提取特征。VGG16 是用于图像特征提取的 CNN,而 LSTM 是用于提取代表序列的时间特征的 RNN。这两者的结合是目前处理这种网络的最先进的技术。然后在组合模型的顶部添加一个带有 dropout 层的多层感知器,以形成深度网络。

更多内容

你可以在网上找到更多 Avi Singh 的相关实验(`https://avisingh599.github.io/deeplearning/visual-qa/`),其中对不同的模型进行了比较,包括一个带有图像 CNN 模型的简单"词袋"模型,一个 LSTM 模型和一个类似本节中的 LSTM + CNN 模型。博客文章还讨论了每个模型的不同训练策略。

除此之外,感兴趣的读者可以在互联网上找到一个很好的图形用户界面(`https://github.com/anujshah1003/VQA-Demo-GUI`),这个图形界面建立在 Avi Singh 的 demo 上面,它允许你交互地加载图像并询问相关问题。YouTube 视频也能在网站 `https://www.youtube.com/watch?v=7FB9PvzOuQY` 上找到。

5.9 利用预训练网络进行视频分类的 6 种方法

对视频进行分类是一个活跃的研究领域,因为处理这种类型的问题需要大量的数据。内存需求经常达到现代 GPU 的极限,可能需要在多台机器上进行分布式的训练。目前学者们正在探索复杂度不断增加的几个方向,来回顾一下。

第一种方法是通过将视频的每一帧视为一幅单独的图像,利用二维 CNN 进行处理。这种方法将视频分类问题简化为图像分类问题。每帧视频图像都有类别输出,并且根据各帧输出的类别,选择频率最高的类别作为视频的分类结果。

第二种方法是创建一个单一的网络,将二维 CNN 与一个 RNN 结合在一起。这个想法是,CNN 将考虑到图像分量,而 RNN 将考虑每个视频的序列信息。这种类型的网络可能非常难以训练,因为要优化的参数数量非常大。

第三种方法是使用三维卷积网络,其中三维卷积网络是二维 CNN 的在 3D 张量(时间,图像宽度,图像高度)上运行的扩展。这种方法是图像分类的另一个自然延伸,但三维卷积网络可能很难训练。

第四种方法基于智能方法的直觉。它们可以用于存储视频中每个帧的离线功能,而不是直接使用 CNN 进行分类。这个想法基于,特征提取可以非常有效地进行迁移学习,如前面章节所示。在提取所有的特征之后,可以将它们作为一组输入传递给 RNN,其将在多个帧中学习序列并输出最终的分类。

第五种方法是第四种方法的简单变体,其中最后一层是 MLP 而不是 RNN。在某些情况下,就计算需求而言,这种方法可以更简单并且成本更低。

第六种方法也是第四种方法的变体,其中特征提取阶段采用三维 CNN 来提取空间和视觉特征,然后将这些特征传递给 RNN 或 MLP。

使用哪种方法取决于具体应用,并没有统一的答案。前三种方法通常计算量更大,而后三种方法计算成本更低,而且性能更好。

在本节中将展示如何使用第六种方法。利用论文 "*Temporal Activity Detection in Untrimmed Videos with Recurrent Neural Networks*"(Montes, Alberto and Salvador, Amaia and Pascual, Santiago and Giro-i-Nieto, Xavier,2016,`https://arxiv.org/abs/1608.08128`)中的实验结果。这项工作旨在解决 ActivityNet 挑战赛中的问题(`http://activity-net.org/challenges/2016/`),重点是从用户生成的视频中识别高层次和目标导向的活动,类似于互联网门户中的活动。面临的挑战是如何在两个不同的任务中生成 200 个活动类别:

- 分类挑战:给定一个长视频,预测视频中的活动标签。
- 检测挑战:给定一个长视频,预测视频中活动的标签和时间范围。

提出的架构由两个阶段组成,如下图所示。第一阶段将视频信息编码成小视频剪辑的单个矢量表示。为了达到这个目的,使用 C3D 网络。C3D 网络使用 3D 卷积来从视频中提取时空特征,这些特征在前面已被分成 16 帧的剪辑。

第二阶段,一旦提取到视频特征,就要对每个片段上的活动进行分类。为了执行这种分类,使用 RNN。具体来说,使用 LSTM 网络,它尝试利用长期相关性,并且执行视频序列的预测。这是一个训练阶段:

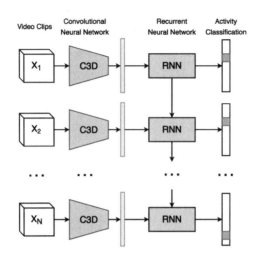

C3D+RNN 示例,见 https://imatge-upc.github.io/activitynet-2016-cvprw/

具体做法

本节简单总结了网站(https://github.com/imatge-upc/activitynet-2016-cvprw/blob/master/misc/nstep_by_step_guide.md)中的结果。

1. 从 git 库中克隆压缩包:

```
git clone https://github.com/imatge-upc/activitynet-2016-cvprw.git
```

2. 下载 ActivityNet v1.3 数据集,大小为 600GB:

```
cd dataset
# This will download the videos on the default directory
sh download_videos.sh username password
# This will download the videos on the directory you specify
sh download_videos.sh username password /path/you/want
```

3. 下载 CNN3d 和 RNN 的预训练权重:

```
cd data/models
sh get_c3d_sports_weights.sh
sh get_temporal_location_weights.sh
```

4. 进行视频分类:

```
python scripts/run_all_pipeline.py -i path/to/test/video.mp4
```

解读分析

如果你对在自己的机器上训练 CNN3D 和 RNN 网络感兴趣,那么可以在互联网上找到本机训练需要使用的特定命令。

目的是提供可用于视频分类的不同方法的高级视图。同样，不是仅有一种方法，而是有多种选择，应该根据具体需求选择最佳方案。

更多内容

CNN-LSTM 体系结构是一个新的 RNN 层，输入变换和循环变换的输入都是卷积的。尽管命名非常相似，但 CNN-LSTM 层不同于 CNN 和 LSTM 的组合。该模型在论文"*Convolutional LSTM Network: A Machine Learning Approach for Precipitation Nowcasting*"（Xingjian Shi, Zhourong Chen, Hao Wang, Dit-Yan Yeung, Wai-kin Wong, Wang-chun Woo, 2015，`https://arxiv.org/abs/1506.04214`）中被提出。2017 年一些人开始使用此模块的视频进行实验，但这仍是一个非常活跃的研究领域。

CHAPTER 6

第 6 章

循环神经网络

6.1 引言

本章将讨论如何使用**循环神经网络**（Recurrent Neural Network，RNN）在保持序列顺序很重要的应用领域中，进行深度学习，我们主要关注文本分析和**自然语言处理**（NLP），但也会看到用于预测比特币价格的序列例子。

很多实时情况都能通过时间序列模型来描述。例如，如果你想写一个文档，单词的顺序很重要，当前的单词肯定取决于以前的单词。如果把注意力放在文字写作上，很明显，一个单词中的下一个字符取决于之前的字符（例如，*The quick brown f...*，下一个字母是 o 的概率很高），如下图所示。关键思想是在给定上下文的情况下产生下一个字符的分布，然后从分布中取样产生下一个候选字符：

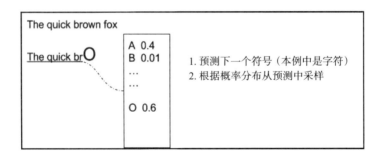

关于"The quick brown fox"句子的预测示例

一个简单的变体是存储多个预测值，并创建一个预测扩展树，如下图所示：

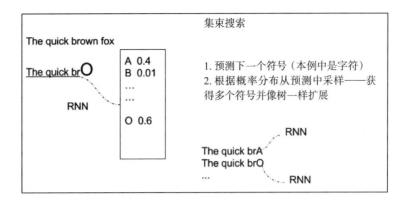

关于"The quick brown fox"句子的预测树示例

基于序列的模型可以用在很多领域中。在音乐中,一首曲子的下一个音符肯定取决于前面的音符,而在视频领域,电影中的下一帧肯定与先前的帧有关。此外,在某些情况下,视频的当前帧、单词、字符或音符不仅仅取决于过去的信号,而且还取决于未来的信号。

基于时间序列的模型可以用 RNN 来描述,其中,时刻 i 输入为 X_i,输出为 Y_i,时刻 $[0, i-1]$ 区间的状态信息被反馈至网络。这种反馈过去状态的思想被循环描述出来,如下图所示:

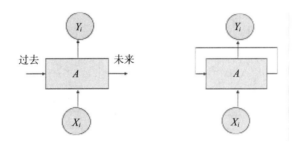

反馈的描述

展开(unfolding)网络可以更清晰地表达循环关系,如下图所示:

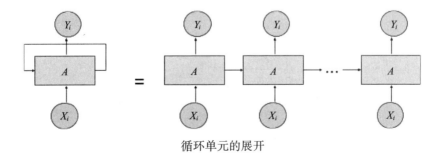

循环单元的展开

最简单的 RNN 单元由简单的 $tanh$ 函数组成,即双曲正切函数,如下图所示:

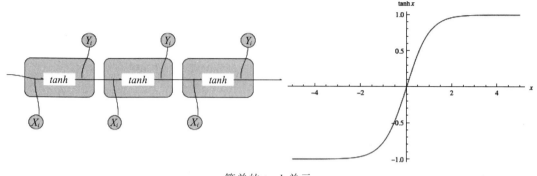

简单的 tanh 单元

梯度消失与梯度爆炸

由于存在两个稳定性问题，训练 RNN 是很困难的。由于反馈环路的缘故，梯度可以很快地发散到无穷大，或者迅速变为 0。如下图所示，在这两种情况下，网络将停止学习任何有用的东西。梯度爆炸的问题可以通过一个简单的策略来解决，就是**梯度裁剪**。梯度消失的问题则难以解决，它涉及更复杂的 RNN 基本单元（例如**长短时记忆（LSTM）网络或门控循环单元（GRU）**）的定义。先来讨论梯度爆炸和梯度裁剪：

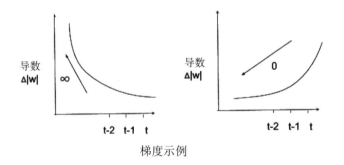

梯度示例

梯度裁剪包括对梯度限定最大值，以使其不能无界增长。如下图所示，该方法提供了一个解决梯度爆炸问题的简单方案：

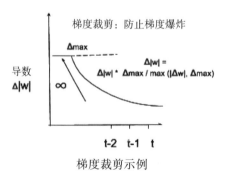

梯度裁剪示例

解决梯度消失需要一个更复杂的记忆模型,它可以有选择地忘记以前的状态,只记住真正重要的状态。如下图所示,将输入以概率 $p \in [0, 1]$ 写入记忆块 M,并乘以输入的权重。以类似的方式,以概率 $p \in [0, 1]$ 读取输出,并乘以输出的权重。再用一个概率来决定要记住或忘记什么:

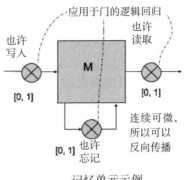

记忆单元示例

长短时记忆网络

长短时记忆网络(LSTM)可以控制何时让输入进入神经元,何时记住之前时序中学到的东西,以及何时让输出传递到下一个时间戳。所有这些决策仅仅基于输入就能自我调整。乍一看,LSTM 看起来很难理解,但事实并非如此。我们用下图来解释它是如何工作的:

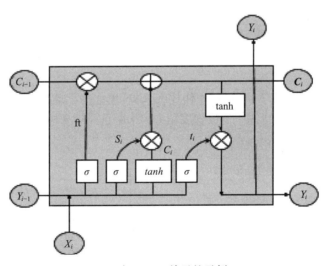

一个 LSTM 单元的示例

首先,需要一个逻辑函数 σ(参考第 2 章)计算出介于 0 和 1 之间的值,并且控制哪个信息片段流经 LSTM 门。请记住,logisitic 函数是可微的,所以它允许反向传播。然后需要

一个运算符 ⊗ 对两个相同维数的矩阵进行点乘产生一个新矩阵，其中新矩阵的第 ij 个元素是两个原始矩阵第 ij 个元素的乘积。同样，需要一个运算符 ⊕ 将两个相同维数的矩阵相加，其中新矩阵的第 ij 个元素是两个原始矩阵第 ij 个元素的和。在这些基本模块中，将 i 时刻的输入 x_i 与前一步的输出 y_i 放在一起。

方程 $f_i = \sigma(W_f \cdot [y_{i-1}, x_i]+b_f)$ 是逻辑回归函数，通过控制激活门 ⊗ 决定前一个单元状态 C_{i-1} 中有多少信息应该传输给下一个单元状态 C_i（W_f 是权重矩阵，b_f 是偏置）。逻辑输出 1 意味着完全保留先前单元状态 C_t-1，输出 0 代表完全忘记 C_{i-1}，输出（0,1）中的数值则代表要传递的信息量。

接着，方程 $\hat{C}_i = \tanh(W_c \cdot [Y_{i-1}, X_i]+b_c)$ 根据当前输入产生新信息，方程 $s_i = \sigma(W_s \cdot [Y_{i-1}, X_i]+b_s)$ 则能控制有多少新信息 $\hat{C}_i$ 通过运算符 ⊕ 被加入到单元状态 C_i 中。

利用运算符 ⊗ 和 ⊕，给出公式 $C_i = f_i * C_{i-1} + s_i * \hat{C}_i$ 对单元状态进行更新。

最后，需要确定当前单元状态的哪些信息输出到 Y_i。很简单，再次采用逻辑回归方程，通过 ⊗ 运算符控制候选值的哪一部分应该输出。在这里有一点需要注意，单元状态是通过 tanh 函数压缩到 [-1, 1]。这部分对应的方程是 $Y_i = t_i * \tanh(C_i)$。

现在我明白这看起来像很多数学理论，但有两个好消息。首先，如果你明白想要达到的目标，那么数学部分就不是那么难；其次，你可以使用 LSTM 单元作为标准 RNN 单元的黑盒替换，并立即解决梯度消失问题。因此你真的不需要知道所有的数学理论，你只需从库中取出 TensorFlow LSTM 并使用它。

门控循环单元和窥孔 LSTM

近年来已经提出了许多 LSTM 的变种模型，其中有两个很受欢迎：窥孔（peephole）LSTM 允许门层查看单元状态，如下图中虚线所示；而门控循环单元（GRU）将隐藏状态和单元状态合并为一个信息通道。

同样，GRU 和窥孔 LSTM 都可以用作标准 RNN 单元的黑盒插件，而不需要知道底层数学理论。这两种单元都可以用来解决梯度消失的问题，并用来构建深度神经网络。

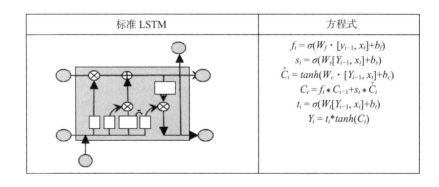

标准 LSTM	方程式
	$f_t = \sigma(W_f \cdot [y_{i-1}, x_i]+b_f)$ $s_i = \sigma(W_s[Y_{i-1}, x_i]+b_s)$ $\check{C}_i = tanh(W_c \cdot [Y_{i-1}, x_i]+b_c)$ $C_i = f_i * C_{i-1} + s_i * \check{C}_i$ $t_i = \sigma(W_t[Y_{i-1}, x_i]+b_t)$ $Y_i = t_i * tanh(C_i)$

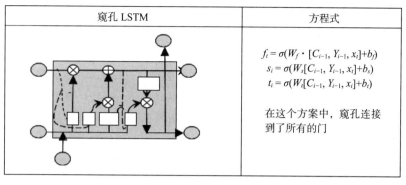

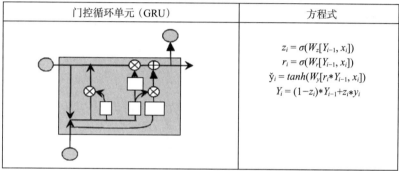

标准 LTSM、窥孔 LTSM、GRU 示例

处理向量序列

真正使 RNN 强大的是它能够处理向量序列,其中 RNN 的输入和输出可以是序列,下图很好地说明了这一点,最左边的例子是一个传统(非递归)网络,后面跟着一个序列输出的 RNN,接着跟着一个序列输入的 RNN,其次跟着序列输入和序列输出不同步的 RNN,最后是序列输入和序列输出同步的 RNN。

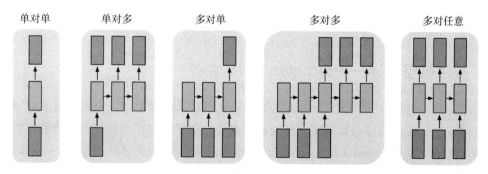

RNN 序列示例(http://karpathy.github.io/2015/05/21/rnn-effectiveness/)

机器翻译是输入序列和输出序列中不同步的一个例子:网络将输入文本作为一个序列

读取，读完全文后输出目标语言。

视频分类是输入序列和输出序列同步的一个例子：视频输入是一系列帧，对于每一帧，在输出中提供分类标签。

如果你想知道更多关于 RNN 的有趣应用，则必读 Andrej Karpathy 的博客，`http://karpathy.github.io/2015/05/21/rnn-effectiveness/`，他训练网络写莎士比亚风格的散文（用 Karpathy 的话说：能勉强承认是莎士比亚的真实样品），写玄幻主题的维基百科文章，写愚蠢不切实际问题的定理证明（用 Karpathy 的话说：更多幻觉代数几何），并写出 Linux 代码片段（用 Karpathy 的话说：模型首先逐字列举了 GNU 许可字符，产生一些宏然后隐藏到代码中）。

以下示例摘自 `http://karpathy.github.io/2015/05/21/rnn-effectiveness/`：

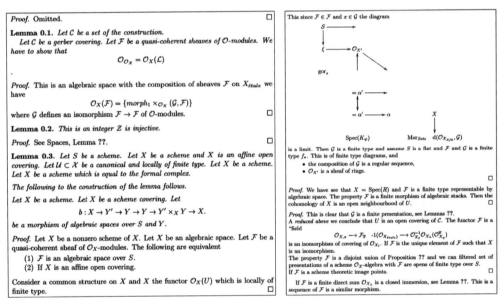

RNN 生成的文本示例

6.2 神经机器翻译——seq2seq RNN 训练

seq2seq 是一类特殊的 RNN，在机器翻译、文本自动摘要和语音识别中有着成功的应用。在本节中，我们将讨论如何实现神经机器翻译，得到类似于谷歌神经机器翻译系统得到的结果（`https://research.googleblog.com/2016/09/a-neural-network-for-machine.html`）。关键是输入一个完整的文本序列，理解整个语义，然后输出翻译结果作为另一个序列。阅读整个序列的想法与以前的架构截然不同，在该架构中，一组固定词汇从一种源语言翻译成目标语言。

本节受到了 Minh-Thang Luong 于 2016 年所写的博士论文"Neural Machine Translation"（https://github.com/lmthang/thesis/blob/master/thesis.pdf）的启发。第一个关键概念是编码器 – 解码器架构，其中编码器将源语句转换为表示语义的向量，然后这个向量通过解码器产生翻译结果。编码器和解码器都是 RNN，它们可以捕捉语言中的长距离依赖关系，例如性别一致性和语法结构，而不必事先知道它们，也不需要跨语言进行 1:1 映射。它能够流利地翻译并且具有强大的功能。

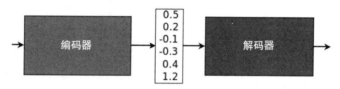

编码器 – 解码器示例（见 https://github.com/lmthang/thesis/blob/master/thesis.pdf）

来看一个 RNN 例子：将 She loves cute cats 翻译成 Elle aime les chats mignons。

有两个 RNN：一个充当编码器，一个充当解码器。源语句 She loves cute cats 后面跟着一个分隔符"-"和目标语句 Elle aime les chats mignon。这两个关联语句被输入给编码器用于训练，并且解码器将产生目标语句 Elle aime les chats mignon。当然，需要大量类似例子来获得良好的训练。

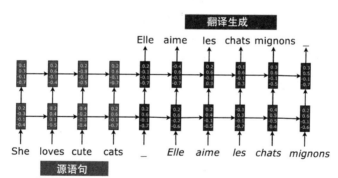

(NMT 序列模型——一个深度循环结构示例，用于将源语句"She loves cute cats"翻译成目标语句"Elle aimel les chats mignons"。解码器侧，前面时序中产生的单词作为输出下一个单词的输入，"_"代表语句的结束)

NMT 的序列模型示例（见 https://github.com/lmthang/thesis/blob/master/thesis.pdf）

现在有一些可以使用的 RNN 变体，具体介绍其中的一些：
- RNN 可以是单向的或双向的，后者将捕捉双向的长时间依赖关系。
- RNN 可以有多个隐藏层，层数的选择对于优化来说至关重要，一方面，更深的网络可以学到更多知识，另一方面，训练需要花费很长时间而且可能会过度拟合。
- RNN 可以具有嵌入层，其将单词映射到嵌入空间中，在嵌入空间中相似单词的映射恰好也非常接近。
- RNN 可以使用简单的重复性单元、LSTM、窥孔 LSTM 或者 GRU。

仍然考虑博士论文"Neural Machine Translation"(https://github.com/lmthang/thesis/blob/master/thesis.pdf),可以使用嵌入层将输入的句子映射到一个嵌入空间。然后,存在两个连接在一起的 RNN——源语言的编码器和目标语言的解码器。如下图所示,有多个隐藏层和两个流动方向:前馈垂直方向连接隐藏层,水平方向是将知识从上一步转移到下一步的循环部分。

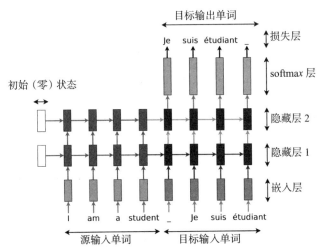

神经机器翻译——一个深度循环结构示例,由 Sutskever 等人于 2014 年提出,将源语句"I am a student"翻译成"Je suis etudiant"。此处,"_"代表语句的结束

神经机器翻译示例(参考 https://github.com/lmthang/thesis/blob/master/thesis.pdf)

本节使用 NMT(Neural Machine Translation,神经机器翻译),这是一个在 TensorFlow 上在线可得的翻译演示包。

准备工作

NMT 可通过 https://github.com/tensorflow/nmt/ 获取,具体代码可通过 GitHub 获取。

具体做法

1. 从 GitHub 克隆 NMT:

```
git clone https://github.com/tensorflow/nmt/
```

2. 下载一个训练数据集。在这个例子中,使用训练集将越南语翻译成英语,其他数据集可以在 https://nlp.stanford.edu/projects/nmt/ 上获得,如德语和捷克语:

```
nmt/scripts/download_iwslt15.sh /tmp/nmt_data
```

3. 参考 https://github.com/tensorflow/nmt/，这里将定义第一个嵌入层，嵌入层将输入、词汇量尺寸 V 和期望的输出尺寸嵌入到空间中。词汇量尺寸 V 中只有最频繁的单词才考虑被嵌入，所有其他单词则被打上 unknown 标签。在本例中，输入是 time-major，这意味着 max time 是第一个输入参数（https://www.tensorflow.org/api_docs/python/tf/nn/dynamic_rnn）：

```
# Embedding
embedding_encoder = variable_scope.get_variable(
"embedding_encoder", [src_vocab_size, embedding_size], ...)
# Look up embedding:
# encoder_inputs: [max_time, batch_size]
# encoder_emb_inp: [max_time, batch_size, embedding_size]
encoder_emb_inp = embedding_ops.embedding_lookup(
embedding_encoder, encoder_inputs)
```

4. 仍然参考 https://github.com/tensorflow/nmt/，这里定义一个简单的编码器，它使用 `tf.nn.rnn_cell.BasicLSTMCell(num_units)` 作为基本的 RNN 单元。虽然很简单，但要注意给定基本 RNN 单元，我们利用 `tf.nn.dynamic_rnn` 构建了 RNN 的（见 https://www.tensorflow.org/api_docs/python/tf/nn/dynamic_rnn）：

```
# Build RNN cell
encoder_cell = tf.nn.rnn_cell.BasicLSTMCell(num_units)

# Run Dynamic RNN
# encoder_outpus: [max_time, batch_size, num_units]
# encoder_state: [batch_size, num_units]
encoder_outputs, encoder_state = tf.nn.dynamic_rnn(
encoder_cell, encoder_emb_inp,
sequence_length=source_sequence_length, time_major=True)
```

5. 定义解码器。首先要有一个基本的 RNN 单元：`tf.nn.rnn_cell.BasicLSTMCell`，以此来创建一个基本的采样解码器 `tf.contrib.seq2seq.BasicDecoder`，将结果输入到解码器 `tf.contrib.seq2seq.dynamic_decode` 中进行动态解码。

```
# Build RNN cell
decoder_cell = tf.nn.rnn_cell.BasicLSTMCell(num_units)
# Helper
helper = tf.contrib.seq2seq.TrainingHelper(
decoder_emb_inp, decoder_lengths, time_major=True)
# Decoder
decoder = tf.contrib.seq2seq.BasicDecoder(
decoder_cell, helper, encoder_state,
output_layer=projection_layer)
# Dynamic decoding
outputs, _ = tf.contrib.seq2seq.dynamic_decode(decoder, ...)
logits = outputs.rnn_output
```

6. 网络的最后一个阶段是 softmax dense 阶段，将最高隐藏状态转换为 logit 向量：

```
projection_layer = layers_core.Dense(
tgt_vocab_size, use_bias=False)
```

7. 定义在训练阶段使用的交叉熵函数和损失：

```
crossent = tf.nn.sparse_softmax_cross_entropy_with_logits(
 labels=decoder_outputs, logits=logits)
train_loss = (tf.reduce_sum(crossent * target_weights) /
 batch_size)
```

8. 定义反向传播所需的步骤，并使用适当的优化器（本例中使用Adam）。请注意，梯度已被剪裁，Adam使用预定义的学习率：

```
# Calculate and clip gradients
params = tf.trainable_variables()
gradients = tf.gradients(train_loss, params)
clipped_gradients, _ = tf.clip_by_global_norm(
 gradients, max_gradient_norm)
# Optimization
optimizer = tf.train.AdamOptimizer(learning_rate)
update_step = optimizer.apply_gradients(
 zip(clipped_gradients, params))
```

9. 运行代码并理解不同的执行步骤。首先，创建训练图，然后开始迭代训练。评价指标是BLEU（bilingual evaluation understudy），这个指标是评估将一种自然语言机器翻译为另一种自然语言的文本质量的标准，质量被认为是算法的结果和人工操作结果的一致性。正如你所看到的，指标值随着时间而增长：

```
python -m nmt.nmt --src=vi --tgt=en --vocab_prefix=/tmp/nmt_data/vocab --
train_prefix=/tmp/nmt_data/train --dev_prefix=/tmp/nmt_data/tst2012 --
test_prefix=/tmp/nmt_data/tst2013 --out_dir=/tmp/nmt_model --
num_train_steps=12000 --steps_per_stats=100 --num_layers=2 --num_units=128
--dropout=0.2 --metrics=bleu
# Job id 0
[...]
# creating train graph ...
num_layers = 2, num_residual_layers=0
cell 0 LSTM, forget_bias=1 DropoutWrapper, dropout=0.2 DeviceWrapper,
device=/gpu:0
cell 1 LSTM, forget_bias=1 DropoutWrapper, dropout=0.2 DeviceWrapper,
device=/gpu:0
cell 0 LSTM, forget_bias=1 DropoutWrapper, dropout=0.2 DeviceWrapper,
device=/gpu:0
cell 1 LSTM, forget_bias=1 DropoutWrapper, dropout=0.2 DeviceWrapper,
device=/gpu:0
start_decay_step=0, learning_rate=1, decay_steps 10000,decay_factor 0.98
[...]
# Start step 0, lr 1, Thu Sep 21 12:57:18 2017
# Init train iterator, skipping 0 elements
global step 100 lr 1 step-time 1.65s wps 3.42K ppl 1931.59 bleu 0.00
global step 200 lr 1 step-time 1.56s wps 3.59K ppl 690.66 bleu 0.00
[...]
global step 9100 lr 1 step-time 1.52s wps 3.69K ppl 39.73 bleu 4.89
global step 9200 lr 1 step-time 1.52s wps 3.72K ppl 40.47 bleu 4.89
global step 9300 lr 1 step-time 1.55s wps 3.62K ppl 40.59 bleu 4.89
[...]
# External evaluation, global step 9000
decoding to output /tmp/nmt_model/output_dev.
```

```
done, num sentences 1553, time 17s, Thu Sep 21 17:32:49 2017.
bleu dev: 4.9
saving hparams to /tmp/nmt_model/hparams
# External evaluation, global step 9000
decoding to output /tmp/nmt_model/output_test.
done, num sentences 1268, time 15s, Thu Sep 21 17:33:06 2017.
bleu test: 3.9
saving hparams to /tmp/nmt_model/hparams
[...]
global step 9700 lr 1 step-time 1.52s wps 3.71K ppl 38.01 bleu 4.89
```

解读分析

所有上述代码已经在 https://github.com/tensorflow/nmt/blob/master/nmt/model.py 上给出。关键是将两个 RNN 打包在一起，第一个是嵌入空间的编码器，将相似的单词映射得很接近，编码器理解训练样例的语义，并产生一个张量作为输出。然后通过将编码器的最后一个隐藏层连接到解码器的初始层可以简单地将该张量传递给解码器。请注意，学习能够进行是因为损失函数基于交叉熵，且 `labels=decoder_outputs`。

如下图所示，代码学习如何翻译，并通过 BLEU 指标的迭代跟踪进度：

TensorBoard 中的 BLEU 指标示例

6.3 神经机器翻译——seq2seq RNN 推理

本节使用前一节的结果将源语言翻译成目标语言。这个想法非常简单：一个源语句作为两个组合的 RNN（编码器 + 解码器）的输入。一旦句子结束，解码器将发出 logit 值，采用贪婪策略输出与最大值相关的单词。例如，单词 *moi* 作为来自解码器的第一个标记被输出，因为这个单词具有最大的 logit 值。此后，单词 *suis* 输出，等等：

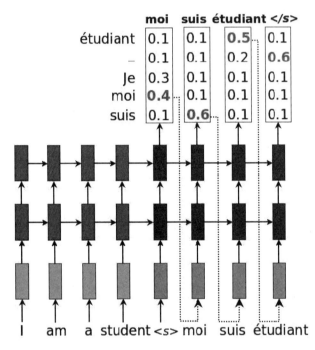

带概率分布的 NMT 序列模型示例（参考 https://github.com/lmthang/thesis/blob/master/thesis.pdf）

解码器的输出有多种策略：
- **贪婪**：输出对应最大 logit 值的单词。
- **采样**：通过对众多 logit 值采样输出单词。
- **集束搜索**：有多个预测，因此创建一个可能结果的扩展树。

具体做法

1. 制定解码器采样的贪婪策略。这很简单，因为可以使用 `tf.contrib.seq2seq.GreedyEmbeddingHelper` 中定义的库，由于不知道目标句子的准确长度，因此这里使用启发式方法将其限制为源语句长度的两倍。

```
# Helper
helper = tf.contrib.seq2seq.GreedyEmbeddingHelper(
    embedding_decoder,
    tf.fill([batch_size], tgt_sos_id), tgt_eos_id)

# Decoder
decoder = tf.contrib.seq2seq.BasicDecoder(
    decoder_cell, helper, encoder_state,
    output_layer=projection_layer)
# Dynamic decoding
outputs, _ = tf.contrib.seq2seq.dynamic_decode(
    decoder, maximum_iterations=maximum_iterations)
translations = outputs.sample_id
maximum_iterations = tf.round(tf.reduce_max(source_sequence_length) * 2)
```

2. 现在可以运行网络，输入一个从未见过的语句（`inference_input_file=/tmp/my_infer_file`），并让网络转换结果（`inference_output_file=/tmp/nmt_model/output_infer`）：

```
python -m nmt.nmt \
  --out_dir=/tmp/nmt_model \
  --inference_input_file=/tmp/my_infer_file.vi \
  --inference_output_file=/tmp/nmt_model/output_infer
```

解读分析

两个 RNN 封装在一起形成编码器–解码器 RNN 网络。解码器发出的 logits 被贪婪策略转换成目标语言的单词。作为一个例子，下面显示了一个从越南语到英语的自动翻译的例子：

- **输入越南语句子**：Khi tôi còn nhỏ , Tôi nghĩ rằng BắcTriều Tiên là đất nước tốt nhất trên thế giới và tôi thường hát bài " Chúng ta chẳng có gì phải ghen tị.
- **输出英语句子**：When I'm a very good , I'm going to see the most important thing about the most important and I'm not sure what I'm going to say.

6.4 你所需要的是注意力—另一个 seq2seq RNN 例子

本节提出了**注意力**（Attention）机制，这是神经网络翻译的最新解决方案。注意力的思想是 2015 年在论文 "*Neural Machine Translation by Jointly Learning to Align and Translate*"（Dzmitry Bahdanau，Kyunghyun Cho 和 Yoshua Bengio，ICLR，2015，`https://arxiv.org/abs/1409.0473`）中提出的，它需要在编码器和解码器 RNN 之间增加额外的连接。事实上，仅将解码器与编码器的最新层连接会存在信息瓶颈，而且不一定能够传递先前编码器层的信息。下图说明了采用注意力机制的方法：

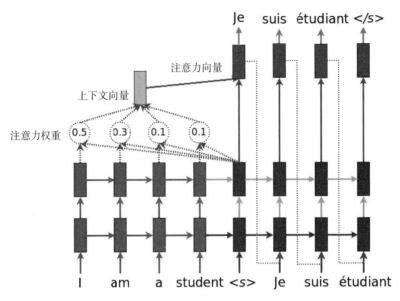

引入注意力模型的 NMT 示例

(参考 https://github.com/lmthang/thesis/blob/master/thesis.pdf)

需要考虑以下三个方面：
- 将当前目标隐藏状态与所有先前的源状态一起使用，以导出注意力权重，用于给先前序列中的信息分配不同的注意力大小。
- 总结注意力权重的结果创建上下文向量。
- 将上下文向量与当前目标隐藏状态相结合以获得注意力向量。

具体做法

1. 通过使用库 `tf.contrib.seq2seq.LuongAttention` 来定义注意力机制，该库实现了文献 "*Effective Approaches to Attention-based Neural Machine Translation*"（Minh-Thang Luong，Hieu Pham 和 Christopher D. Manning，2015）中定义的注意力模型：

```
# attention_states: [batch_size, max_time, num_units]
attention_states = tf.transpose(encoder_outputs, [1, 0, 2])

# Create an attention mechanism
attention_mechanism = tf.contrib.seq2seq.LuongAttention(
 num_units, attention_states,
 memory_sequence_length=source_sequence_length)
```

2. 通过一个注意力包装器，使用所定义的注意力机制作为解码器单元进行封装：

```
decoder_cell = tf.contrib.seq2seq.AttentionWrapper(
 decoder_cell, attention_mechanism,
 attention_layer_size=num_units)
```

3. 运行代码查看结果。可以立即注意到注意力机制在 BLEU 评分方面产生了显著的改善：

```
python -m nmt.nmt \
> --attention=scaled_luong \
> --src=vi --tgt=en \
> --vocab_prefix=/tmp/nmt_data/vocab \
> --train_prefix=/tmp/nmt_data/train \
> --dev_prefix=/tmp/nmt_data/tst2012 \
> --test_prefix=/tmp/nmt_data/tst2013 \
> --out_dir=/tmp/nmt_attention_model \
> --num_train_steps=12000 \
> --steps_per_stats=100 \
> --num_layers=2 \
> --num_units=128 \
> --dropout=0.2 \
> --metrics=bleu
[...]
# Start step 0, lr 1, Fri Sep 22 22:49:12 2017
# Init train iterator, skipping 0 elements
global step 100 lr 1 step-time 1.71s wps 3.23K ppl 15193.44 bleu 0.00
[...]
# Final, step 12000 lr 0.98 step-time 1.67 wps 3.37K ppl 14.64, dev ppl
14.01, dev bleu 15.9, test ppl 12.58, test bleu 17.5, Sat Sep 23 04:35:42
2017
# Done training!, time 20790s, Sat Sep 23 04:35:42 2017.
# Start evaluating saved best models.
[..]
loaded infer model parameters from
/tmp/nmt_attention_model/best_bleu/translate.ckpt-12000, time 0.06s
# 608
src: nhưng bạn biết điều gì không ?
ref: But you know what ?
nmt: But what do you know ?
[...]
# Best bleu, step 12000 step-time 1.67 wps 3.37K, dev ppl 14.01, dev bleu
15.9, test ppl 12.58, test bleu 17.5, Sat Sep 23 04:36:35 2017
```

解读分析

注意力机制是使用编码器 RNN 内部状态获得的信息，并将该信息与解码器的最终状态进行组合的机制，关键思想是可以对源序列中的信息分配不同的注意力。下图的 BLEU 得分显示了应用注意力机制后的优势。

注意到，之前所给出的同样的图中，没有使用注意力机制。

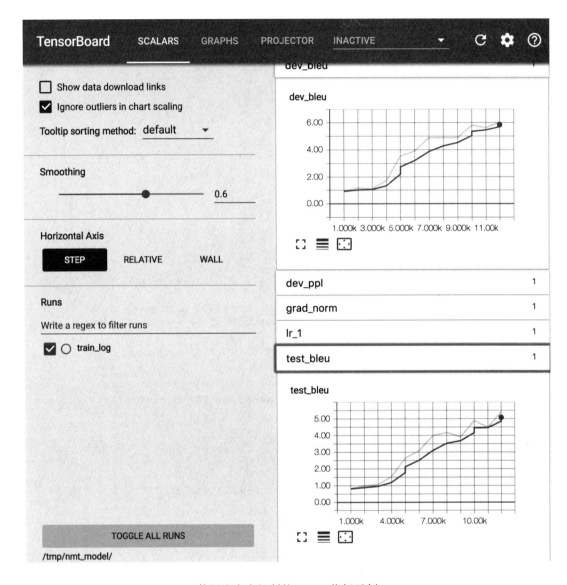

使用注意力机制的 BLEU 指标示例

更多内容

值得注意的是，seq2seq 不仅仅可以用于机器翻译，来看一些例子：

- Lukasz Kaiser 在论文 "Grammar as a Foreign Language"（`https://arxiv.org/abs/1412.7449`）中使用 seq2seq 模型构建语法成分解析器，成分解析树将文本分成子句。树中的非终端是短语类型，终端是句子中的单词，边缘未标记。
- seq2seq 的另一个应用是 SyntaxNet，又名 Parsey McParserFace（一个句法分析

器，`https://research.googleblog.com/2016/05/announcing-syntaxnet-worlds-most.html`），这是许多 NLU 系统上的一个关键组件。在这个系统中输入一个句子，它会自动给句子中的每一个单词打上 POS（Part-of-Speech）标签，用来描述这些词的句法功能，并在依存句法树中呈现。这些句法关系直接涉及句子的潜在含义。

下图给出了这个概念的思想：

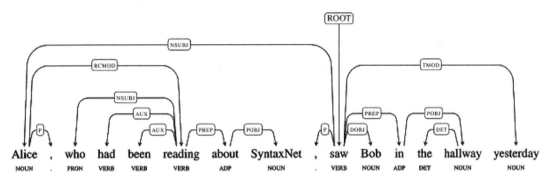

SyntaxNet 示例（参考 `https://research.googleblog.com/2016/05/announcing-syntaxnet-worlds-most.html`）

6.5 使用 RNN 像莎士比亚一样写作

在本节中，我们将学习如何生成类似于莎士比亚风格的文本。核心思想非常简单：以莎士比亚写的真实文本作为输入，并输入到即将要训练的 RNN 中；然后，用训练好的模型来生成新文本，这些文本看起来像是英国最伟大的作家所写的。

> 为了简单起见，这里将使用基于 TensorFlow 运行的框架 TFLearn（`http://tflearn.org/`），这里所使用的例子只是标准版的一部分（参考 `https://github.com/tflearn/tflearn/blob/master/examples/nlp/lstm_generator_shakespeare.py`），所开发的模型是字符级 RNN 语言模型，考虑的序列则是字符序列而不是单词序列。

具体做法

1. 用 `pip` 方式安装 TFLearn：

`pip install -I tflearn`

2. 导入一些有用的模块，并下载莎士比亚写的文本。本例使用的文本位于 `https://raw.githubusercontent.com/tflearn/tflearn.github.io/master/`

resources/shakespeare_input.txt：

```
import os
import pickle
from six.moves import urllib
import tflearn
from tflearn.data_utils import *
path = "shakespeare_input.txt"
char_idx_file = 'char_idx.pickle'
if not os.path.isfile(path):
urllib.request.urlretrieve("https://raw.githubusercontent.com/tflearn/tflea
rn.github.io/master/resources/shakespeare_input.txt", path)
```

3. 将输入文本转换为向量，并通过 `string_to_semi_redundant_sequences()` 返回解析的序列和目标以及关联的字典（函数输出一个元组：包括输入、目标和字典）：

```
maxlen = 25
char_idx = None
if os.path.isfile(char_idx_file):
  print('Loading previous char_idx')
  char_idx = pickle.load(open(char_idx_file, 'rb'))
X, Y, char_idx = \
textfile_to_semi_redundant_sequences(path, seq_maxlen=maxlen, redun_step=3,
pre_defined_char_idx=char_idx)
pickle.dump(char_idx, open(char_idx_file,'wb'))
```

4. 定义由三个LSTM组成的RNN，每个LTSM有512个节点，并返回完整序列而不是仅返回最后一个序列。请注意，使用概率为50%的drop-out模块来连接LSTM模块。最后一层是全连接层，softmax长度等于字典尺寸。损失函数采用 `categorical_crossentropy`，优化器采用Adam：

```
g = tflearn.input_data([None, maxlen, len(char_idx)])
g = tflearn.lstm(g, 512, return_seq=True)
g = tflearn.dropout(g, 0.5)
g = tflearn.lstm(g, 512, return_seq=True)
g = tflearn.dropout(g, 0.5)
g = tflearn.lstm(g, 512)
g = tflearn.dropout(g, 0.5)
g = tflearn.fully_connected(g, len(char_idx), activation='softmax')
g = tflearn.regression(g, optimizer='adam',
loss='categorical_crossentropy',
learning_rate=0.001)
```

5. 现在可以用库函数 `flearn.models.generator.SequenceGenerator(network, dictionary=char_idx,seq_maxlen=maxle,clip_gradients=5.0,checkpoint_path='model_shakespeare')` 生成序列：

```
m = tflearn.SequenceGenerator(g, dictionary=char_idx,
seq_maxlen=maxlen,
clip_gradients=5.0,
checkpoint_path='model_shakespeare')
```

6. 经过50次迭代，从输入文本中选取一个随机序列并生成一个新的文本。温度参数控制所创建序列的多样性；接近于 0 的温度创建的序列看起来就像用于训练的样本，温度越高，结果越多样：

```
for i in range(50):
    seed = random_sequence_from_textfile(path, maxlen)
    m.fit(X, Y, validation_set=0.1, batch_size=128,
          n_epoch=1, run_id='shakespeare')
    print("-- TESTING...")
    print("-- Test with temperature of 1.0 --")
    print(m.generate(600, temperature=1.0, seq_seed=seed))
    print("-- Test with temperature of 0.5 --")
    print(m.generate(600, temperature=0.5, seq_seed=seed))
```

解读分析

当一件新的未知或遗忘的艺术作品需要被鉴定归于某位作者时，就会有著名学者将这件作品与作者的其他作品进行比较。学者们所做的是在作者已知作品的文本序列中寻找共同特征，并希望在鉴定作品中找到相似的特征。

本节的工作方式与之前的相似：RNN 学习莎士比亚作品中的特征，然后这些特征被用来产生新的、从未见过的文本，这些文本很好地代表了最伟大的英国作家的写作风格。

来看执行示例：

```
python shakespeare.py
Loading previous char_idx
Vectorizing text...
Text total length: 4,573,338
Distinct chars  : 67
Total sequences : 1,524,438
---------------------------------
Run id: shakespeare
Log directory: /tmp/tflearn_logs/
```

第一次迭代

此时，网络正在学习一些基本结构，包括虚拟人物（DIA、SURYONT、HRNTLGIPRMAR 和 ARILEN）的对话，但是英语水平还很糟糕，很多单词并不是真正的英文：

```
Training samples: 1371994
Validation samples: 152444
--
Training Step: 10719 | total loss: 2.22092 | time: 22082.057s
| Adam | epoch: 001 | loss: 2.22092 | val_loss: 2.12443 -- iter:
1371994/1371994
-- TESTING...
-- Test with temperature of 1.0 --
'st thou, malice?
If thou caseghough memet oud mame meard'ke. Afs weke wteak, Dy ny wold' as
to of my tho gtroy ard has seve, hor then that wordith gole hie, succ,
caight fom?
DIA:
A gruos ceen, I peey
```

```
by my
Wiouse rat Sebine would.
waw-this afeean.
SURYONT:
Teeve nourterong a oultoncime bucice'is furtutun
Ame my sorivass; a mut my peant?
Am:
Fe, that lercom ther the nome, me, paatuy corns wrazen meas ghomn'ge const
pheale,
As yered math thy vans:
I im foat worepoug and thit mije woml!
HRNTLGIPRMAR:
I'd derfomquesf thiy of doed ilasghele hanckol, my corire-hougangle!
Kiguw troll! you eelerd tham my fom Inow lith a
-- Test with temperature of 0.5 --
'st thou, malice?
If thou prall sit I har, with and the sortafe the nothint of the fore the
fir with with the ceme at the ind the couther hit yet of the sonsee in
solles and that not of hear fore the hath bur.
ARILEN:
More you a to the mare me peod sore,
And fore string the reouck and and fer to the so has the theat end the
dore; of mall the sist he the bot courd wite be the thoule the to nenge ape
and this not the the ball bool me the some that dears,
The be to the thes the let the with the thear tould fame boors and not to
not the deane fere the womour hit muth so thand the e meentt my to the
treers and woth and wi
```

经过几次迭代

此时，网络正在学习对话的正确结构，使写出来的英语看起来更像正确的句子，例如：`Well,there shall the things to need the offer to our heart` 和 `There is not that be so then to the death To make the body and all the mind`：

```
--------------------------------
Training samples: 1371994
Validation samples: 152444
--
Training Step: 64314 | total loss: 1.44823 | time: 21842.362s
| Adam | epoch: 006 | loss: 1.44823 | val_loss: 1.40140 -- iter:
1371994/1371994
--
-- Test with temperature of 0.5 --
in this kind.
THESEUS:
There is not that be so then to the death
To make the body and all the mind.
BENEDICK:
Well, there shall the things to need the offer to our heart,
To not are he with him: I have see the hands are to true of him that I am
not,
The whom in some the fortunes,
Which she were better not to do him?
KING HENRY VI:
I have some a starter, and and seen the more to be the boy, and be such a
```

```
plock and love so say, and I will be his entire,
And when my masters are a good virtues,
That see the crown of our worse,
This made a called grace to hear him and an ass,
And the provest and stand,
```

更多内容

博文 "The Unreasonable Effectiveness of Recurrent Neural Networks" (http://karpathy.github.io/2015/05/21/rnn-effectiveness/) 描述了一套吸引人的字符级语言 RNN 模型，具体包括以下部分：

- 类似本例的莎士比亚文本生成。
- 类似于本例的维基百科文本生成，但是基于不同的训练文本。
- 类似于本例的代数几何（LaTex）文本生成，但是基于不同的训练文本。
- 类似于本例的 Linux 源代码文本生成，但是基于不同的训练文本。
- 类似于本例的婴儿起名文本生成，但是基于不同的训练文本。

6.6 基于 RNN 学习预测比特币价格

本节将介绍如何利用 RNN 预测未来的比特币价格。核心思想是过去观察到的价格时间序列为未来价格提供了一个很好的预估器。经 MIT 授权许可，本节将使用 https://github.com/guillaume-chevalier/seq2seq-signal-prediction 中的代码。给定时间间隔的比特币值通过 https://www.coindesk.com/api/ 的 API 下载，以下是 API 文档的一部分：

> *We offer historical data from our Bitcoin Price Index through the following endpoint:*
> *https://api.coindesk.com/v1/bpi/historical/close.json*
> *By default, this will return the previous 31 days' worth of data. This endpoint accepts the following optional parameters:*
> *?index=[USD/CNY]The index to return data for. Defaults to USD.*
> *?currency=<VALUE>The currency to return the data in, specified in ISO 4217 format. Defaults to USD.*
> *?start=<VALUE>&end=<VALUE> Allows data to be returned for a specific date range. Must be listed as a pair of start and end parameters, with dates supplied in the YYYY-MM-DD format, e.g. 2013-09-01 for September 1st, 2013.*
> *?for=yesterday Specifying this will return a single value for the previous day. Overrides the start/end parameter.*
> *Sample Request:*
> https://api.coindesk.com/v1/bpi/historical/close.json?start=2013-09-01&end=2013-09-05
> *Sample JSON Response:*
> {"bpi":{"2013-09-01":128.2597,"2013-09-02":127.3648,"2013-09-03":127.5915,"2013-0

9-04":120.5738,"2013-09-05":120.5333},"disclaimer":"This data was produced from the CoinDesk Bitcoin Price Index. BPI value data returned as USD.","time":{"updated":"Sep 6, 2013 00:03:00 UTC","updatedISO":"2013-09-06T00:03:00+00:00"}}

具体做法

1. 克隆下面的 GitHub 存储库。这是一个鼓励用户尝试 seq2seq 神经网络架构的项目：

```
git clone
https://github.com/guillaume-chevalier/seq2seq-signal-prediction.git
```

2. 在上述 GitHub 库基础上，考虑使用以下能够下载和标准化比特币历史值（美元或欧元）数据的函数，这些函数在 `dataset.py` 中定义。训练集和测试集根据 80/20 规律分开，因此，20% 的测试数据是最新的比特币值。每个示例都包含 40 个美元（USD）数据点，特征轴/维度上是欧元（EUR）数据。数据根据均值和标准差进行归一化处理，函数 `generate_x_y_data_v4` 生成尺寸为 `batch_size` 的训练数据（或测试数据）的随机样本：

```
def loadCurrency(curr, window_size):
    """
    Return the historical data for the USD or EUR bitcoin value. Is done
with an web API call.
    curr = "USD" | "EUR"
    """
    # For more info on the URL call, it is inspired by :
    # https://github.com/Levino/coindesk-api-node
    r = requests.get(
"http://api.coindesk.com/v1/bpi/historical/close.json?start=2010-07-17&end=2017-03-03&currency={}".format(
            curr
        )
    )
    data = r.json()
    time_to_values = sorted(data["bpi"].items())
    values = [val for key, val in time_to_values]
    kept_values = values[1000:]
    X = []
    Y = []
    for i in range(len(kept_values) - window_size * 2):
        X.append(kept_values[i:i + window_size])
        Y.append(kept_values[i + window_size:i + window_size * 2])
    # To be able to concat on inner dimension later on:
    X = np.expand_dims(X, axis=2)
    Y = np.expand_dims(Y, axis=2)
    return X, Y
def normalize(X, Y=None):
    """
    Normalise X and Y according to the mean and standard
deviation of the X values only.
    """
    # # It would be possible to normalize with last rather than mean, such
as:
    # lasts = np.expand_dims(X[:, -1, :], axis=1)
    # assert (lasts[:, :] == X[:, -1, :]).all(), "{}, {}, {}.
```

第6章 循环神经网络

```python
            {}".format(lasts[:, :].shape, X[:, -1, :].shape, lasts[:, :], X[:, -1, :])
        mean = np.expand_dims(np.average(X, axis=1) + 0.00001, axis=1)
        stddev = np.expand_dims(np.std(X, axis=1) + 0.00001, axis=1)
        # print (mean.shape, stddev.shape)
        # print (X.shape, Y.shape)
        X = X - mean
        X = X / (2.5 * stddev)
        if Y is not None:
            assert Y.shape == X.shape, (Y.shape, X.shape)
            Y = Y - mean
            Y = Y / (2.5 * stddev)
            return X, Y
        return X

def fetch_batch_size_random(X, Y, batch_size):
    """
    Returns randomly an aligned batch_size of X and Y among all examples.
    The external dimension of X and Y must be the batch size
(eg: 1 column = 1 example).
    X and Y can be N-dimensional.
    """
    assert X.shape == Y.shape, (X.shape, Y.shape)
    idxes = np.random.randint(X.shape[0], size=batch_size)
    X_out = np.array(X[idxes]).transpose((1, 0, 2))
    Y_out = np.array(Y[idxes]).transpose((1, 0, 2))
    return X_out, Y_out
X_train = []
Y_train = []
X_test = []
Y_test = []

def generate_x_y_data_v4(isTrain, batch_size):
    """
    Return financial data for the bitcoin.
    Features are USD and EUR, in the internal dimension.
    We normalize X and Y data according to the X only to not
    spoil the predictions we ask for.
    For every window (window or seq_length), Y is the prediction following
X.
    Train and test data are separated according to the 80/20
rule.
    Therefore, the 20 percent of the test data are the most
    recent historical bitcoin values. Every example in X contains
    40 points of USD and then EUR data in the feature axis/dimension.
    It is to be noted that the returned X and Y has the same shape
    and are in a tuple.
    """
    # 40 pas values for encoder, 40 after for decoder's predictions.
    seq_length = 40
    global Y_train
    global X_train
    global X_test
    global Y_test
    # First load, with memoization:
    if len(Y_test) == 0:
        # API call:
        X_usd, Y_usd = loadCurrency("USD",
```

```
            window_size=seq_length)
        X_eur, Y_eur = loadCurrency("EUR",
window_size=seq_length)
        # All data, aligned:
        X = np.concatenate((X_usd, X_eur), axis=2)
        Y = np.concatenate((Y_usd, Y_eur), axis=2)
        X, Y = normalize(X, Y)
        # Split 80-20:
        X_train = X[:int(len(X) * 0.8)]
        Y_train = Y[:int(len(Y) * 0.8)]
        X_test = X[int(len(X) * 0.8):]
        Y_test = Y[int(len(Y) * 0.8):]
    if isTrain:
        return fetch_batch_size_random(X_train, Y_train, batch_size)
    else:
        return fetch_batch_size_random(X_test,  Y_test,  batch_size)
```

3. 生成训练集、验证集和测试集,并定义一些超参数,例如 `batch_size`、`hidden_dim`(RNN 中隐藏神经元的数量)和 `layers_stacked_count`(堆栈循环单元的数量)。另外,定义一些用于微调优化器性能的参数,例如优化器的学习率、迭代次数、优化器模拟退火的 `lr_decay`、优化器的动量以及避免过拟合的 L2 正则化。请注意,GitHub 存储库默认 `batch_size = 5` 和 `nb_iters = 150`,但我设置 `batch_size = 1000` 和 `nb_iters = 100000`,已经获得了更好的结果:

```
from datasets import generate_x_y_data_v4
generate_x_y_data = generate_x_y_data_v4
import tensorflow as tf
import numpy as np
import matplotlib.pyplot as plt
%matplotlib inline
sample_x, sample_y = generate_x_y_data(isTrain=True, batch_size=3)
print("Dimensions of the dataset for 3 X and 3 Y training
examples : ")
print(sample_x.shape)
print(sample_y.shape)
print("(seq_length, batch_size, output_dim)")
print sample_x, sample_y
# Internal neural network parameters
seq_length = sample_x.shape[0]  # Time series will have the same past and
future (to be predicted) lenght.
batch_size = 5  # Low value used for live demo purposes - 100 and 1000
would be possible too, crank that up!
output_dim = input_dim = sample_x.shape[-1]  # Output dimension (e.g.:
multiple signals at once, tied in time)
hidden_dim = 12  # Count of hidden neurons in the recurrent units.
layers_stacked_count = 2  # Number of stacked recurrent cells, on the
neural depth axis.
# Optmizer:
learning_rate = 0.007  # Small lr helps not to diverge during training.
nb_iters = 150  # How many times we perform a training step (therefore how
many times we show a batch).
lr_decay = 0.92  # default: 0.9 . Simulated annealing.
momentum = 0.5  # default: 0.0 . Momentum technique in weights update
lambda_l2_reg = 0.003  # L2 regularization of weights - avoids overfitting
```

4. 将网络定义为由基本 GRU 单元组成的编码器-解码器。网络由 `layers_stacked_count=2` 个 RNN 组成，使用 TensorBoard 对网络进行可视化。请注意，`hidden_dim=12` 是循环单元中隐藏的神经元：

```
tf.nn.seq2seq = tf.contrib.legacy_seq2seq
tf.nn.rnn_cell = tf.contrib.rnn
tf.nn.rnn_cell.GRUCell = tf.contrib.rnn.GRUCell
tf.reset_default_graph()
# sess.close()
sess = tf.InteractiveSession()
with tf.variable_scope('Seq2seq'):
    # Encoder: inputs
    enc_inp = [
        tf.placeholder(tf.float32, shape=(None, input_dim), name="inp_{}".format(t))
           for t in range(seq_length)
    ]
    # Decoder: expected outputs
    expected_sparse_output = [
        tf.placeholder(tf.float32, shape=(None, output_dim), name="expected_sparse_output_".format(t))
           for t in range(seq_length)
    ]
    # Give a "GO" token to the decoder.
    # You might want to revise what is the appended value "+ enc_inp[:-1]".
    dec_inp = [ tf.zeros_like(enc_inp[0], dtype=np.float32, name="GO") ] + enc_inp[:-1]
    # Create a `layers_stacked_count` of stacked RNNs (GRU cells here).
    cells = []
    for i in range(layers_stacked_count):
        with tf.variable_scope('RNN_{}'.format(i)):
            cells.append(tf.nn.rnn_cell.GRUCell(hidden_dim))
            # cells.append(tf.nn.rnn_cell.BasicLSTMCell(...))
    cell = tf.nn.rnn_cell.MultiRNNCell(cells)
    # For reshaping the input and output dimensions of the seq2seq RNN:
    w_in = tf.Variable(tf.random_normal([input_dim, hidden_dim]))
    b_in = tf.Variable(tf.random_normal([hidden_dim], mean=1.0))
    w_out = tf.Variable(tf.random_normal([hidden_dim, output_dim]))
    b_out = tf.Variable(tf.random_normal([output_dim]))
reshaped_inputs = [tf.nn.relu(tf.matmul(i, w_in) + b_in) for i in enc_inp]
# Here, the encoder and the decoder uses the same cell, HOWEVER,
    # the weights aren't shared among the encoder and decoder, we have two
    # sets of weights created under the hood according to that function's
def.
    dec_outputs, dec_memory = tf.nn.seq2seq.basic_rnn_seq2seq(
        enc_inp,
        dec_inp,
        cell
    )
output_scale_factor = tf.Variable(1.0, name="Output_ScaleFactor")
    # Final outputs: with linear rescaling similar to batch norm,
    # but without the "norm" part of batch normalization hehe.
    reshaped_outputs = [output_scale_factor*(tf.matmul(i, w_out) + b_out)
for i in dec_outputs]
```

```
    # Merge all the summaries and write them out to /tmp/bitcoin_logs (by
default)
    merged = tf.summary.merge_all()
    train_writer = tf.summary.FileWriter('/tmp/bitcoin_logs',
sess.graph)
```

5. 运行 TensorBoard 并可视化由 RNN 编码器和 RNN 解码器组成的网络：

```
tensorboard --logdir=/tmp/bitcoin_logs
```

以下是代码的流程：

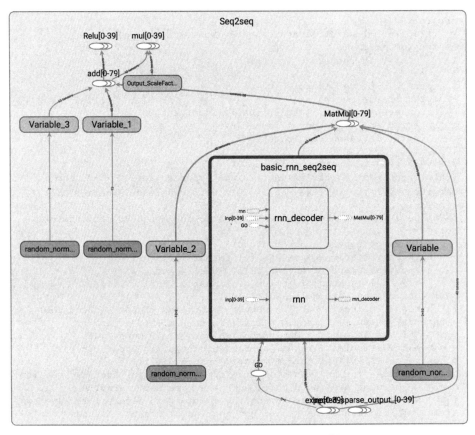

TensorBoard 中的比特币价格预测示例

6. 定义一个 L2 正则化损失函数，以避免过度拟合并具有更好的泛化能力。优化器选择 RMSprop，其中 `learning_rate`、`decay` 和 `momentum` 的值在第 3 步中已给出：

```
# Training loss and optimizer
with tf.variable_scope('Loss'):
    # L2 loss
    output_loss = 0
    for _y, _Y in zip(reshaped_outputs, expected_sparse_output):
```

```
        output_loss += tf.reduce_mean(tf.nn.l2_loss(_y - _Y))
    # L2 regularization (to avoid overfitting and to have a  better
generalization capacity)
    reg_loss = 0
    for tf_var in tf.trainable_variables():
        if not ("Bias" in tf_var.name or "Output_" in tf_var.name):
            reg_loss += tf.reduce_mean(tf.nn.l2_loss(tf_var))
    loss = output_loss + lambda_l2_reg * reg_loss

with tf.variable_scope('Optimizer'):
    optimizer = tf.train.RMSPropOptimizer(learning_rate, decay=lr_decay,
momentum=momentum)
    train_op = optimizer.minimize(loss)
```

7. 生成训练数据并在数据集的 `batch_size` 示例上运行优化程序，为批量训练做好准备。同样，从数据集的 `batch_size` 示例生成测试数据，为测试做好准备。训练运行迭代次数为 `nb_iters+1`，并每训练 10 次迭代来测试一次结果：

```
def train_batch(batch_size):
    """
    Training step that optimizes the weights
    provided some batch_size X and Y examples from the dataset.
    """
    X, Y = generate_x_y_data(isTrain=True, batch_size=batch_size)
    feed_dict = {enc_inp[t]: X[t] for t in range(len(enc_inp))}
    feed_dict.update({expected_sparse_output[t]: Y[t] for t in
range(len(expected_sparse_output))})
    _, loss_t = sess.run([train_op, loss], feed_dict)
    return loss_t

def test_batch(batch_size):
    """
    Test step, does NOT optimizes. Weights are frozen by not
    doing sess.run on the train_op.
    """
    X, Y = generate_x_y_data(isTrain=False, batch_size=batch_size)
    feed_dict = {enc_inp[t]: X[t] for t in range(len(enc_inp))}
    feed_dict.update({expected_sparse_output[t]: Y[t] for t in
range(len(expected_sparse_output))})
    loss_t = sess.run([loss], feed_dict)
    return loss_t[0]

# Training
train_losses = []
test_losses = []
sess.run(tf.global_variables_initializer())

for t in range(nb_iters+1):
    train_loss = train_batch(batch_size)
    train_losses.append(train_loss)
    if t % 10 == 0:
        # Tester
        test_loss = test_batch(batch_size)
        test_losses.append(test_loss)
        print("Step {}/{}, train loss: {}, \tTEST loss: {}".format(t,
nb_iters, train_loss, test_loss))
```

```
print("Fin. train loss: {}, \tTEST loss: {}".format(train_loss, test_loss))
```

8. 将 `n_predictions` 测试结果可视化，`nb_predictions` 取 5，预测值用黄色圆点实际值用蓝色 × 符号表示。请注意，预测从直方图中的最后一个蓝点开始，可以看出，即使是这个简单的模型也是相当准确的：

```
# Test
nb_predictions = 5
print("Let's visualize {} predictions with our signals:".format(nb_predictions))
X, Y = generate_x_y_data(isTrain=False, batch_size=nb_predictions)
feed_dict = {enc_inp[t]: X[t] for t in range(seq_length)}
outputs = np.array(sess.run([reshaped_outputs], feed_dict)[0])
for j in range(nb_predictions):
    plt.figure(figsize=(12, 3))
    for k in range(output_dim):
        past = X[:,j,k]
        expected = Y[:,j,k]
        pred = outputs[:,j,k]
        label1 = "Seen (past) values" if k==0 else "_nolegend_"
        label2 = "True future values" if k==0 else "_nolegend_"
        label3 = "Predictions" if k==0 else "_nolegend_"
        plt.plot(range(len(past)), past, "o--b", label=label1)
        plt.plot(range(len(past), len(expected)+len(past)), expected, "x--b", label=label2)
        plt.plot(range(len(past), len(pred)+len(past)), pred, "o--y", label=label3)
    plt.legend(loc='best')
    plt.title("Predictions v.s. true values")
    plt.show()
```

结果如下：

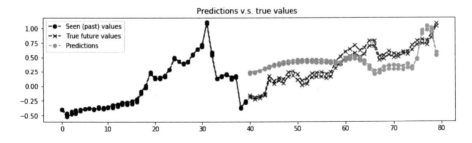

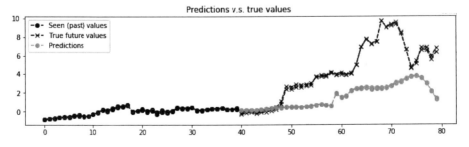

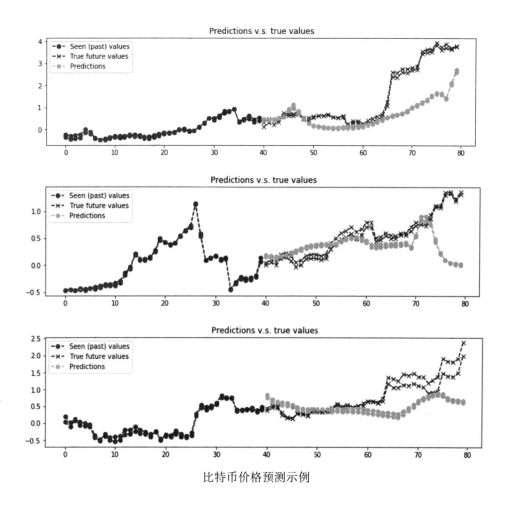

比特币价格预测示例

解读分析

比特币价格的预测是使用一个基于 GRU 基本单元的 RNN 组成的编码器 – 解码器。RNN 非常擅长学习序列，事实上即使是只有两层和 12 个 GRU 单元的简单模型，比特币价格预测也是相当准确的。当然，这个预测代码并不是鼓励投资比特币，而只是讨论深度学习方法。而且，为了确认是否存在数据过度拟合的情况，需要进行更多的实验。

更多内容

预测股票市场是一个很酷的 RNN 应用，并有一些使用方便的软件包可用，例如：

- Drnns 预测为了实现深度 RNN，使用了来自 Kaggle "股市预测"数据集的"每日新闻"中的 Keras 神经网络库。该数据集的任务是将当前和过去的新闻标题作为特征，预测道琼斯指数的走势，开源代码可在 https://github.com/jvpoulos/

drnns-prediction 中找到。
- Michael Luk 写了一篇有趣的博文（`https://sflscientific.com/data-science-blog/2017/2/10/predicting-stock-volume-with-lstm`），如何使用 RNN 预测可口可乐股票成交量。
- Jakob Aungiers 也写了一篇有趣的博文（`http://www.jakob-aungiers.com/articles/a/LSTM-Neural-Network-for-Time-Series-Prediction`），主题是关于 LSTM 神经网络进行时间序列预测的。

6.7 多对一和多对多的 RNN 例子

本节通过提供 RNN 映射的各种例子来总结本章已经讨论过的 RNN 知识。为了简单起见，将采用 Keras 编写如下图所示的一对一、一对多、多对一和多对多映射：

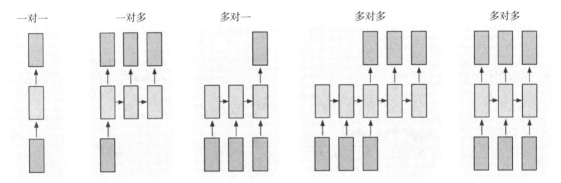

RNN 序列示例（参考 `http://karpathy.github.io/2015/05/21/rnn-effectiveness/`）

具体做法

1. 如果你想创建一对一的映射，这不是 RNN 而是一个全连接层。假设已经定义了一个模型，并且你想添加全连接网络，那么这在 Keras 中很容易实现：

```
model = Sequential()
model.add(Dense(output_size, input_shape=input_shape))
```

2. 如果要创建一对多映射，可以使用 `RepeatVector(...)` 来实现。请注意，`return_sequences` 是一个布尔值，用于决定是返回输出序列中的最后一个输出还是全部序列：

```
model = Sequential()
model.add(RepeatVector(number_of_times,input_shape=input_shape))
model.add(LSTM(output_size, return_sequences=True))
```

3. 如果你想创建**多对一**映射，可以用下面的 LSTM 代码片段来实现：

```
model = Sequential()
model.add(LSTM(1, input_shape=(timesteps, data_dim)))
```

4. 如果要创建**多对多**映射，在输入和输出长度与循环步数匹配时，可以使用以下 LSTM 代码片段来实现：

```
model = Sequential()
model.add(LSTM(1, input_shape=(timesteps, data_dim),
return_sequences=True))
```

解读分析

Keras 允许你轻松地编写各种形式的 RNN，包括一对一、一对多、多对一以及多对多的映射，上面的例子正说明了用 Keras 实现它们非常容易。

CHAPTER 7
第 7 章

无监督学习

到目前为止，讨论的所有模型都是基于监督式学习模式的，训练数据集包含输入及其标签。本章主要介绍无监督学习。

7.1 引言

在机器学习中，有三种不同的学习模式：监督学习、无监督学习和强化学习。

监督学习，也称为有导师学习，网络输入包括数据和相应的输出标签信息。例如，在 MNIST 数据集中，手写数字的每个图像都有一个标签，代表图片中的数字值。

强化学习，也称为评价学习，不给网络提供期望的输出，但空间会提供给出一个奖惩的反馈，当输出正确时，给网络奖励，当输出错误时就惩罚网络。

无监督学习，也称为无导师学习，在网络的输入中没有相应的输出标签信息，网络接收输入，但既没有提供期望的输出，也没有提供来自环境的奖励，神经网络要在这种情况下学习输入数据中的隐藏结构。无监督学习非常有用，因为现存的大多数数据是没有标签的，这种方法可以用于诸如模式识别、特征提取、数据聚类和降维等任务。在本章和下一章中，主要学习基于无监督学习的机器学习和神经网络相关技术。

7.2 主成分分析

主成分分析（Principal Component Analysis, PCA）是一种多变量统计方法，它是最常用的降维方法之一，通过正交变换将一组可能存在相关性的变量数据转换为一组线性不相关的变量，转换后的变量被称为**主成分**。可以使用两种方法进行 PCA—**特征分解**或**奇异值分解**（SVD）。

准备工作

PCA 将 n 维输入数据缩减为 r 维，其中 $r < n$。简单地说，PCA 实质上是一个基变换，

使得变换后的数据有最大的方差，也就是通过对坐标轴的旋转和坐标原点的平移使得其中一个轴（主轴）与数据点之间的方差最小，坐标转换后去掉高方差的正交轴，得到降维数据集。这里使用 SVD 方法进行 PCA 降维，假定有 $p \times n$ 维数据样本 X，共有 p 个样本，每行是 n 维，$p \times n$ 实矩阵可以分解为：

$$X = U \Sigma V^T$$

这里，正交阵 U 的维数是 $p \times n$，正交阵 V 的维数是 $n \times n$（正交阵满足：$UU^T = V^T V = 1$），Σ 是 $n \times n$ 的对角阵。接下来，将 Σ 分割成 r 列，记作 Σ_r；利用 U 和 V 便能够得到降维数据点 Y_r：

$$Y_r = U \Sigma_r$$

本节代码改编自链接 `https://github.com/eliorc/Medium/blob/master/PCA-tSNE-AE.ipynb` 中的代码。

具体做法

1. 导入所需的模块，除了 TensorFlow，还需要 `numpy` 进行基本的矩阵计算，用 `matplotlib`、`mpl_toolkit` 和 `seaborn` 绘制图形：

```
import tensorflow as tf
import numpy as np
import matplotlib.pyplot as plt
from mpl_toolkits.mplot3d import Axes3D
import seaborn as sns
%matplotlib inline
```

2. 加载数据集，此处使用常用的 MNIST 数据集：

```
from tensorflow.examples.tutorials.mnist import input_data
mnist = input_data.read_data_sets("MNIST_data/")
```

3. 定义类 `TF_PCA`，此类初始化如下：

```
def __init__(self, data,  dtype=tf.float32):
        self._data = data
        self._dtype = dtype
        self._graph = None
        self._X = None
        self._u = None
        self._singular_values = None
        self._siqma = None
```

4. 定义 `fit` 函数计算输入数据的 SVD。定义计算图，以此计算奇异值和正交矩阵 U，`self._X` 以占位符的形式读入数据 `self.data`，`tf.svd` 以递减顺序返回形状为 [..., p] 的奇异值 s（`singular_values`），然后使用 `tf.diag` 将奇异值转换为对角矩阵：

```
def fit(self):
        self._graph = tf.Graph()
        with self._graph.as_default():
            self._X = tf.placeholder(self._dtype, shape=self._data.shape)
```

```
        # Perform SVD
        singular_values, u, _ = tf.svd(self._X)
        # Create sigma matrix
        sigma = tf.diag(singular_values)
    with tf.Session(graph=self._graph) as session:
        self._u, self._singular_values, self._sigma =
session.run([u, singular_values, sigma], feed_dict={self._X:
self._data})
```

5. 现在有了 sigma 矩阵、正交矩阵 U 和奇异值，下面定义 reduce 函数来计算降维数据。该方法需要 n_dimensions 和 keep_info 两个输入参数之一，n_dimensions 参数表示在降维数据中保持的维数，keep_info 参数表示保留信息的百分比（0.8 意味着保持 80% 的原始数据）。该方法创建一个计算图，对 sigma 矩阵进行分割并计算降维数据集 Y_r：

```
def reduce(self, n_dimensions=None, keep_info=None):
    if keep_info:
        # Normalize singular values
        normalized_singular_values = self._singular_values /
sum(self._singular_values)
        # information per dimension
        info = np.cumsum(normalized_singular_values)
# Get the first index which is above the given information
threshold
        it = iter(idx for idx, value in enumerate(info) if value
>= keep_info)
        n_dimensions = next(it) + 1
    with self.graph.as_default():
        # Cut out the relevant part from sigma
        sigma = tf.slice(self._sigma, [0, 0],
[self._data.shape[1], n_dimensions])
        # PCA
        pca = tf.matmul(self._u, sigma)

    with tf.Session(graph=self._graph) as session:
        return session.run(pca, feed_dict={self._X:
self._data})
```

6. TF_PCA 类已经准备就绪，下面会将 MNIST 的每个输入数据从维度为 784（28×28）减小到每个维度为 3。在这里为了对比效果只保留了 10% 的信息，但通常情况下需要保留大约 80% 的信息：

```
tf_pca.fit()
pca = tf_pca.reduce(keep_info=0.1)   # The reduced dimensions
dependent upon the % of information
print('original data shape', mnist.train.images.shape)
print('reduced data shape', pca.shape)
```

代码输出如下：

```
original data shape (55000, 784)
reduced data shape (55000, 3)
```

7. 绘制三维空间中的 55000 个数据点：

```
Set = sns.color_palette("Set2", 10)
color_mapping = {key:value for (key,value) in enumerate(Set)}
colors = list(map(lambda x: color_mapping[x], mnist.train.labels))
fig = plt.figure()
ax = Axes3D(fig)
ax.scatter(pca[:, 0], pca[:, 1],pca[:, 2], c=colors)
```

Out[5]: <mpl_toolkits.mplot3d.art3d.Path3DCollection at 0x1d4cfe4f9e8>

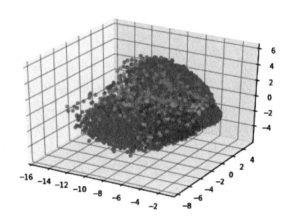

解读分析

前面的代码对 MNIST 图像进行了降维操作。原图的大小为 28×28，利用 PCA 方法把尺寸压缩得更小。通常在图像处理中经常用到降维操作，因为太大的图像尺寸包含大量的冗余数据。

更多内容

TensorFlow 中的 embeddings 技术可以实现从对象到向量的映射，TensorBoard 中的 Embedding Projector 可以交互式地对模型的 embeddings 进行可视化，并提供了三种降维的方法：PCA、t-SNE 和自定义方式，可以使用 Embeddings Projector 来得到与上面类似的结果。这需要从 tensorflow.contrib.tensorboard.plugins 中导入 projector，并且通过简单的三个步骤就可以完成。

1. 加载数据：

```
mnist = input_data.read_data_sets('MNIST_data')
images = tf.Variable(mnist.test.images, name='images')
```

2. 新建一个 metadata 文件（用制表符分隔的 .tsv 文件）：

```
with open(metadata, 'w') as metadata_file:
    for row in mnist.test.labels:
        metadata_file.write('%d\n' % row)
```

3. 将 embeddings 保存在 `Log_DIR` 中：

```python
with tf.Session() as sess:
    saver = tf.train.Saver([images])

    sess.run(images.initializer)
    saver.save(sess, os.path.join(LOG_DIR, 'images.ckpt'))

    config = projector.ProjectorConfig()
    # One can add multiple embeddings.
    embedding = config.embeddings.add()
    embedding.tensor_name = images.name
    # Link this tensor to its metadata file (e.g. labels).
    embedding.metadata_path = metadata
    # Saves a config file that TensorBoard will read during startup.
    projector.visualize_embeddings(tf.summary.FileWriter(LOG_DIR), config)
```

现在就可以使用 TensorBoard 查看 embeddings 了，通过命令行 `tensorboard --logdir=log`，在 Web 浏览器中打开 TensorBoard，然后进入 EMBEDDINGS 选项卡。下图显示的就是使用 PCA 方法运算的前三个主成分为轴的 TensorBoard 投影：

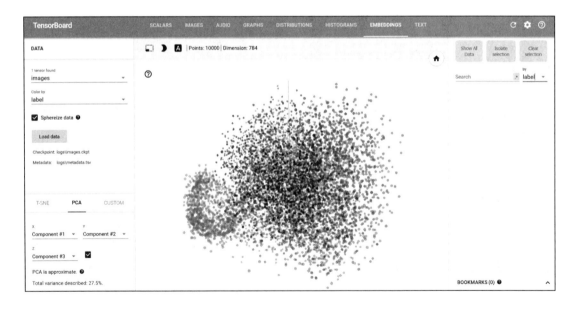

拓展阅读

- `https://arxiv.org/abs/1404.1100`
- `http://www.cs.otago.ac.nz/cosc453/student_tutorials/principal_components.pdf`
- `http://mplab.ucsd.edu/tutorials/pca.pdf`

- http://projector.tensorflow.org/

7.3 k 均值聚类

顾名思义，k 均值聚类是一种对数据进行聚类的技术，即将数据分割成指定数量的几个类，这是一种无监督学习方法，揭示数据的内在性质及规律。还记得哈利波特故事中的分院帽吗？那就是聚类，将新学生（无标签）分成四类：格兰芬多、拉文克拉、赫奇帕奇和斯特莱林。

人是非常擅长分类的，聚类算法试图让计算机也具备这种类似的能力，聚类技术很多，例如层次法、贝叶斯法和划分法。k 均值聚类属于划分聚类方法，将数据分成 k 个簇，每个簇有一个中心，称为**质心**，k 值需要给定。

k 均值聚类算法的工作原理如下：

1. 随机选择 k 个数据点作为初始质心（聚类中心）。
2. 将每个数据点划分给距离最近的质心，衡量两个样本数据点的距离有多种不同的方法，最常用的是欧氏距离。
3. 重新计算每个簇的质心作为新的聚类中心，使其总的平方距离达到最小。
4. 重复第 2 步和第 3 步，直到收敛。

准备工作

使用 TensorFlow 的 Estimator 类 `KmeansClustering` 来实现 k 均值聚类，具体实现可参考 https://github.com/tensorflow/tensorflow/blob/r1.3/tensorflow/contrib/learn/python/learn/estimators/kmeans.py，可以直接进行 k 均值聚类和推理。根据 TensorFlow 文档，`KmeansClustering` 类对象可以使用以下 `__init__` 方法进行实例化：

```
__init__(
num_clusters,
model_dir=None,
initial_clusters=RANDOM_INIT,
distance_metric=SQUARED_EUCLIDEAN_DISTANCE,
random_seed=0,
use_mini_batch=True,
mini_batch_steps_per_iteration=1,
kmeans_plus_plus_num_retries=2,
relative_tolerance=None,
config=None
)
```

TensorFlow 文档对这些参数的定义如下：

num_clusters：要训练的簇数。

model_dir：保存模型结果和日志文件的目录。

initial_clusters：指定如何对簇初始化，取值请参阅 clustering_ops. kmeans。
distance_metric：聚类的距离度量方式，取值请参阅 clustering_ops. kmeans。
random_seed：Python 中的整数类型，用于初始化质心的伪随机序列发生器的种子。
use_mini_batch：如果为 true，运行算法时分批处理数据，否则一次使用全部数据集。
mini_batch_steps_per_iteration：经过指定步数后将计算的簇中心更新回原数据。更多详细信息参见 clustering_ops.py。
kmeans_plus_plus_num_retries：对于在 kmeans ++ 方法初始化过程中采样的每个点，该参数指定在选择最优值之前从当前分布中提取的附加点数。如果指定了负值，则使用试探法对 O（log（num_to_sample））个附加点进行抽样。
relative_tolerance：相对误差，在每一轮迭代之间若损失函数的变化小于这个值则停止计算。有一点要注意就是，如果将 use_mini_batch 设置为 True，程序可能无法正常工作。
配置：请参阅 Estimator。

TensorFlow 支持将欧氏距离和余弦距离作为质心的度量，`KmeansClustering` 类提供了多种交互方法。在这里使用 `fit()`、`clusters()` 和 `predict_clusters_idx()` 方法：

```
fit(
 x=None,
 y=None,
 input_fn=None,
 steps=None,
 batch_size=None,
 monitors=None,
 max_steps=None
)
```

根据 TensorFlow 文档描述，需要给 `fit()` 提供 `input_fn()` 函数，`cluster` 方法返回簇质心，predict_cluster_idx 方法返回得到簇的索引。

具体做法

1. 与以前一样，从加载必要的模块开始，这里需要 TensorFlow、NumPy 和 Matplotlib。这里使用鸢尾花卉数据集，该数据集分为三类，每类都是指一种鸢尾花卉，每类有 50 个实例。可以从 https://archive.ics.uci.edu/ml/datasets/iris 上下载 .csv 文件，也可以使用 sklearn 库的数据集模块（scikit-learn）来加载数据：

```
import numpy as np
import tensorflow as tf
import matplotlib.pyplot as plt
from matplotlib.colors import ListedColormap
# dataset Iris
from sklearn import datasets

%matplotlib inline
```

2. 加载数据集：

```
# import some data to play with
iris = datasets.load_iris()
x = iris.data[:, :2] # we only take the first two features.
y = iris.target
```

3. 绘出数据集查看一下：

```
# original data without clustering
plt.scatter(hw_frame[:,0], hw_frame[:,1])
plt.xlabel('Sepia Length')
plt.ylabel('Sepia Width')
```

代码输出如下：

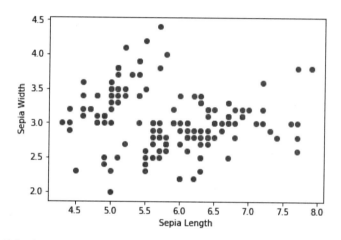

4. 可以看到数据中并没有明显可见的分类。定义 `input_fn` 来给 `fit()` 方法输入数据，函数返回一个 TensorFlow 常量，用来指定 x 的值和维度，类型为 `float`。

```
def input_fn():
    return tf.constant(np.array(x), tf.float32, x.shape),None
```

5. 开始使用 `KmeansClustering` 类，分为 3 类，设置 `num_clusters=3`。通常情况下事先并不知道最优的聚类数量，在这种情况下，常用的方法是采用**肘部法则**（elbow method）来估计聚类数量：

```
kmeans = tf.contrib.learn.KMeansClustering(num_clusters=3,
    relative_tolerance=0.0001, random_seed=2)
kmeans.fit(input_fn=input_fn)
```

6. 使用 `clusters()` 方法找到这些簇，使用 `predict_cluster_idx()` 方法为每个输入点计算分配的簇索引：

```
clusters = kmeans.clusters()
assignments = list(kmeans.predict_cluster_idex(input_fn=input_fn))
```

7. 对创建的簇进行可视化操作，创建一个包装函数 `ScatterPlot`，它将每个点的 X

和 Y 值与每个数据点的簇和簇索引对应起来：

```
def ScatterPlot(X, Y, assignments=None, centers=None):
 if assignments is None:
  assignments = [0] * len(X)
 fig = plt.figure(figsize=(14,8))
 cmap = ListedColormap(['red', 'green', 'blue'])
 plt.scatter(X, Y, c=assignments, cmap=cmap)
 if centers is not None:
  plt.scatter(centers[:, 0], centers[:, 1], c=range(len(centers)),
  marker='+', s=400, cmap=cmap)
 plt.xlabel('Sepia Length')
 plt.ylabel('Sepia Width')
```

使用下面的函数画出簇：

```
ScatterPlot(x[:,0], x[:,1], assignments, clusters)
```

结果如下：

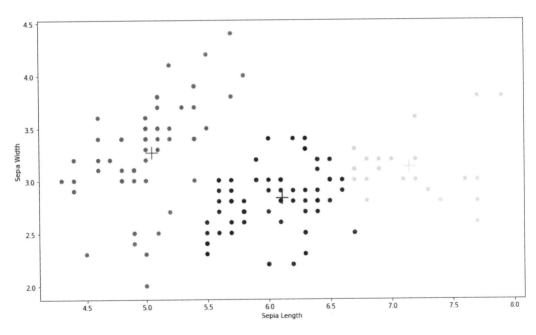

其中"+"号代表三个簇的质心。

解读分析

上面的案例中使用 TensorFlow Estimator 的 k 均值聚类进行了聚类，这里是提前知道簇的数目，因此设置 num_clusters=3。但是在大多数情况下，数据没有标签，我们也不知道有多少簇存在，这时候可以使用肘部法则确定簇的最佳数量。肘部法则选择簇数量的原则是减少距离的平方误差和（SSE），随着簇数量 k 的增加，SSE 是逐渐减小的，直到 SSE =

0，当 k 等于数据点的数量时，每个点都是自己的簇。这里想要的是一个较小的 k 值，而且 SSE 也较小。在 TensorFlow 中，可以使用 `KmeansClustering` 类中定义的 `score()` 方法计算 SSE，该方法返回所有样本点距最近簇的距离之和：

```
sum_distances = kmeans.score(input_fn=input_fn, steps=100)
```

对于鸢尾花卉数据，如果针对不同的 k 值绘制 SSE，能够看到 $k=3$ 时，SSE 的变化是最大的；之后变化趋势减小，因此肘部 k 值可设置为 3：

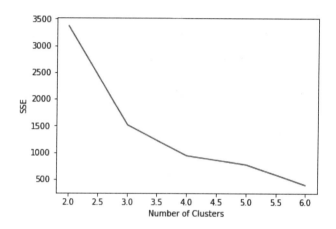

更多内容

k 均值聚类因其简单、快速、强大而被广泛应用，当然它也有不足之处，最大的不足就是用户必须指定簇的数量；其次，算法不保证全局最优；再次，对异常值非常敏感。

拓展阅读

- Kanungo, Tapas, et al. *An efficient k-means clustering algorithm: Analysis and implementation*. IEEE transactions on pattern analysis and machine intelligence 24.7 (2002): 881-892.

- Ortega, Joaquín Pérez, et al. *Research issues on k-means algorithm: An experimental trial using matlab*. CEUR Workshop Proceedings: Semantic Web and New Technologies.

- http://home.deib.polimi.it/matteucc/Clustering/tutorial_html/kmeans.html

- Chen, Ke. *On coresets for k-median and k-means clustering in metric and euclidean spaces and their applications*. SIAM Journal on Computing 39.3 (2009): 923-947.

- https://en.wikipedia.org/wiki/Determining_the_number_of_clusters_in_a_data_set

7.4 自组织映射

自组织映射（SOM）网络也被称为 **Kohonen 网络**或者**胜者独占单元**（WTU），是受人脑特征启发而提出的一种非常特殊的神经网络。在大脑中，不同的感官输入以拓扑顺序的方式呈现。与其他神经网络不同，SOM 神经元之间并不是通过权重相互连接的，相反，它们能够影响彼此的学习。SOM 最重要的特点是神经元以拓扑方式表示所学到的输入信息。

在 SOM 中，神经元通常放置在（一维或二维）网格的节点处，更高的维数也可以用，但实际很少使用。网格中的每个神经元都可以通过权重矩阵连接到所有输入单元。下图中有 3×4（12）个神经元和 7 个输入，为了清晰只画出了所有输入连接到一个神经元的权重向量，在这种情况下，每个神经元将拥有 7 个权值，因此权重矩阵维数为 12×7。

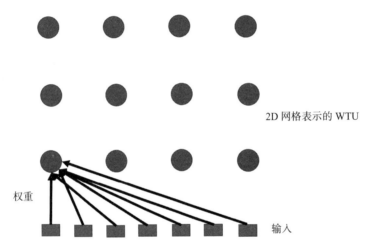

2D 网格表示的 WTU

权重

输入

SOM 通过竞争机制进行学习，可以认为它是 PCA 的非线性推广，因此 SOM 可以像 PCA 一样用于降维。

准备工作

先来了解它是如何工作的，第一步，用随机值或从输入中随机采样对连接权重进行初始化，网格中的每个神经元都被赋予一个位置。数据输入后，测量输入向量（X）和所有神经元权向量（W）之间的距离，与输入数据距离最小的神经元为胜者（WTU），距离度量如下：

$$d_j = \sqrt{\sum_{i=1}^{N}(W_{ji} - x_i)^2}$$

其中，d_j 是神经元 j 的权重与输入 X 之间的距离，最小距离的神经元是胜者。

第二步，调整获胜神经元及其邻域神经元的权重，以确保如果下一次是相同的输入，则胜者还是同一个神经元。网络采用邻域函数 $\Lambda(r)$ 确定哪些邻域神经元权重需要修改，通常使用高斯墨西哥帽函数作为邻域函数，数学表达式如下：

$$\Lambda(r) = e^{-\frac{d^2}{2\sigma^2}}$$

其中，σ 是随时间变化的神经元影响半径，d 是距离获胜神经元的距离：

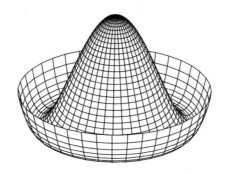

邻域函数的一个重要特性是它的半径随时间而减小，这样刚开始时较多邻域神经元权重被修改，但是随着网络的学习，最终只有少量的神经元的权重被修改（有时只有一个或没有）。权重的改变由下式计算：

$$dW = \eta\Lambda(X-W)$$

按照这个方法继续处理输入，重复执行给定的迭代次数。在迭代过程中利用一个与迭代次数相关的因子来减少学习率和影响半径。

具体做法

1. 首先导入必需的模块：

```
import tensorflow as tf
import numpy as np
import matplotlib.pyplot as plt
%matplotlib inline
```

2. 声明类 WTU 来执行 SOM 的功能。实例化时，参数 $m\times n$ 为二维 SOM 网格的大小，`dim` 为输入数据的维数，`num_interations` 为迭代次数。

```
def __init__(self, m, n, dim, num_iterations, eta = 0.5, sigma = None):
    """
    m x n : The dimension of 2D lattice in which neurons are arranged
    dim : Dimension of input training data
    num_iterations: Total number of training iterations
    eta : Learning rate
    sigma: The radius of neighbourhood function.
    """
    self._m = m
    self._n = n
    self._neighbourhood = []
```

```
        self._topography = []
        self._num_iterations = int(num_iterations)
        self._learned = False
```

3. **__int__** 方法中还定义了计算图和会话。

4. 如果没有提供 `sigma` 值，则采用默认值，通常是 SOM 网格最大维数的一半：

```
if sigma is None:
    sigma = max(m,n)/2.0 # Constant radius
else:
    sigma = float(sigma)
```

5. 在计算图中先定义权重矩阵变量和输入，然后计算获胜者并更新获胜者和它近邻神经元的权重。由于 SOM 具有拓扑映射，此处通过增加操作来获取神经元的拓扑定位。

```
        self._graph = tf.Graph()

# Build Computation Graph of SOM
 with self._graph.as_default():
# Weight Matrix and the topography of neurons
        self._W = tf.Variable(tf.random_normal([m*n, dim], seed = 0))
        self._topography = 
tf.constant(np.array(list(self._neuron_location(m, n))))

        # Placeholders for training data
        self._X = tf.placeholder('float', [dim])
        # Placeholder to keep track of number of iterations
        self._iter = tf.placeholder('float')

        # Finding the Winner and its location
        d = tf.sqrt(tf.reduce_sum(tf.pow(self._W - tf.stack([self._X
            for i in range(m*n)]),2),1))
        self.WTU_idx = tf.argmin(d,0)
        slice_start = tf.pad(tf.reshape(self.WTU_idx,
[1]),np.array([[0,1]]))
        self.WTU_loc = tf.reshape(tf.slice(self._topography,
slice_start,          [1,2]), [2])
        # Change learning rate and radius as a function of iterations
        learning_rate = 1 - self._iter/self._num_iterations
        _eta_new = eta * learning_rate
        _sigma_new = sigma * learning_rate
        # Calculating Neighbourhood function
        distance_square = tf.reduce_sum(tf.pow(tf.subtract(
            self._topography, tf.stack([self.WTU_loc for i in range(m *
n)])), 2), 1)
        neighbourhood_func = tf.exp(tf.negative(tf.div(tf.cast(
distance_square, "float32"), tf.pow(_sigma_new, 2))))
        # multiply learning rate with neighbourhood func
        eta_into_Gamma = tf.multiply(_eta_new, neighbourhood_func)
        # Shape it so that it can be multiplied to calculate dW
        weight_multiplier = tf.stack([tf.tile(tf.slice(
eta_into_Gamma, np.array([i]), np.array([1])), [dim])
for i in range(m * n)])
        delta_W = tf.multiply(weight_multiplier,
tf.subtract(tf.stack([self._X for i in range(m * n)]),self._W))
        new_W = self._W + delta_W
        self._training = tf.assign(self._W,new_W)
```

```
# Initialize All variables
init = tf.global_variables_initializer()
self._sess = tf.Session()
self._sess.run(init)
```

6.在类中定义 `fit()` 方法,执行默认计算图中声明的训练操作,同时还计算质心网格:

```
def fit(self, X):
    """
    Function to carry out training
    """
    for i in range(self._num_iterations):
        for x in X:
            self._sess.run(self._training, feed_dict= {self._X:x,
self._iter: i})


    # Store a centroid grid for easy retreival
    centroid_grid = [[] for i in range(self._m)]
    self._Wts = list(self._sess.run(self._W))
    self._locations = list(self._sess.run(self._topography))
    for i, loc in enumerate(self._locations):
        centroid_grid[loc[0]].append(self._Wts[i])
    self._centroid_grid = centroid_grid

    self._learned = True
```

7.定义一个函数来获取二维网格中获胜神经元的索引和位置:

```
def winner(self, x):
    idx = self._sess.run([self.WTU_idx,self.WTU_loc], feed_dict =
{self._X:x})
    return idx
```

8.定义一些辅助函数来执行网格中神经元的二维映射,并将输入向量映射到二维网格中的相关神经元中:

```
def _neuron_location(self,m,n):
    """
    Function to generate the 2D lattice of neurons
    """
    for i in range(m):
        for j in range(n):
            yield np.array([i,j])
def get_centroids(self):
    """
    Function to return a list of 'm' lists, with each inner
list containing the 'n' corresponding centroid locations    as 1-D
NumPy arrays.
    """
    if not self._learned:
        raise ValueError("SOM not trained yet")
    return self._centroid_grid

def map_vects(self, X):
    """
    Function to map each input vector to the relevant neuron
```

```
in the lattice
    """
    if not self._learned:
        raise ValueError("SOM not trained yet")
    to_return = []
    for vect in X:
        min_index = min([i for i in range(len(self._Wts))],
        key=lambda x: np.linalg.norm(vect -
self._Wts[x]))
        to_return.append(self._locations[min_index])
return to_return
```

9. 现在我们的 WTU 类已经可以使用了,从 `.csv` 文件中读取数据并将其标准化:

```
def normalize(df):
    result = df.copy()
    for feature_name in df.columns:
        max_value = df[feature_name].max()
        min_value = df[feature_name].min()
        result[feature_name] = (df[feature_name] - min_value) /
(max_value - min_value)
    return result

# Reading input data from file
import pandas as pd
df = pd.read_csv('colors.csv') # The last column of data file is a
label
data = normalize(df[['R', 'G', 'B']]).values
name = df['Color-Name'].values
n_dim = len(df.columns) - 1

# Data for Training
colors = data
color_names = name
```

10. 使用 WTU 类进行降维,并将其排列在一个拓扑图中:

```
som = WTU(30, 30, n_dim, 400, sigma=10.0)
som.fit(colors)

# Get output grid
image_grid = som.get_centroids()

# Map colours to their closest neurons
mapped = som.map_vects(colors)

# Plot
plt.imshow(image_grid)
plt.title('Color Grid SOM')
for i, m in enumerate(mapped):
    plt.text(m[1], m[0], color_names[i], ha='center', va='center',
bbox=dict(facecolor='white', alpha=0.5, lw=0))
```

最后绘图如下:

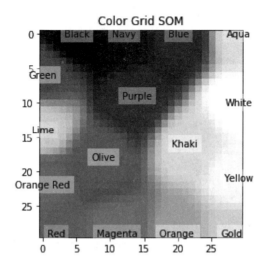

解读分析

SOM 是一个计算密集型网络，因此对于大型数据集并不实用，不过，该算法很容易理解，很容易发现输入数据之间的相似性。因此被广泛用于图像分割和自然语言处理的单词相似性映射中。

拓展阅读

- 一篇用通俗易懂的语言解释 SOM 的很棒的博客文章：http://www.ai-junkie.com/ann/som/som1.html
- SOM 的简介：https://en.wikipedia.org/wiki/Self-organizing_map
- Kohonen 关于 SOM 的开创性论文："The self-organizing map", Neurocomputing 21.1 (1998): 1-6：https://pdfs.semanticscholar.org/8c6a/aea3159e9f49283de252d0548b337839ca6f.pdf

7.5 受限玻尔兹曼机

受限玻尔兹曼机（RBM）是一个两层神经网络，第一层被称为**可见层**，第二层被称为**隐藏层**，因为网络只有两层，所以又被称为**浅层神经网络**。该模型最早由 Paul Smolensky 于 1986 年提出（他称其为 Harmony 网络），此后 Geoffrey Hinton 在 2006 年提出了**对比散度**（Contrastive Divergence，CD）方法对 RBM 进行训练。可见层中的每个神经元与隐藏层中的所有神经元都相连接，但是同一层的神经元之间无连接，所有的神经元输出状态只有两种。

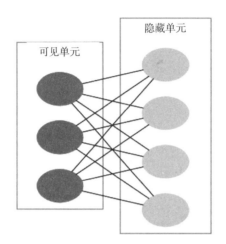

来源:Qwertyus-Own, CCBY-SA3.0, 参考 `https://commons.wikimedia.org/w/index.php?curid=22717044`

RBM 可以用于降维、特征提取和协同过滤,RBM 的训练可以分成三部分:正向传播、反向传播和比较。

准备工作

下面看看 RBM 的表达式。

正向传播:可见层(V)已知,利用权重(W)和偏置(c)采样出隐藏层(h_0),根据下式的随机概率(σ 是随机概率),隐藏单元开启或关闭:

$$p(h_i/v_o) = \sigma\left(V^\mathrm{T}W + c\right)_i$$

反向传播:反过来,隐藏层 h_0 已知,通过相同的权重 W 采样出可见层,但是偏置 c 不同,以此重建输入。采样概率为:

$$p(v_i/h_o) = \sigma\left(W^\mathrm{T}h_o + b\right)_i$$

这两个传递过程重复 k 步或直到收敛,研究表明,$k = 1$ 就已经能给出很好的结果,所以此处设置 $k = 1$。

RBM 模型是一个基于能量的模型,对于一组给定的状态(可见向量 V 和隐藏向量)可构造能量函数:

$$E(v,h) = -b^\mathrm{T}V - c^\mathrm{T}h - V^\mathrm{T}Wh$$

与每个可见向量 V 相关联的是自由能量,一个单独配置的能量,要想与其他含有 V 的配置的能量相等,则:

$$F(v) = -b^\mathrm{T}V - \sum_{j \in \mathrm{hidden}} \log\left(1 + e^{cj + V^\mathrm{T}W}\right)$$

使用对比发散度目标函数，即 Mean($F(V_{\text{original}})$)−Mean($F(V_{\text{constructed}})$)，则权重的变化由下式给出：

$$dW = \eta \left(\{V^\mathrm{T} h\}_{\text{input}} - \{V^\mathrm{T} h\}_{\text{reconstructed}} \right)$$

其中，η 是学习率，偏置 b 和 c 也存在类似表达式。

具体做法

1. 导入模块：

```
import tensorflow as tf
import numpy as np
from tensorflow.examples.tutorials.mnist import input_data
import matplotlib.pyplot as plt
%matplotlib inline
```

2. 编写一个类 RBM 来完成主要工作。__init__ 中将构建完整的计算图、正向传播、反向传播和目标函数，此处使用 TensorFlow 内置的优化器来更新权重和偏置：

```
class RBM(object):
    def __init__(self, m, n):
        """
        m: Number of neurons in visible layer
        n: number of neurons in hidden layer
        """
        self._m = m
        self._n = n
        # Create the Computational graph
        # Weights and biases
        self._W = tf.Variable(tf.random_normal(shape=(self._m,self._n)))
        self._c = tf.Variable(np.zeros(self._n).astype(np.float32))
#bias for hidden layer
        self._b = tf.Variable(np.zeros(self._m).astype(np.float32))
#bias for Visible layer
        # Placeholder for inputs
        self._X = tf.placeholder('float', [None, self._m])
        # Forward Pass
        _h = tf.nn.sigmoid(tf.matmul(self._X, self._W) + self._c)
        self.h = tf.nn.relu(tf.sign(_h - tf.random_uniform(tf.shape(_h))))
        #Backward pass
        _v = tf.nn.sigmoid(tf.matmul(self.h, tf.transpose(self._W)) + self._b)
        self.V = tf.nn.relu(tf.sign(_v - tf.random_uniform(tf.shape(_v))))
        # Objective Function
        objective = tf.reduce_mean(self.free_energy(self._X)) - tf.reduce_mean(self.free_energy(self.V))
        self._train_op = tf.train.GradientDescentOptimizer(1e-3).minimize(objective)
```

```
        # Cross entropy cost
        reconstructed_input = self.one_pass(self._X)
        self.cost =
tf.reduce_mean(tf.nn.sigmoid_cross_entropy_with_logits(
labels=self._X, logits=reconstructed_input))
```

3. 在 RBM 类中定义 fit() 方法，在 __init__ 中声明所有的操作之后，训练操作就是简单地在会话中调用 train_op。我们使用批量训练：

```
    def fit(self, X, epochs = 1, batch_size = 100):
        N, D = X.shape
        num_batches = N // batch_size
        obj = []
        for i in range(epochs):
            #X = shuffle(X)
            for j in range(num_batches):
                batch = X[j * batch_size: (j * batch_size +
batch_size)]
                _, ob = self.session.run([self._train_op,self.cost
], feed_dict={self._X: batch})
                if j % 10 == 0:
                    print('training epoch {0} cost
{1}'.format(j,ob))
                obj.append(ob)
        return obj
```

4. 定义其他辅助函数计算 logit 误差，并从网络中返回重建的图像：

```
def set_session(self, session):
    self.session = session

def free_energy(self, V):
    b = tf.reshape(self._b, (self._m, 1))
    term_1 = -tf.matmul(V,b)
    term_1 = tf.reshape(term_1, (-1,))
    term_2 = -tf.reduce_sum(tf.nn.softplus(tf.matmul(V,self._W) +
        self._c))
    return term_1 + term_2

def one_pass(self, X):
    h = tf.nn.sigmoid(tf.matmul(X, self._W) + self._c)
    return tf.matmul(h, tf.transpose(self._W)) + self._b

def reconstruct(self,X):
    x = tf.nn.sigmoid(self.one_pass(X))
    return self.session.run(x, feed_dict={self._X: X})
```

5. 加载 MNIST 数据集：

```
mnist = input_data.read_data_sets("MNIST_data/", one_hot=True)
trX, trY, teX, teY = mnist.train.images, mnist.train.labels,
mnist.test.images, mnist.test.labels
```

6. 在 MNIST 数据集上训练 RBM：

```
Xtrain = trX.astype(np.float32)
Xtest = teX.astype(np.float32)
_, m = Xtrain.shape
rbm = RBM(m, 100)
```

```
#Initialize all variables
init = tf.global_variables_initializer()
with tf.Session() as sess:
    sess.run(init)
    rbm.set_session(sess)
    err = rbm.fit(Xtrain)
    out = rbm.reconstruct(Xest[0:100])   # Let us reconstruct Test
Data
```

7. 损失函数随着训练次数的变化如下：

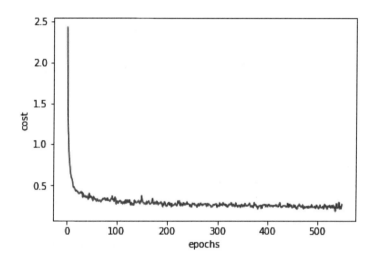

解读分析

由于能够重建图像，RBM 可用于从现有数据中生成更多的数据。通过下面的画图代码对比原始和重建的 MNIST 图像：

```
row, col = 2, 8
idx = np.random.randint(0, 100, row * col // 2)
f, axarr = plt.subplots(row, col, sharex=True, sharey=True, figsize=(20,4))
for fig, row in zip([Xtest_noisy,out], axarr):
    for i,ax in zip(idx,row):
        ax.imshow(fig[i].reshape((28, 28)), cmap='Greys_r')
        ax.get_xaxis().set_visible(False)
        ax.get_yaxis().set_visible(False)
```

结果如下：

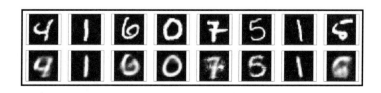

拓展阅读

- Smolensky, Paul. *Information processing in dynamical systems: Foundations of harmony theory*. No. CU-CS-321-86. COLORADO UNIV AT BOULDER DEPT OF COMPUTER SCIENCE, 1986. (`Link`)

- Salakhutdinov, Ruslan, Andriy Mnih, and Geoffrey Hinton. *Restricted Boltzmann machines for collaborative filtering*. Proceedings of the 24th international conference on Machine learning. ACM, 2007.(`Link`)

- Hinton, Geoffrey. *A practical guide to training restricted Boltzmann machines*. Momentum 9.1 (2010): 926.(`Link`)

- 如果你对数学感兴趣，那么这是一个很好的教程：`http://deeplearning.net/tutorial/rbm.html#rbm`。

7.6 基于 RBM 的推荐系统

网络零售商利用推荐系统向顾客推荐产品，例如，亚马逊会告诉你购买这个产品的其他顾客对哪些产品感兴趣，Netflix 根据你观看的内容以及其他有同样兴趣的 Netflix 用户观看过的内容向你推荐电视剧和电影。这些推荐系统都是基于协同过滤进行推荐的，在协同过滤中，系统根据用户过去的行为建立一个模型。这里利用上一节中的 RBM 模型，建立一个基于协同过滤的推荐系统用于推荐电影，这里的主要问题就是大多数用户不会评价所有的产品/电影，大部分数据是缺失的。如果有 M 个产品和 N 个用户，那么需要建立一个 $N \times M$ 的矩阵，矩阵中包括用户已知的评分，未知则置为 0。

准备工作

使用协同过滤构建推荐系统，首先准备数据。此处使用从 `https://grouplens.org/datasets/movielens/` 获取的电影数据为例，该数据由 `movies.dat` 和 `ratings.dat` 两个 .dat 文件组成，`movies.dat` 文件包含三列，分别为 3883 部电影的影片 ID、标题和类型，`ratings.dat` 文件包含四列，分别为用户 ID、影片 ID、评分和时长。现在合并这两个数据文件来构建一个矩阵，其中，每个用户都拥有对所有 3883 个电影的评分。但是用户通常不会对所有电影进行评分，所以只有某些电影的评分是非零的（需要标准化），而其余的则是零，为 0 的这部分将不会对隐藏层产生影响。

具体做法

1. 使用上一节中创建的 RBM 类。先定义 RBM 网络，可见单元的数量是电影的数量 3883 (`movies_df` 是从 `movies.dat` 文件读取的数据结构)：

```
m = len(movies_df)  # Number of visible units
n = 20   # Number of Hidden units
recommender = rbm.RBM(m,n)
```

2. 使用 Pandas 的 merge 和 groupby 命令创建一个列表 *trX*，存放 1000 名左右用户的用户影片评分，注意这里进行了标准化，列表大小是 1000×3883，用来训练 RBM：

```
Xtrain = np.array(trX)
init = tf.global_variables_initializer()
with tf.Session() as sess:
 sess.run(init)
 recommender.set_session(sess)
 err = recommender.fit(Xtrain, epochs=10)
```

3. 随着训练次数的增加，cross-logit 误差在减小，如下图所示：

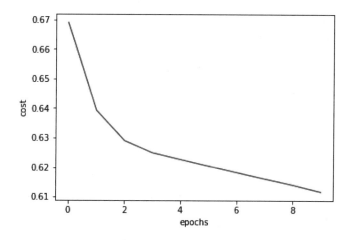

4. 网络训练好后，对随机选择的一个用户（例如 ID 为 150）进行电影推荐：

```
user_index = 150
x = np.array([Xtrain[user_index, :]])
init = tf.global_variables_initializer()
with tf.Session() as sess:
 sess.run(init)
 recommender.set_session(sess)
 out = recommender.reconstruct(x.astype(np.float32))
```

5. 可以看到这个用户的推荐结果，推荐分数如下：

Out[13]:

	MovieID	Title	Genre	List_Index	Recommendation_Score
0	1	Toy Story (1995)	Animation|Children's|Comedy	0	0.560319
1	2	Jumanji (1995)	Adventure|Children's|Fantasy	1	0.699984
2	3	Grumpier Old Men (1995)	Comedy|Romance	2	0.751833
3	4	Waiting to Exhale (1995)	Comedy|Drama	3	0.894270
4	5	Father of the Bride Part II (1995)	Comedy	4	0.238415

更多内容

Geoffrey Hinton 教授带领多伦多大学团队赢得了求解最优协同过滤的 Netflix 竞赛，他们的算法使用 RBM 预测用户对电影的评分（https://en.wikipedia.org/wiki/Netflix_Prize）。他们的工作详情可以从他们的论文中找到，网址为 http://www.cs.toronto.edu/~hinton/absps/netflixICML.pdf。

一个 RBM 隐藏单元的输出可以被送到另一个 RBM 的可见单元，重复这个过程就形成了堆叠 RBM，堆叠 RBM 中的每一个 RBM 都是独立训练的，无视其他 RBM 的存在。深度信任网络（DBN）就是由一系列 RBM 堆叠而成，DBN 可以使用有监督或无监督的方法进行训练，下一节中将了解更多信息。

7.7 用 DBN 进行情绪检测

本节中首先介绍如何利用堆叠 RBM 创建 DBN，然后利用 DBN 检测情绪。有趣的地方在于这里采用了两种不同的学习模式：首先，使用无监督学习逐一预训练 RBM，然后用监督学习训练 MLP 层。

准备工作

使用上一节中创建的 RBM 类，唯一的改动就是训练后不需要重建图像。此处堆叠 RBM 将数据一直向前传递到 DBN 的最后一层——MLP 层。类中不再使用 `reconstruct()` 函数，并增加了 `rbm_output()` 函数：

```
def rbm_output(self,X):
    x = tf.nn.sigmoid(tf.matmul(X, self._W) + self._c)
    return self.session.run(x, feed_dict={self._X: X})
```

数据使用 Kaggle 面部表情识别数据，可从地址 https://www.kaggle.com/c/challenges-in-representation-learning-facial-expression-recognition-challenge 获得，数据的描述如下：

该数据集由 48×48 像素的面部灰度图像组成，面部已经过自动调整，大致居中，并在每张图像中占据大致相同的空间。任务是根据面部表情显示的情绪将每个人脸分类成七个类别之一（0 = 生气，1 = 厌恶，2 = 恐惧，3 = 快乐，4 = 悲伤，5 = 惊奇，6 = 平静）。

train.csv 包含两列："情绪"和"像素"，"情绪"列包含从 0 到 6（包括 0 和 6）的数字代码，代表图像中的情绪，"像素"列是用引号括起来的字符串，字符串的内容是按行定序的用空格分隔的像素值。test.csv 则只包含"像素"列，任务就是预测情绪列。

训练集由 28709 个样本组成，最后的测试集由 3589 个样本组成，用来评比出比赛的冠军。

这个数据集由 Pierre-Luc Carrier 和 Aaron Courville 编写，是一个正在进行的研究项目的一部分，他们慷慨地为研讨会组织者提供了数据集的初步版本供本次比赛使用。

完整的数据在 `fer2013.csv` 文件中，从其中分离出训练、验证和测试数据：

```
data = pd.read_csv('data/fer2013.csv')
tr_data = data[data.Usage == "Training"]
test_data = data[data.Usage == "PublicTest"]
mask = np.random.rand(len(tr_data)) < 0.8
train_data = tr_data[mask]
val_data = tr_data[~mask]
```

需要对数据进行预处理，将像素和情绪标签分开。为此，构建两个函数，函数 `dense_to_one_hot()` 为标签执行独热编码，函数 `preprocess_data()` 将每个像素分隔成一个数组，利用这两个函数生成训练、验证和测试数据集的输入特征和标签：

```
def dense_to_one_hot(labels_dense, num_classes):
    num_labels = labels_dense.shape[0]
    index_offset = np.arange(num_labels) * num_classes
    labels_one_hot = np.zeros((num_labels, num_classes))
    labels_one_hot.flat[index_offset + labels_dense.ravel()] = 1
    return labels_one_hot
def preprocess_data(dataframe):
    pixels_values = dataframe.pixels.str.split(" ").tolist()
    pixels_values = pd.DataFrame(pixels_values, dtype=int)
    images = pixels_values.values
    images = images.astype(np.float32)
    images = np.multiply(images, 1.0/255.0)
    labels_flat = dataframe["emotion"].values.ravel()
    labels_count = np.unique(labels_flat).shape[0]
    labels = dense_to_one_hot(labels_flat, labels_count)
    labels = labels.astype(np.uint8)
    return images, labels
```

使用前面代码中定义的函数获得训练所需格式的数据。情绪检测 DBN 模型的构建原则基本与这篇关于 MNIST 的论文相似：`https://www.cs.toronto.edu/~hinton/absps/fastnc.pdf`。

具体做法

1. 导入依赖库 TensorFlow、NumPy 和 Pandas 来读取 `.csv` 文件，还需导入 Matplolib：

```
import tensorflow as tf
import numpy as np
import pandas as pd
import matplotlib.pyplot as plt
```

2. 使用辅助函数 `preprocess_data()` 获得训练、验证和测试数据：

```
X_train, Y_train = preprocess_data(train_data)
X_val, Y_val = preprocess_data(val_data)
X_test, Y_test = preprocess_data(test_data)
```

3. 大致查看一下数据，绘制均值图像，并输出训练、验证和测试数据集中的图像数量：

```
# Explore Data
mean_image = X_train.mean(axis=0)
std_image = np.std(X_train, axis=0)
print("Training Data set has {} images".format(len(X_train)))
print("Validation Data set has {} images".format(len(X_val)))
print("Test Data set has {} images".format(len(X_test)))
plt.imshow(mean_image.reshape(48,48), cmap='gray')
```

得到结果如下:

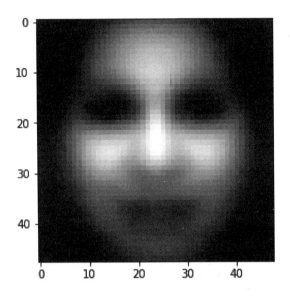

4. 查看训练样本中的图像及其标签:

```
classes = 
['angry','disgust','fear','happy','sad','surprise','neutral']
num_classes = len(classes)
samples_per_class = 7
for y,cls in enumerate(classes):
    idxs = np.flatnonzero(np.argmax(Y_train, axis =1) == y)
    idxs = np.random.choice(idxs, samples_per_class, 
replace=False)
    for i, idx in enumerate(idxs):
        plt_idx = i * num_classes + y + 1
        plt.subplot(samples_per_class, num_classes, plt_idx)
        plt.imshow(X_train[idx].reshape(48,48), cmap='gray')
#pixel height and width
        plt.axis('off')
        if i == 0:
            plt.title(cls)
plt.show()
```

效果图如下:

5. 定义堆叠 RBM 的结构,每层的 RBM 将前一个 RBM 的输出作为其输入:

```
RBM_hidden_sizes = [1500, 700, 400] #create 4 layers of RBM with
size 1500, 700, 400 and 100
#Set input as training data
inpX = X_train
#Create list to hold our RBMs
rbm_list = []
#Size of inputs is the number of inputs in the training set
input_size = inpX.shape[1]
#For each RBM we want to generate
for i, size in enumerate(RBM_hidden_sizes):
    print ('RBM: ',i,' ',input_size,'->', size)
    rbm_list.append(RBM(input_size, size))
    input_size = size
```

其中包含了三个 RBM:第一个 RBM 带有 2304(48×48)个输入和 1500 个隐藏单元,第二个 RBM 带有 1500 个输入和 700 个隐藏单元,第三个 RBM 带有 700 个输入和 400 个隐藏单元。

6. 逐层训练每个 RBM,训练方法为**贪婪训练**。在原论文中,对 MNIST 进行训练时每个 RBM 的训练次数是 30,此处增加训练次数应该能提高网络的性能:

```
# Greedy wise training of RBMs
init = tf.global_variables_initializer()
for rbm in rbm_list:
    print ('New RBM:')
    #Train a new one
    with tf.Session() as sess:
        sess.run(init)
        rbm.set_session(sess)
        err = rbm.fit(inpX, 5)
        inpX_n = rbm.rbm_output(inpX)
        print(inpX_n.shape)
        inpX = inpX_n
```

7. 定义 DBN 类,用三层 RBM 和两个 MLP 层来构建 DBN,RBM 层的权重是从预训练的 RBM 中加载的。同时也声明了 DBN 训练和预测的方法;网络利用最小化均方损失函数

的方法进行微调:

```
class DBN(object):

    def __init__(self, sizes, X, Y, eta = 0.001, momentum = 0.0, epochs = 10, batch_size = 100):
        #Initialize hyperparameters
        self._sizes = sizes
        print(self._sizes)
        self._sizes.append(1000) # size of the first FC layer
        self._X = X
        self._Y = Y
        self.N = len(X)
        self.w_list = []
        self.c_list = []
        self._learning_rate = eta
        self._momentum = momentum
        self._epochs = epochs
        self._batchsize = batch_size
        input_size = X.shape[1]
        #initialization loop
        for size in self._sizes + [Y.shape[1]]:
            #Define upper limit for the uniform distribution range
            max_range = 4 * math.sqrt(6. / (input_size + size))

            #Initialize weights through a random uniform distribution
            self.w_list.append(
                np.random.uniform( -max_range, max_range, [input_size, size]).astype(np.float32))

            #Initialize bias as zeroes
            self.c_list.append(np.zeros([size], np.float32))
            input_size = size


        # Build DBN
        #Create placeholders for input, weights, biases, output
        self._a = [None] * (len(self._sizes) + 2)
        self._w = [None] * (len(self._sizes) + 1)
        self._c = [None] * (len(self._sizes) + 1)
        self._a[0] = tf.placeholder("float", [None, self._X.shape[1]])
        self.y = tf.placeholder("float", [None, self._Y.shape[1]])

        #Define variables and activation function
        for i in range(len(self._sizes) + 1):
            self._w[i] = tf.Variable(self.w_list[i])
            self._c[i] = tf.Variable(self.c_list[i])
        for i in range(1, len(self._sizes) + 2):
            self._a[i] = tf.nn.sigmoid(tf.matmul(self._a[i - 1], self._w[i - 1]) + self._c[i - 1])

        #Define the cost function
        cost = tf.reduce_mean(tf.nn.softmax_cross_entropy_with_logits(labels=self.y, logits= self._a[-1]))
        #cost = tf.reduce_mean(tf.square(self._a[-1] - self.y))
```

```python
            #Define the training operation (Momentum Optimizer minimizing the Cost function)
            self.train_op = tf.train.AdamOptimizer(learning_rate=self._learning_rate).minimize(cost)

            #Prediction operation
            self.predict_op = tf.argmax(self._a[-1], 1)
    #load data from rbm
    def load_from_rbms(self, dbn_sizes,rbm_list):
            #Check if expected sizes are correct
            assert len(dbn_sizes) == len(self._sizes)

            for i in range(len(self._sizes)):
                #Check if for each RBN the expected sizes are correct
                assert dbn_sizes[i] == self._sizes[i]

            #If everything is correct, bring over the weights and biases
            for i in range(len(self._sizes)-1):
                self.w_list[i] = rbm_list[i]._W
                self.c_list[i] = rbm_list[i]._c


    def set_session(self, session):
        self.session = session

    #Training method
    def train(self, val_x, val_y):
        #For each epoch
        num_batches = self.N // self._batchsize

        batch_size = self._batchsize
        for i in range(self._epochs):
            #For each step
            for j in range(num_batches):
                batch = self._X[j * batch_size: (j * batch_size + batch_size)]
                batch_label = self._Y[j * batch_size: (j * batch_size + batch_size)]

                self.session.run(self.train_op, feed_dict={self._a[0]: batch, self.y: batch_label})

                for j in range(len(self._sizes) + 1):
                    #Retrieve weights and biases
                    self.w_list[j] = sess.run(self._w[j])
                    self.c_list[j] = sess.run(self._c[j])

                train_acc = np.mean(np.argmax(self._Y, axis=1) == self.session.run(self.predict_op, feed_dict={self._a[0]: self._X, self.y: self._Y}))

                val_acc = np.mean(np.argmax(val_y, axis=1) == self.session.run(self.predict_op, feed_dict={self._a[0]: val_x, self.y: val_y}))
```

```
                        print (" epoch " + str(i) + "/" + str(self._epochs) +
" Training Accuracy: " +   str(train_acc) + " Validation Accuracy: "
+ str(val_acc))

    def predict(self, X):
        return self.session.run(self.predict_op,
feed_dict={self._a[0]: X})
```

8. 实例化 DBN 对象并进行训练,预测测试数据的标签:

```
nNet = DBN(RBM_hidden_sizes, X_train, Y_train, epochs = 80)
with tf.Session() as sess:
    #Initialize Variables
    sess.run(tf.global_variables_initializer())
    nNet.set_session(sess)
    nNet.load_from_rbms(RBM_hidden_sizes,rbm_list)
    nNet.train(X_val, Y_val)
    y_pred = nNet.predict(X_test)
```

解读分析

RBM 使用无监督学习来学习出模型的隐藏特征,然后将预训练的 RBM 与全连接层一起进行微调。

这里的准确度在很大程度上取决于图像表示。在前面的章节中没有进行图像处理,只有缩放比例为 0 到 1 之间的灰度图像,但是如果按照链接 http://deeplearning.net/wp-content/uploads/2013/03/dlsvm.pdf 中所陈述的增加图像处理,则精确度会进一步提高。因此在 `preprocess_data` 函数中将图像与 100.0 / 255.0 相乘,并添加如下几行代码:

```
std_image = np.std(X_train, axis=0)
X_train = np.divide(np.subtract(X_train,mean_image), std_image)
X_val = np.divide(np.subtract(X_val,mean_image), std_image)
X_test = np.divide(np.subtract(X_test,mean_image), std_image)
```

更多内容

在前面的例子中,没有进行预处理时,三个数据集的精确度大约只有 40%,添加预处理后,训练数据的精确度增加到 90%,不过对于验证和测试集,仍然只能达到约 45% 的精确度。

很多方法可以用来改善结果,比如,在代码中使用的数据集是仅有 22000 个图像的 Kaggle 数据集,如果你观察这些图像,会发现增加只过滤面部的步骤会改善结果;另一种策略是增大而不是缩小隐藏层,如在论文 https://www.cs.swarthmore.edu/~meeden/cs81/s14/papers/KevinVincent.pdf 中提到的。

在识别情绪方面的另一个成功的改进是使用面部关键点而不是整个面部进行训练,参见 http://cs229.stanford.edu/proj2010/McLaughlinLeBayanbat-RecognizingEmotionsWithDeepBeliefNets.pdf。

利用前面的例子,你可以灵活使用这些改进并探索如何提高性能,愿 GPU 助你一臂之力!

CHAPTER 8

第 8 章

自动编码机

自动编码机是前馈非循环神经网络，是一种无监督机器学习方法，具有非常好的提取数据特征表示的能力，它是深层置信网络的重要组成部分，在图像重构、聚类、机器翻译等方面有着广泛的应用。在本章中，你将学习和实现自动编码机的不同改进，并最终学习如何使用层叠自动编码机。

8.1 引言

自动编码机也被称为 diabolo 网络或 autoassociator，最初是在 20 世纪 80 年代由 Hinton 和 PDP 小组提出的[1]。这是一种前馈神经网络，没有任何反馈，通过无监督学习方式进行学习。就像第 3 章中的多层感知机一样，也使用 BP 算法进行学习，但最大的不同在于，自动编码机的目标是重构一样的输入。

可以将自动编码机看作由两个级联网络组成，第一个网络是一个编码器，负责接收输入 x，并将输入通过函数 h 变换为信号 y：

$$y = h(x)$$

第二个网络将编码的信号 y 作为其输入，通过函数 f 得到重构的信号 r：

$$r = f(y) = f(h(x))$$

定义误差 e 为原始输入 x 与重构信号 r 之差，$e = x - r$，网络训练的目标是减少**均方误差**（MSE），同 MLP 一样，误差被反向传播回隐藏层。下图中显示了自动编码机的结构，用颜色区分编码器和解码器。自动编码机可以进行权值共享，即解码器和编码器的权值彼此互为转置，这样可以加快网络学习的速度，因为训练参数的数量减少了，但同时降低了网络的灵活程度。自动编码机与第 7 章中介绍的 RBM 非常类似，本质区别在于，自动编码机中神经元的状态是确定性的，而 RBM 中神经元的状态是概率性的。

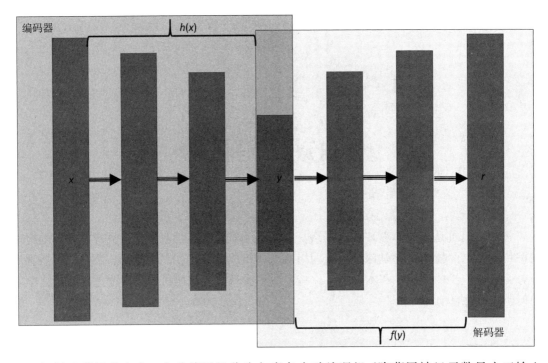

根据隐藏层的大小，自动编码机分为欠完备自动编码机（隐藏层神经元数量小于输入层神经元数量）和过完备自动编码机（隐藏层神经元数量大于输入层神经元数量）。根据对损失函数的约束条件，又可以分为：稀疏自动编码机、去噪自动编码机和卷积自动编码机，本章将学习这些自动编码机的不同版本，并使用 TensorFlow 来实现它们。

自动编码机的一个非常好的应用是降维[2]，而且自动编码机能产生比 PCA 更好的结果。自动编码机也可用于特征提取[3]、文档检索[2]、分类和异常检测。

拓展阅读

- Rumelhart, David E., Geoffrey E. Hinton, and Ronald J. Williams. Learning internal representations by error propagation. No. ICS-8506. California Univ San Diego La Jolla Inst for Cognitive Science, 1985. (http://www.cs.toronto.edu/~fritz/absps/pdp8.pdf)

- Hinton, Geoffrey E., and Ruslan R. Salakhutdinov. *Reducing the dimensionality of data with neural networks*, science 313.5786 (2006): 504-507. (https://pdfs.semanticscholar.org/7d76/b71b700846901ac4ac119403aa737a285e36.pdf)

- Masci, Jonathan, et al. *Stacked convolutional auto-encoders for hierarchical feature extraction*. Artificial Neural Networks and Machine Learning–ICANN 2011 (2011): 52-59. (https://www.researchgate.net/profile/Jonathan_Masci/publication/221078713_Stacked_Convolutional_Auto-Encoders_for_Hierarchical_Feature_Extraction/links/0deec518b9c6ed4634000000/Stacked-Convolutional-Auto-Encoders-for-Hierarchical-Feature-Extraction.pdf)

- Japkowicz, Nathalie, Catherine Myers, and Mark Gluck. *A novelty detection approach to classification*. IJCAI. Vol. 1. 1995. (http://www.ijcai.org/Proceedings/95-1/Papers/068.pdf)

8.2 标准自动编码机

由 Hinton 提出的标准自动编码机只有一个隐藏层，隐藏层中神经元的数量少于输入（和输出）层中神经元的数量，这会压缩网络中的信息，因此可以将隐藏层看作是一个压缩层，限定保留的信息。自动编码机的学习包括在隐藏层上对输入信号进行压缩表示，然后在输出层尽可能地复现原始输入：

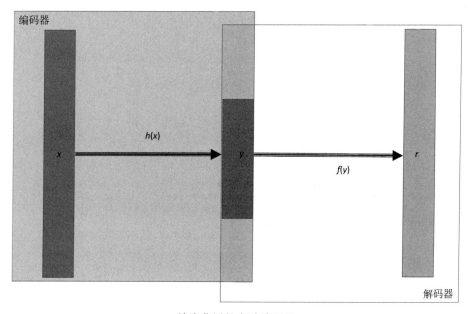

单隐藏层的自动编码机

准备工作

本节利用自动编码机进行图像重构，将利用 MNIST 数据训练自动编码机，并使用它来重构测试图像。

具体做法

1. 导入所有必要的模块：

```
import tensorflow as tf
import numpy as np
```

```
from tensorflow.examples.tutorials.mnist import input_data
import matplotlib.pyplot as plt
%matplotlib inline
```

2. 从 TensorFlow 中获取 MNIST 数据,这里要注意的一点是,标签并没有进行独热编码,因为并没有使用标签来训练网络。自动编码机是通过无监督学习进行训练的。

```
mnist = input_data.read_data_sets("MNIST_data/")
trX, trY, teX, teY = mnist.train.images, mnist.train.labels,
mnist.test.images, mnist.test.labels
```

3. 声明 `AutoEncoder` 类,使用 `init` 方法初始化自动编码机的权重、偏置和占位符,也可以在 `init` 方法中构建全部的计算图。还需要定义编码器、解码器、`set_session`(会话建立)和 `fit` 方法。此处构建的自动编码机使用简单的均方误差作为损失函数,使用 `AdamOptimizer` 进行优化。

```
class AutoEncoder(object):
    def __init__(self, m, n, eta = 0.01):
        """
        m: Number of neurons in input/output layer
        n: number of neurons in hidden layer
        """
        self._m = m
        self._n = n
        self.learning_rate = eta
        # Create the Computational graph
        
        # Weights and biases
        self._W1 = tf.Variable(tf.random_normal(shape=(self._m,self._n)))
        self._W2 = tf.Variable(tf.random_normal(shape=(self._n,self._m)))
        self._b1 = tf.Variable(np.zeros(self._n).astype(np.float32)) #bias for hidden layer
        self._b2 = tf.Variable(np.zeros(self._m).astype(np.float32)) #bias for output layer
        
        # Placeholder for inputs
        self._X = tf.placeholder('float', [None, self._m])
        
        self.y = self.encoder(self._X)
        self.r = self.decoder(self.y)
        error = self._X - self.r
        
        self._loss = tf.reduce_mean(tf.pow(error, 2))
        self._opt =
        tf.train.AdamOptimizer(self.learning_rate).minimize(self._loss)
    
    
    def encoder(self, x):
        h = tf.matmul(x, self._W1) + self._b1
        return tf.nn.sigmoid(h)
    
    
    def decoder(self, x):
        h = tf.matmul(x, self._W2) + self._b2
        return tf.nn.sigmoid(h)
```

```
def set_session(self, session):
self.session = session


def reduced_dimension(self, x):
h = self.encoder(x)
return self.session.run(h, feed_dict={self._X: x})

def reconstruct(self,x):
h = self.encoder(x)
r = self.decoder(h)
return self.session.run(r, feed_dict={self._X: x})



def fit(self, X, epochs = 1, batch_size = 100):
N, D = X.shape
num_batches = N // batch_size

obj = []
for i in range(epochs):
#X = shuffle(X)
for j in range(num_batches):
    batch = X[j * batch_size: (j * batch_size + batch_size)]
    _, ob = self.session.run([self._opt,self._loss],
feed_dict={self._X: batch})
    if j % 100 == 0 and i % 100 == 0:
        print('training epoch {0} batch {2} cost {1}'.format(i,ob,
j))
obj.append(ob)
return obj
```

为便于使用，此处还定义了两个辅助函数，`reduced_dimension`给出编码器网络的输出，`reconstruct`给出重构的测试图像的输出。

4. 训练时将输入数据转换为`float`型，初始化所有变量并运行会话。在计算时，目前只是测试自动编码机的重构能力：

```
Xtrain = trX.astype(np.float32)
Xtest = teX.astype(np.float32)
_, m = Xtrain.shape

autoEncoder = AutoEncoder(m, 256)

#Initialize all variables
init = tf.global_variables_initializer()
with tf.Session() as sess:
    sess.run(init)
    autoEncoder.set_session(sess)
    err = autoEncoder.fit(Xtrain, epochs=10)
    out = autoEncoder.reconstruct(Xtest[0:100])
```

5. 绘制误差在训练周期中的变化图，验证网络的均方误差在训练时是否得到优化，对于一个好的训练，误差应该随着训练周期的增加而减少：

```
plt.plot(err)
plt.xlabel('epochs')
plt.ylabel('cost')
```

图示如下:

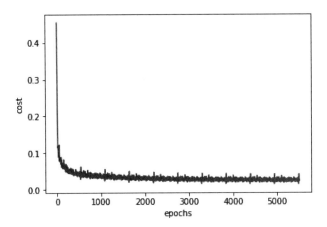

可以看到,随着网络的学习,损失/成本是下降的,当训练周期达到 5000 时,几乎是在一条线上振荡,这意味着进一步增加训练周期将不再有用。如果现在还想要改进模型,应该调整学习率、批量大小和优化器等超参数。

6. 观察重构的图像,对比原始图像和自动编码机生成的重构图像。

```
# Plotting original and reconstructed images
row, col = 2, 8
idx = np.random.randint(0, 100, row * col // 2)
f, axarr = plt.subplots(row, col, sharex=True, sharey=True,
figsize=(20,4))
for fig, row in zip([Xtest,out], axarr):
    for i,ax in zip(idx,row):
        ax.imshow(fig[i].reshape((28, 28)), cmap='Greys_r')
        ax.get_xaxis().set_visible(False)
        ax.get_yaxis().set_visible(False)
```

得到以下结果:

解读分析

有意思的是,在前面的代码中,维数从输入的 784 降到了 256,但是网络仍然可以重构原始图像。将自动编码机性能与 RBM 进行对比,其中隐藏层维数相等:

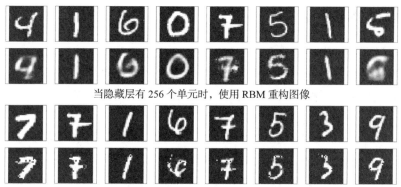

当隐藏层有 256 个单元时，使用 RBM 重构图像

当隐藏层有 256 个单元时，使用自动编码机重构图像

可以看到，由自动编码机重构的图像比 RBM 重构的图像要清晰得多。原因在于自动编码机中有更多的权重（从隐藏层到解码器输出层的权重）被训练。自动编码机学到的细节更多，即使两者都将信息压缩到相同的尺寸，其性能也优于 RBM。

更多内容

像 PCA 一样，自动编码机也可以用于降维，但 PCA 只能进行线性变换，而自动编码机可以使用非线性激活函数，从而在其中引入非线性变换。下图是 Hinton 的论文"Reducing the dimensionality of data with Neural Networks"复现的结果，图 A 显示 PCA 的结果，图 B 是由 RBM 堆叠的自动编码机（每层节点为 784-1000-500-250-2）的结果：

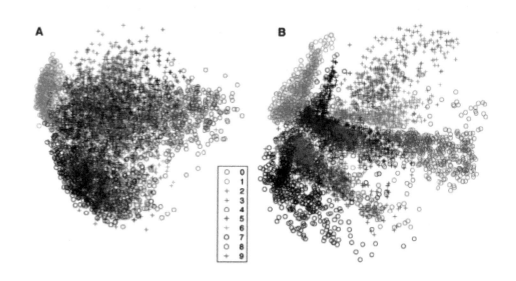

正如稍后会看到的，使用堆叠自动编码机时，每个自动编码机最初会独立进行预训练，然后会对整个网络进行微调以获得更好的性能。

8.3 稀疏自动编码机

上一节中的自动编码机更像是一个识别网络,只是简单重构了输入。而重点应是在像素级重构图像,施加的唯一约束是隐藏层单元的数量。而有趣的是,像素级重构并不能保证网络将从数据集中学习抽象特征,但是可以通过添加更多的约束确保网络从数据集中学习抽象特征。

在稀疏自动编码机中,重构误差中添加了一个稀疏惩罚,用来限定任何时刻的隐藏层中并不是所有单元都被激活。如果 m 是输入模式的总数,那么可以定义一个参数 ρ_hat(细节可以查看 Andrew Ng 的教案:`https://web.stanford.edu/class/cs294a/sparseAutoencoder_2011new.pdf`),用来表示每个隐藏层单元的行为(平均激活多少次)。基本的想法是让约束值 ρ_hat 等于稀疏参数 ρ。具体实现时在原始损失函数中增加表示稀疏性的正则项,损失函数如下:

```
loss = Mean squared error + Regularization for sparsity parameter
```

(损失 = 均方误差 + 稀疏参数正则项)

如果 ρ_hat 偏离 ρ,那么正则项将惩罚网络,一个常规的实现方法是衡量 ρ 和 ρ_hat 之间的 Kullback-Leiber (KL) 散度。

准备工作

在开始之前,先来看一下 KL 散度 D_{KL} 的概念,它是衡量两个分布之间差异的非对称度量,在本节中,两个分布是 ρ 和 ρ_hat。当 ρ 和 ρ_hat 相等时,KL 散度是零,否则会随着两者差异的增大而单调增加,KL 散度的数学表达式如下:

$$D_{KL}\left(\rho \| \hat{\rho}_j\right) = \rho\log\frac{\rho}{\hat{\rho}_j} + (1-\rho)\log\frac{1-\rho}{1-\rho_j}$$

下面是 ρ=0.3 时的 KL 的散度 D_{KL} 的变化图,从图中可以看到,当 ρ_hat=0.3 时,D_{KL}=0;而在 0.3 两侧都会单调递增:

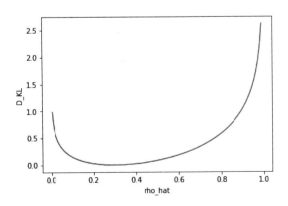

具体做法

1. 导入必要的模块：

```
import tensorflow as tf
import numpy as np
from tensorflow.examples.tutorials.mnist import input_data
import matplotlib.pyplot as plt
%matplotlib inline
```

2. 从 TensorFlow 示例加载 MNIST 数据集：

```
mnist = input_data.read_data_sets("MNIST_data/")
trX, trY, teX, teY = mnist.train.images, mnist.train.labels,
mnist.test.images, mnist.test.labels
```

3. 定义 `SparseAutoEncoder` 类，除了引入 KL 散度损失之外，它与前面的自动编码机类非常相似：

```
def kl_div(self, rho, rho_hat):
  term2_num = tf.constant(1.)- rho
  term2_den = tf.constant(1.) - rho_hat
  kl = self.logfunc(rho,rho_hat) + self.logfunc(term2_num, term2_den)
  return kl

def logfunc(self, x1, x2):
  return tf.multiply( x1, tf.log(tf.div(x1,x2)))
```

将 KL 约束条件添加到损失函数中，如下所示：

```
alpha = 7.5e-5
kl_div_loss = tf.reduce_sum(self.kl_div(0.02,
tf.reduce_mean(self.y,0)))
loss = self._loss + alpha * kl_div_loss
```

其中，`alpha` 是稀疏约束的权重。该类的完整代码如下所示：

```
class SparseAutoEncoder(object):
  def __init__(self, m, n, eta = 0.01):
    """
    m: Number of neurons in input/output layer
    n: number of neurons in hidden layer
    """
    self._m = m
    self._n = n
    self.learning_rate = eta

    # Create the Computational graph

    # Weights and biases
    self._W1 = tf.Variable(tf.random_normal(shape=(self._m,self._n)))
    self._W2 = tf.Variable(tf.random_normal(shape=(self._n,self._m)))
    self._b1 = tf.Variable(np.zeros(self._n).astype(np.float32)) #bias for hidden layer
    self._b2 = tf.Variable(np.zeros(self._m).astype(np.float32)) #bias for output layer
```

```python
# Placeholder for inputs
self._X = tf.placeholder('float', [None, self._m])

self.y = self.encoder(self._X)
self.r = self.decoder(self.y)
error = self._X - self.r

self._loss = tf.reduce_mean(tf.pow(error, 2))
alpha = 7.5e-5
kl_div_loss = tf.reduce_sum(self.kl_div(0.02,
tf.reduce_mean(self.y,0)))
loss = self._loss + alpha * kl_div_loss
self._opt =
tf.train.AdamOptimizer(self.learning_rate).minimize(loss)

def encoder(self, x):
h = tf.matmul(x, self._W1) + self._b1
return tf.nn.sigmoid(h)


def decoder(self, x):
h = tf.matmul(x, self._W2) + self._b2
return tf.nn.sigmoid(h)

def set_session(self, session):
self.session = session


def reduced_dimension(self, x):
h = self.encoder(x)
return self.session.run(h, feed_dict={self._X: x})

def reconstruct(self,x):
h = self.encoder(x)
r = self.decoder(h)
return self.session.run(r, feed_dict={self._X: x})

def kl_div(self, rho, rho_hat):
term2_num = tf.constant(1.)- rho
term2_den = tf.constant(1.) - rho_hat
kl = self.logfunc(rho,rho_hat) + self.logfunc(term2_num,
term2_den)
return kl

def logfunc(self, x1, x2):
return tf.multiply( x1, tf.log(tf.div(x1,x2)))



def fit(self, X, epochs = 1, batch_size = 100):
N, D = X.shape
num_batches = N // batch_size

obj = []
for i in range(epochs):
    #X = shuffle(X)
```

```
        for j in range(num_batches):
            batch = X[j * batch_size: (j * batch_size + batch_size)]
            _, ob = self.session.run([self._opt,self._loss],
feed_dict={self._X: batch})
            if j % 100 == 0:
                print('training epoch {0} batch {2} cost
{1}'.format(i,ob, j))
obj.append(ob)
 return obj
```

4. 声明 `SparseAutoEncoder` 类的一个对象，调用 *fit()* 训练，然后计算重构的图像：

```
Xtrain = trX.astype(np.float32)
Xtest = teX.astype(np.float32)
_, m = Xtrain.shape
sae = SparseAutoEncoder(m, 256)
#Initialize all variables
init = tf.global_variables_initializer()
with tf.Session() as sess:
 sess.run(init)
 sae.set_session(sess)
 err = sae.fit(Xtrain, epochs=10)
 out = sae.reconstruct(Xtest[0:100])
```

5. 重构损失均方误差随网络学习的变化图：

```
plt.plot(err)
plt.xlabel('epochs')
plt.ylabel('Reconstruction Loss (MSE)')
```

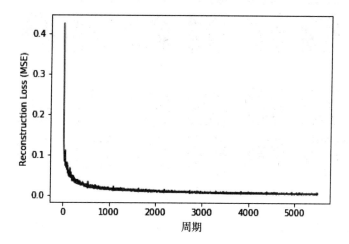

6. 查看重构的图像：

```
# Plotting original and reconstructed images
row, col = 2, 8
idx = np.random.randint(0, 100, row * col // 2)
f, axarr = plt.subplots(row, col, sharex=True, sharey=True,
figsize=(20,4))
for fig, row in zip([Xtest,out], axarr):
```

```
for i,ax in zip(idx,row):
    ax.imshow(fig[i].reshape((28, 28)), cmap='Greys_r')
    ax.get_xaxis().set_visible(False)
    ax.get_yaxis().set_visible(False)
```

结果如下:

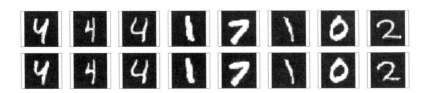

解读分析

必须注意到,稀疏自动编码机的主要代码与标准自动编码机完全相同,稀疏自动编码机只有一个主要变化——增加了 KL 散度损失以确保隐藏(瓶颈)层的稀疏性。如果将两者的重构结果进行比较,则可以看到即使隐藏层中的单元数量相同,稀疏自动编码机也比标准自动编码机好很多:

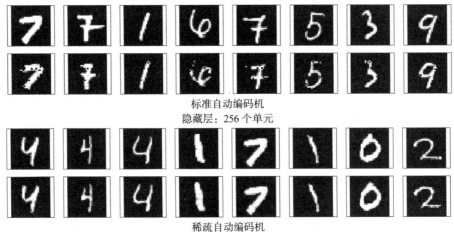

标准自动编码机
隐藏层: 256 个单元

稀疏自动编码机
隐藏层: 256 个单元

在 MNIST 数据集上,标准自动编码机训练后的重构损失是 0.022,而稀疏自动编码机是 0.006,由此可见稀疏自动编码机对数据的内在表示学习得更好一些。

更多内容

输入图像的内在表示是通过权重存储的,可视化网络学习的权重可见,以下分别是标准和稀疏自动编码机的编码层权重,从中可以看到,在标准自动编码机中,许多隐藏层单元的权重非常大,这表明它们是超载的。

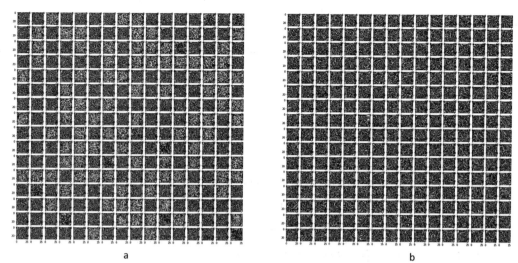

标准自动编码机(a)和稀疏自动编码机(b)的编码层权重

拓展阅读

- http://web.engr.illinois.edu/~hanj/cs412/bk3/KL-divergence.pdf
- https://en.wikipedia.org/wiki/Kullback%E2%80%93Leibler_divergence

8.4 去噪自动编码机

前两节中探讨的两个自动编码机属于欠完备自动编码机，因为隐藏层的维度比输入（输出）层低。去噪自动编码机属于过完备自动编码机，隐藏层的维数大于输入层时效果会更好。

去噪自动编码机从受损（噪声）输入中学习，它向编码器网络提供有噪声的输入，然后将解码器的重构图像与原始输入进行比较，这就会"教会"网络去学习如何对输入去噪。不再只是进行像素比较，为了去噪，它也会学习相邻像素的信息。

准备工作

去噪自动编码机也具有 KL 散度惩罚项，它不同于稀疏自动编码机的主要有两个方面，首先，隐藏层的单元数 n_hidden 大于输入层的单元数 m，即 n_hidden>m；其次，编码器的输入是受损输入，要做到这一点，这里构造了一个给输入添加噪声的受损函数：

```
def corruption(x, noise_factor = 0.3): #corruption of the input
    noisy_imgs = x + noise_factor * np.random.randn(*x.shape)
    noisy_imgs = np.clip(noisy_imgs, 0., 1.)
    return noisy_imgs
```

具体做法

1. 导入必要的模块，导入 TensorFlow 和 numpy 处理输入数据，导入 matplotlib 绘图，等等：

```
import tensorflow as tf
import numpy as np
from tensorflow.examples.tutorials.mnist import input_data
import matplotlib.pyplot as plt
import math
%matplotlib inline
```

2. 加载来自 TensorFlow 示例的数据。本章所有示例都使用了标准的 MNIST 数据库，这样不同自动编码机之间就有了对比的基准。

```
mnist = input_data.read_data_sets("MNIST_data/")
trX, trY, teX, teY = mnist.train.images, mnist.train.labels,
mnist.test.images, mnist.test.labels
```

3. 定义类 `DenoisingAutoEncoder`，与前面的 `SparseAutoEncoder` 类相似。在这里，定义一个受噪声影响的图像的占位符，将这个噪声图像送入编码器；重构误差是原始清晰图像和解码器输出之间的差异，这里保留了稀疏惩罚项，因此，`fit` 函数参数包括原始图像和噪声图像。

```
class DenoisingAutoEncoder(object):
def __init__(self, m, n, eta = 0.01):
"""
m: Number of neurons in input/output layer
n: number of neurons in hidden layer
"""
self._m = m
self._n = n
self.learning_rate = eta

# Create the Computational graph

# Weights and biases
self._W1 = tf.Variable(tf.random_normal(shape=(self._m,self._n)))
self._W2 = tf.Variable(tf.random_normal(shape=(self._n,self._m)))
self._b1 = tf.Variable(np.zeros(self._n).astype(np.float32)) #bias for hidden layer
self._b2 = tf.Variable(np.zeros(self._m).astype(np.float32)) #bias for output layer

# Placeholder for inputs
self._X = tf.placeholder('float', [None, self._m])

self._X_noisy = tf.placeholder('float', [None, self._m])

self.y = self.encoder(self._X_noisy)
self.r = self.decoder(self.y)
error = self._X - self.r

self._loss = tf.reduce_mean(tf.pow(error, 2))
```

```python
#self._loss = tf.reduce_mean(tf.nn.sigmoid_cross_entropy_with_logits(labels
=self._X, logits = self.r))
alpha = 0.05
kl_div_loss = tf.reduce_sum(self.kl_div(0.02, tf.reduce_mean(self.y,0)))
loss = self._loss + alpha * kl_div_loss
self._opt = tf.train.AdamOptimizer(self.learning_rate).minimize(loss)


def encoder(self, x):
h = tf.matmul(x, self._W1) + self._b1
return tf.nn.sigmoid(h)


def decoder(self, x):
h = tf.matmul(x, self._W2) + self._b2
return tf.nn.sigmoid(h)
def set_session(self, session):
self.session = session

def reconstruct(self,x):
h = self.encoder(x)
r = self.decoder(h)
return self.session.run(r, feed_dict={self._X: x})

def kl_div(self, rho, rho_hat):
term2_num = tf.constant(1.)- rho
term2_den = tf.constant(1.) - rho_hat
kl = self.logfunc(rho,rho_hat) + self.logfunc(term2_num, term2_den)
return kl

def logfunc(self, x1, x2):
return tf.multiply( x1, tf.log(tf.div(x1,x2)))

def corrupt(self,x):
return x * tf.cast(tf.random_uniform(shape=tf.shape(x),
minval=0,maxval=2),tf.float32)

def getWeights(self):
return self.session.run([self._W1, self._W2,self._b1, self._b2])

def fit(self, X, Xorg, epochs = 1, batch_size = 100):
N, D = X.shape
num_batches = N // batch_size

obj = []
for i in range(epochs):
#X = shuffle(X)
for j in range(num_batches):
batch = X[j * batch_size: (j * batch_size + batch_size)]
batchO = Xorg[j * batch_size: (j * batch_size + batch_size)]
_, ob = self.session.run([self._opt,self._loss], feed_dict={self._X:
batchO, self._X_noisy: batch})
if j % 100 == 0:
print('training epoch {0} batch {2} cost {1}'.format(i,ob, j))
obj.append(ob)
return obj
```

也可以为自动编码机对象添加噪声，若如此，将使用类中定义的受损方法：self._X_noisy=self.corrupt(self._X)*0.3+self._X*(1-0.3)，此时 fit 函数修改如下：

```
def fit(self, X, epochs = 1, batch_size = 100):
    N, D = X.shape
    num_batches = N // batch_size
    obj = []
    for i in range(epochs):
        #X = shuffle(X)
        for j in range(num_batches):
            batch = X[j * batch_size: (j * batch_size + batch_size)]
            _, ob = self.session.run([self._opt,self._loss], feed_dict={self._X: batch})
            if j % 100 == 0:
                print('training epoch {0} batch {2} cost {1}'.format(i,ob, j))
            obj.append(ob)
    return obj
```

4. 使用前面定义的受损函数来生成一个噪声图像，并提供给会话：

```
n_hidden = 800
Xtrain = trX.astype(np.float32)
Xtrain_noisy = corruption(Xtrain).astype(np.float32)
Xtest = teX.astype(np.float32)
#noise = Xtest * np.random.randint(0, 2, Xtest.shape).astype(np.float32)
Xtest_noisy = corruption(Xtest).astype(np.float32) #Xtest * (1-0.3)+ noise *(0.3)
_, m = Xtrain.shape

dae = DenoisingAutoEncoder(m, n_hidden)

#Initialize all variables
init = tf.global_variables_initializer()
with tf.Session() as sess:
    sess.run(init)
    dae.set_session(sess)
    err = dae.fit(Xtrain_noisy, Xtrain, epochs=10)
    out = dae.reconstruct(Xtest_noisy[0:100])
    W1, W2, b1, b2 = dae.getWeights()
    red = dae.reduced_dimension(Xtrain)
```

5. 随着网络的学习，重构损失在减少：

```
plt.plot(err)
plt.xlabel('epochs')
plt.ylabel('Reconstruction Loss (MSE)')
```

效果图如下：

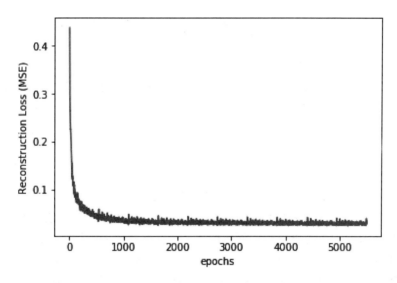

6. 当将来自测试数据集中的噪声图像提供给训练好的网络时,可进行图像重构:

```
# Plotting original and reconstructed images
row, col = 2, 8
idx = np.random.randint(0, 100, row * col // 2)
f, axarr = plt.subplots(row, col, sharex=True, sharey=True, figsize=(20,4))
for fig, row in zip([Xtest_noisy,out], axarr):
 for i,ax in zip(idx,row):
  ax.imshow(fig[i].reshape((28, 28)), cmap='Greys_r')
  ax.get_xaxis().set_visible(False)
  ax.get_yaxis().set_visible(False)
```

得到以下结果:

拓展阅读

- https://cs.stanford.edu/people/karpathy/convnetjs/demo/autoencoder.html
- http://blackecho.github.io/blog/machine-learning/2016/02/29/denoising-autoencoder-tensorflow.html

8.5 卷积自动编码机

研究人员发现,卷积神经网络(CNN)之所以在处理图像上有优势,是因为可以提取隐藏在图像中的空间信息，因此很自然地想到如果可以使用 CNN 构造编码器和解码器网络，会比其他自动编码机工作得更好，因此产生了卷积自动编码机（CAE）。第 4 章中解释了卷积和最大池化的概念，本节将以此为基础来理解卷积自动编码机是如何工作的。

CAE 的编码器和解码器都是 CNN 网络，编码器的卷积网络学习将输入编码为一组信号，然后解码器 CNN 尝试重构来自自动编码机的输入。其中 CNN 作为通用特征提取器进行工作，学习如何最好地捕捉输入特征。

准备工作

在第 4 章中已经了解到，随着卷积层的添加，传递到下一层的空间尺寸信息在减小，但是在自动编码机中，重构图像的大小和深度应与输入图片相同，这意味着解码器应该以某种方式调整图像大小和卷积来重构原始图像。转置卷积层能够增加空间尺寸和卷积，在 TensorFlow 中通过 `tf.nn.conv2d_transpose` 即可实现，但是转置卷积层会导致最终图像中出现伪影。Augustus Odena 等人 [1] 表明使用最近邻或双线性插值（上采样）紧跟着一个卷积层的方式可以避免这些伪影，他们采用最近邻差值（`tf.image.resize_images`）实现，最终取得了非常好的结果，在这里将使用相同的方法。

具体做法

1. 导入必要的模块：

```
import tensorflow as tf
import numpy as np
from tensorflow.examples.tutorials.mnist import input_data
import matplotlib.pyplot as plt
import math
%matplotlib inline
```

2. 加载输入数据：

```
mnist = input_data.read_data_sets("MNIST_data/")
trX, trY, teX, teY = mnist.train.images, mnist.train.labels,
mnist.test.images, mnist.test.labels
```

3. 定义网络参数，同时也计算每个最大池化层的输出空间维度，这些信息在解码器网络中对图像进行上采样：

```
# Network Parameters
h_in, w_in = 28, 28 # Image size height and width
k = 3 # Kernel size
p = 2 # pool
s = 2 # Strides in maxpool
filters = {1:32,2:32,3:16}
activation_fn=tf.nn.relu
# Change in dimensions of image after each MaxPool
h_l2, w_l2 = int(np.ceil(float(h_in)/float(s))) ,
int(np.ceil(float(w_in)/float(s))) # Height and width: second
encoder/decoder layer
```

```
h_l3, w_l3 = int(np.ceil(float(h_l2)/float(s))) ,
   int(np.ceil(float(w_l2)/float(s))) # Height and width: third
   encoder/decoder layer
```

4. 为输入(噪声图像)和目标(对应的清晰图像)创建占位符:

```
X_noisy = tf.placeholder(tf.float32, (None, h_in, w_in, 1), name='inputs')
X = tf.placeholder(tf.float32, (None, h_in, w_in, 1), name='targets')
```

5. 建立编码器和解码器网络:

```
### Encoder
conv1 = tf.layers.conv2d(X_noisy, filters[1], (k,k), padding='same',
activation=activation_fn)
# Output size h_in x w_in x filters[1]
maxpool1 = tf.layers.max_pooling2d(conv1, (p,p), (s,s), padding='same')
# Output size h_l2 x w_l2 x filters[1]
conv2 = tf.layers.conv2d(maxpool1, filters[2], (k,k), padding='same',
activation=activation_fn)
# Output size h_l2 x w_l2 x filters[2]
maxpool2 = tf.layers.max_pooling2d(conv2,(p,p), (s,s), padding='same')
# Output size h_l3 x w_l3 x filters[2]
conv3 = tf.layers.conv2d(maxpool2,filters[3], (k,k), padding='same',
activation=activation_fn)
# Output size h_l3 x w_l3 x filters[3]
encoded = tf.layers.max_pooling2d(conv3, (p,p), (s,s), padding='same')
# Output size h_l3/s x w_l3/s x filters[3] Now 4x4x16

### Decoder
upsample1 = tf.image.resize_nearest_neighbor(encoded, (h_l3,w_l3))
# Output size h_l3 x w_l3 x filters[3]
conv4 = tf.layers.conv2d(upsample1, filters[3], (k,k), padding='same',
activation=activation_fn)
# Output size h_l3 x w_l3 x filters[3]
upsample2 = tf.image.resize_nearest_neighbor(conv4, (h_l2,w_l2))
# Output size h_l2 x w_l2 x filters[3]
conv5 = tf.layers.conv2d(upsample2, filters[2], (k,k), padding='same',
activation=activation_fn)
# Output size h_l2 x w_l2 x filters[2]
upsample3 = tf.image.resize_nearest_neighbor(conv5, (h_in,w_in))
# Output size h_in x w_in x filters[2]
conv6 = tf.layers.conv2d(upsample3, filters[1], (k,k), padding='same',
activation=activation_fn)
# Output size h_in x w_in x filters[1]

logits = tf.layers.conv2d(conv6, 1, (k,k) , padding='same',
activation=None)

# Output size h_in x w_in x 1
decoded = tf.nn.sigmoid(logits, name='decoded')

loss = tf.nn.sigmoid_cross_entropy_with_logits(labels=X, logits=logits)
cost = tf.reduce_mean(loss)
opt = tf.train.AdamOptimizer(0.001).minimize(cost)
```

6. 建立会话:

```
sess = tf.Session()
```

7. 根据给定输入调整模型:

```
epochs = 10
batch_size = 100
# Set's how much noise we're adding to the MNIST images
noise_factor = 0.5
sess.run(tf.global_variables_initializer())
err = []
for i in range(epochs):
 for ii in range(mnist.train.num_examples//batch_size):
 batch = mnist.train.next_batch(batch_size)
 # Get images from the batch
 imgs = batch[0].reshape((-1, h_in, w_in, 1))
 
 # Add random noise to the input images
 noisy_imgs = imgs + noise_factor * np.random.randn(*imgs.shape)
 # Clip the images to be between 0 and 1
 noisy_imgs = np.clip(noisy_imgs, 0., 1.)
 
 # Noisy images as inputs, original images as targets
 batch_cost, _ = sess.run([cost, opt], feed_dict={X_noisy: noisy_imgs,X: imgs})
 err.append(batch_cost)
 if ii%100 == 0:
 print("Epoch: {0}/{1}... Training loss {2}".format(i, epochs, batch_cost))
```

8. 网络学习误差如下:

```
plt.plot(err)
plt.xlabel('epochs')
plt.ylabel('Cross Entropy Loss')
```

效果图如下:

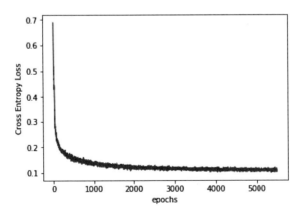

9. 重构的图像。

```
fig, axes = plt.subplots(rows=2, cols=10, sharex=True, sharey=True,
figsize=(20,4))
```

```
in_imgs = mnist.test.images[:10]
noisy_imgs = in_imgs + noise_factor * np.random.randn(*in_imgs.shape)
noisy_imgs = np.clip(noisy_imgs, 0., 1.)
reconstructed = sess.run(decoded, feed_dict={X_noisy:
noisy_imgs.reshape((10, 28, 28, 1))})
for images, row in zip([noisy_imgs, reconstructed], axes):
 for img, ax in zip(images, row):
 ax.imshow(img.reshape((28, 28)), cmap='Greys_r')
 ax.get_xaxis().set_visible(False)
 ax.get_yaxis().set_visible(False)
```

下图是前面代码的输出:

10. 关闭会话:

```
sess.close()
```

解读分析

这是一个去噪 CAE,可以看到,与仅由一个隐藏层组成的简单去噪自动编码机相比,去噪 CAE 对图像去噪效果更好。

更多内容

CAE 也被研究人员用于语义分割,可以参考一篇有趣的文章 "*Segnet: a Deep Convolutional Encoder-Decoder Architecture for Image Segmentation*"(Badrinayanan 等人,2015,`https://arxiv.org/pdf/1511.00561.pdf`),该网络使用 VGG16 的卷积层作为编码器网络,并且包含相应的解码器结构,每个编码器对应一个解码器网络。解码器对从对应编码器接收到的最大池化输入特征映射进行了非线性上采样。

拓展阅读

- `https://distill.pub/2016/deconv-checkerboard/`
- `https://pgaleone.eu/neural-networks/2016/11/24/convolutional-autoencoders/`
- `https://arxiv.org/pdf/1511.00561.pdf`
- `https://github.com/arahusky/Tensorflow-Segmentation`

8.6 堆叠自动编码机

到目前为止介绍的自动编码机(除了 CAE)仅由单层编码器和单层解码器组成。编码器和解码器网络也可能有多层,使用更深的编码器和解码器网络可以使自动编码机表示更复杂的特征,将一个编码器提取的特征作为输入传递到下一个编码器,这种结构被称为堆叠自动编码机(或者**深度自动编码机**)。堆叠自动编码机可以作为一个网络进行训练,训练目标是最小化重构误差;也可以首先使用之前学习的无监督方法对每个编码器/解码器网络进行预训练,然后对整个网络进行微调。有人指出,通过预训练(逐层贪婪训练),效果会更好。

准备工作

本节将使用逐层贪婪方法来训练堆叠自动编码机,为了降低训练难度,这里使用共享权重,因此相应的编码器/解码器权重将是彼此的转置。

具体做法

1. 导入所有必要的模块:

```python
import tensorflow as tf
import numpy as np
from tensorflow.examples.tutorials.mnist import input_data
import matplotlib.pyplot as plt
%matplotlib inline
```

2. 加载数据集:

```python
mnist = input_data.read_data_sets("MNIST_data/")
trX, trY, teX, teY = mnist.train.images, mnist.train.labels, mnist.test.images, mnist.test.labels
```

3. 定义类 `StackedAutoencoder`。`__init__` 方法包括一个列表,其中包含每个自动编码机中的诸多神经元,从第一个输入自动编码机和学习率开始。由于每层都有不同的输入和输出维度,因此选择一个字典数据结构来表示每层的权重、偏置和输入:

```python
class StackedAutoEncoder(object):
    def __init__(self, list1, eta = 0.02):
        """
        list1: [input_dimension, hidden_layer_1, ....,hidden_layer_n]
        """
        N = len(list1)-1
        self._m = list1[0]
        self.learning_rate = eta

        # Create the Computational graph
        self._W = {}
        self._b = {}
        self._X = {}
        self._X['0'] = tf.placeholder('float', [None, list1[0]])

        for i in range(N):
            layer = '{0}'.format(i+1)
            print('AutoEncoder Layer {0}: {1} --> {2}'.format(layer, list1[i], list1[i+1]))
```

```
    self._W['E' + layer] = tf.Variable(tf.random_normal(shape=(list1[i],
list1[i+1])),name='WtsEncoder'+layer)
    self._b['E'+ layer] =
tf.Variable(np.zeros(list1[i+1]).astype(np.float32),name='BiasEncoder'+laye
r)
    self._X[layer] = tf.placeholder('float', [None, list1[i+1]])
    self._W['D' + layer] = tf.transpose(self._W['E' + layer]) # Shared weights
    self._b['D' + layer] =
tf.Variable(np.zeros(list1[i]).astype(np.float32),name='BiasDecoder' +
layer)
    # Placeholder for inputs
    self._X_noisy = tf.placeholder('float', [None, self._m])
```

4. 建立计算图, 在预训练时为每个自动编码机定义优化参数, 当上一个自动编码机的输出作为当前自动编码机的输入时, 为其定义重构损失, 为此定义了方法 `pretrain` 和 `one_pass`, 分别为每个自动编码机返回编码器的训练操作和输出:

```
    self.train_ops = {}
    self.out = {}


    for i in range(N):
    layer = '{0}'.format(i+1)
    prev_layer = '{0}'.format(i)
    opt = self.pretrain(self._X[prev_layer], layer)
    self.train_ops[layer] = opt
    self.out[layer] = self.one_pass(self._X[prev_layer], self._W['E'+layer],
    self._b['E'+layer], self._b['D'+layer])
```

5. 在计算图中对整个堆叠自动编码机进行微调, 这里使用类方法 `encoder` 和 `decoder` 来实现:

```
    self.y = self.encoder(self._X_noisy,N) #Encoder output
    self.r = self.decoder(self.y,N) # Decoder ouput

    optimizer = tf.train.AdamOptimizer(self.learning_rate)
    error = self._X['0'] - self.r # Reconstruction Error

    self._loss = tf.reduce_mean(tf.pow(error, 2))
    self._opt = optimizer.minimize(self._loss)
```

6. 定义类方法 `fit`, 对每个自动编码机执行批量预训练, 然后进行微调。在训练时使用正常的输入; 在微调时使用受损输入。这使得可以用堆叠自动编码机从噪声输入中进行重构:

```
    def fit(self, Xtrain, Xtr_noisy, layers, epochs = 1, batch_size = 100):
    N, D = Xtrain.shape
    num_batches = N // batch_size
    X_noisy = {}
    X = {}
    X_noisy['0'] = Xtr_noisy
    X['0'] = Xtrain
    for i in range(layers):
    Xin = X[str(i)]
    print('Pretraining Layer ', i+1)
```

```
    for e in range(5):
     for j in range(num_batches):
      batch = Xin[j * batch_size: (j * batch_size + batch_size)]
      self.session.run(self.train_ops[str(i+1)], feed_dict= {self._X[str(i)]:
batch})
    print('Pretraining Finished')
    X[str(i+1)] = self.session.run(self.out[str(i+1)], feed_dict =
{self._X[str(i)]: Xin})

    obj = []
    for i in range(epochs):
     for j in range(num_batches):
      batch = Xtrain[j * batch_size: (j * batch_size + batch_size)]
      batch_noisy = Xtr_noisy[j * batch_size: (j * batch_size + batch_size)]
      _, ob = self.session.run([self._opt,self._loss], feed_dict={self._X['0']:
batch, self._X_noisy: batch_noisy})
      if j % 100 == 0 :
    print('training epoch {0} batch {2} cost {1}'.format(i,ob, j))
    obj.append(ob)
    return obj
```

7. 各个类方法如下：

```
def encoder(self, X, N):
 x = X
 for i in range(N):
  layer = '{0}'.format(i+1)
  hiddenE = tf.nn.sigmoid(tf.matmul(x, self._W['E'+layer]) +
self._b['E'+layer])
  x = hiddenE
 return x

def decoder(self, X, N):
 x = X
 for i in range(N,0,-1):
  layer = '{0}'.format(i)
  hiddenD = tf.nn.sigmoid(tf.matmul(x, self._W['D'+layer]) +
self._b['D'+layer])
  x = hiddenD
 return x

def set_session(self, session):
 self.session = session

def reconstruct(self,x, n_layers):
 h = self.encoder(x, n_layers)
 r = self.decoder(h, n_layers)
 return self.session.run(r, feed_dict={self._X['0']: x})

def pretrain(self, X, layer ):
 y = tf.nn.sigmoid(tf.matmul(X, self._W['E'+layer]) + self._b['E'+layer])
 r =tf.nn.sigmoid(tf.matmul(y, self._W['D'+layer]) + self._b['D'+layer])

 # Objective Function
 error = X - r # Reconstruction Error
  loss = tf.reduce_mean(tf.pow(error, 2))
 opt = tf.train.AdamOptimizer(.001).minimize(loss, var_list =
```

```
        [self._W['E'+layer],self._b['E'+layer],self._b['D'+layer]])
    return opt

def one_pass(self, X, W, b, c):
    h = tf.nn.sigmoid(tf.matmul(X, W) + b)
    return h
```

8. 使用之前定义的 **corruption** 函数为图像加入噪声，最后创建一个 **StackAutoencoder** 对象并对其进行训练：

```
Xtrain = trX.astype(np.float32)
Xtrain_noisy = corruption(Xtrain).astype(np.float32)
Xtest = teX.astype(np.float32)
Xtest_noisy = corruption(Xtest).astype(np.float32)
_, m = Xtrain.shape

list1 = [m, 500, 50] # List with number of neurons in Each hidden layer, starting from input layer
n_layers = len(list1)-1
autoEncoder = StackedAutoEncoder(list1)

#Initialize all variables
init = tf.global_variables_initializer()

with tf.Session() as sess:
    sess.run(init)
    autoEncoder.set_session(sess)
    err = autoEncoder.fit(Xtrain, Xtrain_noisy, n_layers, epochs=30)
    out = autoEncoder.reconstruct(Xtest_noisy[0:100],n_layers)
```

9. 随着堆叠自动编码机的微调，重构误差不断降低。可以看到，由于进行了预训练，重构损失是从非常低的水平开始的：

```
plt.plot(err)
plt.xlabel('epochs')
plt.ylabel('Fine Tuning Reconstruction Error')
```

效果图如下：

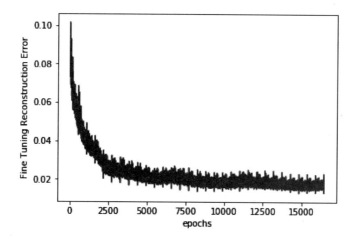

10. 现在来测试一下层叠自动编码机的性能。下图是存在噪声的测试图像以及去噪后的手写图像：

解读分析

堆叠自动编码机的实验表明，预训练应该用低的学习率，这确保了在微调期间有更好的收敛性和性能。

更多内容

本章介绍的内容都是关于自动编码机的，自动编码机目前仅仅用于降维和信息检索，并引起了广泛的兴趣。首先它是无监督的，其次是因为它们可以和 FCN 一起使用。自动编码机可以帮助解决维度爆炸的魔咒，研究人员也已经证实可将其用于分类和异常检测。

拓展阅读

- 一篇很好的堆叠自动编码机的文章，网址为 http://ufldl.stanford.edu/wiki/index.php/Stacked_Autoencoders。
- Schwenk, Holger. "The diabolo classifier." Neural Computation 10.8 (1998): 2175-2200.
- Sakurada, Mayu, and Takehisa Yairi. "Anomaly detection using autoencoders with nonlinear dimensionality reduction." Proceedings of the MLSDA 2014 2nd Workshop on Machine Learning for Sensory Data Analysis. ACM, 2014.
- 堆叠自动编码机的 TensorBoard 可视化和实。现：https://github.com/cmgreen210/TensorFlowDeepAutoencoder。

CHAPTER 9

第 9 章

强化学习

这章将介绍**强化学习**,这是一个有着巨大研究前景的算法。

9.1 引言

2016 年 3 月,谷歌公司 DeepMind 团队的 AlphaGo 以 4 比 1 战胜第 18 届世界围棋冠军李世石,这是一场具有历史意义的比赛。让电脑学会下围棋是一件十分困难的事情,它有

208,168,199,381,979,984,699,478,633,344,862,770,286,522,
453,884,530,548,425,639,456,820,927,419,612,738,015,378,
525,648,451,698,519,643,907,259,916,015,628,128,546,089,
888,314,427, 129,715,319,317,557,736,620,397,247,064,840,935

种可能的落子位置。在围棋中获胜不可能只靠简单的蛮力,它需要技巧、创造力,以及类似职业棋手的直觉。

通过融合深度强化学习网络和最先进的树搜索算法,AlphaGo 实现了这一创举。本章将介绍强化学习以及强化学习的算法案例。

第一个问题就是什么是强化学习,它与前几章介绍的监督学习和无监督学习有什么区别?

喂养过宠物的人都知道,想要训练宠物,最有效的方法就是当它做得好的时候奖励它,做得不好的时候惩罚它。**强化学习**就是一种类似的学习算法。神经网络算法采取一系列**动作**(a),它将会引起一系列与环境有关的状态 (s) 变化,相应地,它就可以得到奖励或者惩罚。

以一只狗为例,狗是这里的主体,狗主动采取动作,那么对应地做出反应,比如扔给它一块骨头作为奖赏。

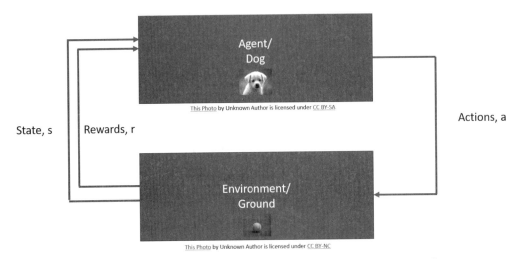

改编自 Sutton 和 Barto 的 "Reinforcement Learning:an Introduction"

> 我们的大脑也有一组位于前脑底部的皮层下核，称为**基础神经节**。根据神经系统科学中的论述，基础神经节负责动作的选择，也就是说，它负责给出在任一给定时应该从几种可执行动作中选择执行哪一个动作。

算法的目的是最大化奖励和减少惩罚，在这个决策过程中存在众多困难，其中最重要的就是如何最大化未来回报，也就是未来信度分配问题。算法会根据某些策略（π）决定它的动作，同时，也会根据与环境的交互来学习该策略（π）。策略学习算法很多，在本章中将对其中几种进行研究，不同的策略学习算法都会通过试错的过程来学习得到最优策略（π*），这其中必须要与环境进行交互。在这里选用提供了众多不同环境的 OpenAI Gym 作为交互的环境。

本章只介绍强化学习的基本概念，并且假设你已经熟知马尔可夫决策过程、折扣因子以及价值函数（状态值和动作值）。

这里定义一次迭代为游戏的一次运行，比如数独游戏的一次运行。通常，强化学习算法都需要多次迭代，来学习能够得到最大化奖励的最优策略。

通过强化学习可以让智能体在没有游戏先验知识的基础上学会打游戏，甚至在游戏中战胜人类。

9.2 学习 OpenAI Gym

使用 OpenAI Gym 作为实践环境，这是一个可以用来研究和比较强化学习算法的开源工具包，包含了各种可用来训练和研究新的强化学习算法的模拟环境。

准备工作

首先需要安装 OpenAI Gym，最简洁的方法是使用 `pip install gym`。OpenAI Gym 提供了多种环境，比如 Atari、棋盘游戏以及 2D 或 3D 游戏引擎等。在 Windows 上的最小安装只支持算法基本环境，如 toy_text 和 classic_control 这几种。如果你想研究其他环境，需要安装更多依赖项，如 OS X 和 Ubuntu 系统支持完整版本。详细的说明可以在 OpenAI Gym 的 GitHub 链接（https://github.com/openai/gym#installing-dependencies-for-specific-environments）中阅读。

具体做法

1. OpenAI Gym 提供了一个统一的环境接口，智能体可以通过三种基本方法：重置、执行和回馈与环境交互。重置操作会重置环境并返回观测值；执行操作会在环境中执行一个时间步长，并返回观测值、奖励、状态和信息；回馈操作会回馈环境的一个帧，比如弹出交互窗口。

2. 使用 OpenAI Gym 首先需要将其载入：

```
import gym
```

3. 构造一个初始环境：

```
env_name = 'Breakout-v3'
env = gym.make(env_name)
```

4. 通过重置来启动环境：

```
obs = env.reset()
```

5. 显示环境的基本情况：

```
print(obs.shape)
```

6. 可以执行的动作可以通过指令 `actions=env.action_space` 获得，比如 Breakout-v4 有四种可能的动作：等待、发射、向左和向右。动作总数可以通过指令 `env.action_space.n` 来获得。

7. 定义智能体从四种可能的动作中随机选择一种动作来执行：

```
def random_policy(n):
    action = np.random.randint(0,n)
    return action
```

8. 使用 `obs`、`rewards`、`done`、`info=env.step(action)` 让智能体继续随机执行 1000 步：

```
for step in range(1000): # 1000 steps max
 action = random_policy(env.action_space.n)
 obs, reward, done, info = env.step(action)
env.render()
if done:
    img = env.render(mode='rgb_array')
```

```
plt.imshow(img)
plt.show()
print("The game is over in {} steps".format(step))
break
```

观测值 obs 将环境信息传给智能体，在这里是一个 210×160×3 大小的彩色图片信息。每一步执行过程中，智能体都会得到一个 0 或 1 的奖励（reward），在 OpenAI Gym 中，奖励范围是 [-inf,inf]。游戏结束时，环境会返回 done=True。信息（info）在调试时非常有用，但智能体得不到这个信息。env.render() 指令会显示一个窗口提示当前环境的状态，当调用这个指令时，可以通过窗口看到算法是如何学习和执行的。在训练过程中，为了节约时间建议注释掉这条指令。

9. 最后，关闭环境：

```
env.close()
```

解读分析

上述代码实现了智能体如何从四个动作中随机选择其中一个：

另一个需要注意的是，在这个环境中，动作空间是离散的，观测空间是 Box 类型的。在 OpenAI Gym 中提到的动作空间和观测空间的离散和 Box 是指它们允许的数值。离散空间是一个非负数值区间，在这里为（0,1,2,3）；而观测空间是一个 n 维的盒子，比如 Pac-Man 中任何合理的观测都是一个 210× 160 × 3 的数组。

更多内容

OpenAI Gym 里包含很多环境，这都是它们社区的积极贡献。若要获取所有环境列表，可以运行如下代码（来自 https://github.com/openai/gym）：

```
from gym import envs
env_ids = [spec.id for spec in envs.registry.all()]
print("Total Number of environments are", len(env_ids))
for env_id in sorted(env_ids):
    print(env_id)
```

目前，OpenAI Gym 里共包含 777 种不同的环境，下图是早先使用相同随机算法的 Pac-Man 游戏的图像。

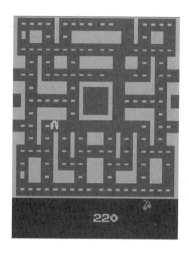

拓展阅读

- 更多不同环境的详细信息，请访问 https://gym.openai.com/envs。
- 了解 Wiki 维护的更多环境的介绍，请访问 https://github.com/openai/gym/wiki。
- 安装说明和依赖包的详细信息，请访问 https://github.com/openai/gym。

9.3 用神经网络智能体玩 Pac-Man 游戏

本节使用一个简单的神经网络智能体玩 Pac-Man 游戏，首先创建一组随机权重和偏置，之后让它进行游戏，然后选择玩最长时间的智能体，因为它的策略很有可能是最优决策。

准备工作

这里的智能体不学习任何策略，而是依赖初始权重进行决策，也就是固定策略。智能体根据神经网络算出的概率来选择动作，这个决策完全基于对当前环境的观测。

这里采用全连接神经网络实现这个过程，神经网络的输入由环境的观测空间决定，输出神经元的数目取决于可能的离散动作数。Pac-Man 包含九个动作——等待、右转、左转、向上、向下、向左移动、向右移动、向上移动和向下移动，所以神经网络有 9 个输出神经元。

具体做法

1. 导入模块，也导入 gym，以便使用它提供的不同环境：

```
import gym
import numpy as np
import tensorflow as tf
import matplotlib.pyplot as plt
```

2. 定义 `RlAgent` 类。该部分主要包含三个方法，`__init__` 方法初始化神经网络的大小并创建计算图，这里使用 `TensorFlow` 里的 `tf.multinomial` 函数计算采取哪种动作，该函数计算网络的 9 个输出神经元的 sigmoid 值，网络会根据概率大小选择动作。`predit` 函数返回神经网络预测的动作结果，`get_weights` 函数输出获胜智能体的权重和偏置：

```
class RlAgent(object):
 def __init__(self,m,n,ini=False,W=None, b=None ):
    self._graph = tf.Graph()
    with self._graph.as_default():
      self._X = tf.placeholder(tf.float32,shape=(1,m))
      if ini==False:
          self.W = tf.Variable(tf.random_normal([m,n]),trainable=False)
          self.bias = tf.Variable(tf.random_normal([1,n]),trainable=False)
      else:
          self.W = W
          self.bias = b
      out = tf.nn.sigmoid(tf.matmul(self._X,self.W)+ self.bias)
      self._result = tf.multinomial(out,1)
      init = tf.global_variables_initializer()

    self._sess = tf.Session()
    self._sess.run(init)

 def predict(self, X):
    action = self._sess.run(self._result, feed_dict= {self._X: X})
    return action

 def get_weights(self):
    W, b = self._sess.run([self.W, self.bias])
    return W, b
```

3. 定义辅助函数 `play_one_episode` 进行一轮游戏：

```
def play_one_episode(env, agent):
    obs = env.reset()
    img_pre = preprocess_image(obs)
    done = False
    t = 0
    while not done and t < 10000:
        env.render()  # This can be commented to speed up
        t += 1
        action = agent.predict(img_pre)
        #print(t,action)
        obs, reward, done, info = env.step(action)
        img_pre = preprocess_image(obs)
        if done:
```

```
            break
    return t
```

4.`play_multiple_episodes` 函数创建一个智能体，让这个智能体多玩几次游戏并返回平均持续时间：

```
def play_multiple_episodes(env, T,ini=False, W=None, b=None):
    episode_lengths = np.empty(T)
    obs = env.reset()
    img_pre = preprocess_image(obs)
    if ini== False:
        agent = RlAgent(img_pre.shape[1],env.action_space.n)
    else:
        agent = RlAgent(img_pre.shape[1],env.action_space.n,ini, W, b)
    for i in range(T):
        episode_lengths[i] = play_one_episode(env, agent)
    avg_length = episode_lengths.mean()
    print("avg length:", avg_length)
    if ini == False:
        W, b = agent.get_weights()
    return avg_length, W, b
```

5.`random_search` 函数调用 `play_multiple_episodes` 函数，每次调用时，都会用一组新的随机权重和偏置实例化一个新的智能体。最后，从这些随机初始化的神经网络智能体中选出结果最好的作为优胜者：

```
def random_search(env):
    episode_lengths = []
    best = 0
    for t in range(10):
        print("Agent {} reporting".format(t))
        avg_length, wts, bias = play_multiple_episodes(env, 10)
        episode_lengths.append(avg_length)
        if avg_length > best:
            best_wt = wts
            best_bias = bias
            best = avg_length
    return episode_lengths, best_wt, best_bias
```

6.每执行一个步骤，环境都会返回一个观测值。观测值具有三个颜色通道，并且需要预处理，这里将之转换为灰度，增强对比度，整形成行向量后，输入神经网络中：

```
def preprocess_image(img):
    img = img.mean(axis =2) # to grayscale
    img[img==150] = 0  # Bring about a better contrast
    img = (img - 128)/128 - 1 # Normalize image from -1 to 1
    m,n = img.shape
    return img.reshape(1,m*n)
```

7.神经网络智能体依次实例化并执行，对比后选出最优智能体。出于计算效率的考虑，这里只运行 10 个智能体，每个玩 10 个游戏，将游戏时间最长的结果视为最佳决策：

```
if __name__ == '__main__':
    env_name = 'Breakout-v0'
```

```
#env_name = 'MsPacman-v0'
env = gym.make(env_name)
episode_lengths, W, b = random_search(env)
plt.plot(episode_lengths)
plt.show()
print("Final Run with best Agent")
play_multiple_episodes(env,10, ini=True, W=W, b=b)
```

运行结果如下：

可以看到，智能体通过这种随机初始化的方法，可以实现游戏平均持续时间达到 615.5。

9.4 用 Q learning 玩 Cart-Pole 平衡游戏

如在引言中所讨论的，有一个由状态 s 描述的环境（$s \in S$，S 是所有可能状态的集合），一个能够执行动作 a 的 agent（$a \in A$，A 是所有可能动作的集合），智能体的动作致使智能体从一个状态转移到另外一个状态。智能体的行为会得到奖励，而智能体的目标就是最大化奖励。在 Q learning 中，智能体计算能够最大化奖励 R 的状态-动作组合，以此学习要采取的动作（策略 π），在选择动作时，智能体不仅要考虑当前的奖励，还要尽量考虑未来的奖励：

$$Q: S \times A \to R$$

智能体从任意初始状态 Q 开始，选择一个动作 a 并得到奖励 r，然后更新状态为 s'（主要受过去的状态 s 和动作 a 的影响），新的 Q 值为：

$$Q(s,a) = (1-\alpha)Q(s,a) + \alpha\left[r + \gamma \max_{a'} Q(s',a')\right]$$

其中，α 是学习率，γ 是折扣因子。第一项保留 Q 的旧值，第二项对 Q 值进行更新估计（包括当前奖励和未来动作的折扣奖励），这会导致在结果状态不满意时降低 Q 值，从而确保智能体在下一次处于此状态时不会选择相同的动作。类似地，当对当前状态满意时，对应的

Q 值将增加。

Q learning 的最简单实现包括维护和更新一个状态 - 动作值的对应表，表格大小为 $N \times M$，其中 N 是所有可能状态的数量，M 是所有可能动作的数量。对于大多数环境来说，这个表格会相当大，表格越大，搜索所需的时间越长，存储表格所需的内存越多，因此该方案并不可行。在本节中将使用 Q learning 的 NN 实现，神经网络被用作函数逼近器来预测值函数（Q），NN 的输出节点等于可能动作的数量，它们的输出表示相应动作的值函数。

准备工作

本节将训练一个线性神经网络来实践 `CartPole-v0` 环境（https://github.com/openai/gym/wiki/CartPole-v0），目标是平衡小车上的杆子，观测状态由 4 个连续的参数组成：推车位置 [-2.4, 2.4]，车速 [-∞, ∞]，杆子角度 [~ -41.8º, ~ 41.8º] 与杆子末端速度 [-∞, ∞]。通过向左或向右推车能够实现平衡，所以动作空间由两个动作组成，下图就是 `CartPole-v0` 环境空间：

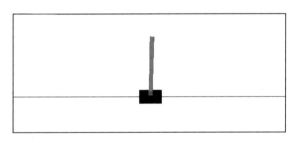

对于 Q learning，需要找到一种方法来量化连续的观测状态值。这里使用 `FeatureTransformer` 类来实现，首先生成观测空间中的 20000 个随机样本，然后用 scikit 的 `StandardScaler` 类将样本标准化，`RBFSampler` 用不同的方差来覆盖观测空间不同的部分。`FeatureTransformer` 类是用随机的观测空间样本实例化的，然后用 `fit_transform` 函数训练 `RBFSampler`。

上述过程执行结束之后，调用 `transform` 方法将连续观测空间转换为特征表示。

```
class FeatureTransformer:
 def __init__(self, env):
    obs_examples = np.random.random((20000, 4))
    print(obs_examples.shape)
    scaler = StandardScaler()
    scaler.fit(obs_examples)

    # Used to converte a state to a featurizes represenation.
    # We use RBF kernels with different variances to cover different parts
of the space
    featurizer = FeatureUnion([
        ("cart_position", RBFSampler(gamma=0.02, n_components=500)),
        ("cart_velocity", RBFSampler(gamma=1.0, n_components=500)),
        ("pole_angle", RBFSampler(gamma=0.5, n_components=500)),
```

```
            ("pole_velocity", RBFSampler(gamma=0.1, n_components=500))
            ])
        feature_examples = 
featurizer.fit_transform(scaler.transform(obs_examples))
        print(feature_examples.shape)

        self.dimensions = feature_examples.shape[1]
        self.scaler = scaler
        self.featurizer = featurizer

    def transform(self, observations):
        scaled = self.scaler.transform(observations)
        return self.featurizer.transform(scaled)
```

具体做法

1. 导入必要的模块。除了常用的 TensorFlow、Numpy 和 Matplotlib 外，还需导入 Gym 并从 scikit 导入一些类：

```
import numpy as np
import tensorflow as tf
import gym
import matplotlib.pyplot as plt
from sklearn.pipeline import FeatureUnion
from sklearn.preprocessing import StandardScaler
from sklearn.kernel_approximation import RBFSampler
```

2. 在 Q learning 中使用神经网络作为函数逼近器来估计值函数。定义一个线性 `NeuralNetwork` 类，把转换后的观测空间作为输入，并预测 Q 估计值。由于有两种可能的动作，需要两个不同的神经网络对象来获得预测的状态－动作值。类中包括训练单个神经网络和预测输出的方法：

```
class NeuralNetwork:
 def __init__(self, D):
        eta = 0.1
        self.W = tf.Variable(tf.random_normal(shape=(D, 1)), name='w')
        self.X = tf.placeholder(tf.float32, shape=(None, D), name='X')
        self.Y = tf.placeholder(tf.float32, shape=(None,), name='Y')
        # make prediction and cost
        Y_hat = tf.reshape(tf.matmul(self.X, self.W), [-1])
        err = self.Y - Y_hat
        cost = tf.reduce_sum(tf.pow(err,2))

        # ops we want to call later
        self.train_op = 
tf.train.GradientDescentOptimizer(eta).minimize(cost)
        self.predict_op = Y_hat

        # start the session and initialize params
        init = tf.global_variables_initializer()
        self.session = tf.Session()
        self.session.run(init)

    def train(self, X, Y):
```

```
        self.session.run(self.train_op, feed_dict={self.X: X, self.Y:
Y})

    def predict(self, X):
        return self.session.run(self.predict_op, feed_dict={self.X: X})
```

3. 下一个重要的类是 `Agent` 类，使用 `NeuralNetwork` 类创建智能体。实例化的智能体有两个线性神经网络，每个有 2000 个输入神经元和 1 个输出神经元。(实质上，这意味着智能体有 2 个神经元，每个神经元有 2000 个输入，因为神经网络的输入层不做任何处理。) `Agent` 类中定义了预测两个神经网络的输出和更新两个神经网络权重的方法。在这里，训练阶段时智能体使用 ε 贪婪策略进行探索，在每一步中，智能体可以选择具有最高 Q 值的动作或随机选择一个动作，具体取决于 epsilon（eps）的值，ε 在训练过程中不断衰减，因此，初始时智能体会做大量的随机动作（exploration，探索），但是随着训练的进行，具有最大 Q 值的动作被采用（exploitation，利用）。这就是所谓的 Exploration - Exploitation 平衡：允许智能体探索随机动作，这能够让智能体尝试新的随机动作并从中学习：

```
class Agent:
 def __init__(self, env, feature_transformer):
  self.env = env
  self.agent = []
  self.feature_transformer = feature_transformer
  for i in range(env.action_space.n):
   model = NeuralNetwork(feature_transformer.dimensions)
   self.agent.append(model)
 def predict(self, s):
  X = self.feature_transformer.transform([s])
  return np.array([m.predict(X)[0] for m in self.agent])

 def update(self, s, a, G):
  X = self.feature_transformer.transform([s])
  self.agent[a].train(X, [G])

 def sample_action(self, s, eps):
  if np.random.random() < eps:
   return self.env.action_space.sample()
  else:
   return np.argmax(self.predict(s))
```

4. 定义一个函数来运行一个步骤，类似于之前使用过的 `play_one` 函数，但现在使用 Q learning 来更新智能体的权重。用 `env.reset()` 重置环境来开始这个步骤，然后直到游戏完成（最大迭代次数以确保程序结束）。像以前一样，智能体基于当前的观测状态（`obs`）选择一个动作并在环境中执行（`env.step(action)`）。不同的是，基于先前的状态和采取动作后的状态，神经网络用 $G = r + \gamma max_{a'} Q(s', a')$ 更新权重，从而可以预测出与动作相对应的准确期望值。为了获得更好的稳定性，此处修改了奖励——每当杆子落下时，智能体将得到 -400 的奖励，否则每一步都会得到 +1 的奖励：

```
def play_one(env, model, eps, gamma):
 obs = env.reset()
```

```python
    done = False
    totalreward = 0
    iters = 0
    while not done and iters < 2000:
        action = model.sample_action(obs, eps)
        prev_obs = obs
        obs, reward, done, info = env.step(action)
        env.render()    # Can comment it to speed up.

    if done:
        reward = -400

    # update the model
    next = model.predict(obs)
    assert(len(next.shape) == 1)
    G = reward + gamma*np.max(next)
    model.update(prev_obs, action, G)

    if reward == 1:
        totalreward += reward
    iters += 1
```

5. 所有的函数和类已经准备好，现在定义智能体和环境（本例中是 `'CartPole-v0'`）。该智能体总共进行 1000 次游戏，并通过价值函数与环境交互来学习：

```python
if __name__ == '__main__':
    env_name = 'CartPole-v0'
    env = gym.make(env_name)
    ft = FeatureTransformer(env)
    agent = Agent(env, ft)
    gamma = 0.97

    N = 1000
    totalrewards = np.empty(N)
    running_avg = np.empty(N)
    for n in range(N):
        eps = 1.0 / np.sqrt(n + 1)
        totalreward = play_one(env, agent, eps, gamma)
        totalrewards[n] = totalreward
        running_avg[n] = totalrewards[max(0, n - 100):(n + 1)].mean()
        if n % 100 == 0:
            print("episode: {0}, total reward: {1} eps: {2} avg reward (last 100): {3}".format(n, totalreward, eps, running_avg[n]), )

    print("avg reward for last 100 episodes:", totalrewards[-100:].mean())
    print("total steps:", totalrewards.sum())

    plt.plot(totalrewards)
    plt.xlabel('episodes')
    plt.ylabel('Total Rewards')
    plt.show()

    plt.plot(running_avg)
```

```
plt.xlabel('episodes')
plt.ylabel('Running Average')
plt.show()
env.close()
```

```
episode: 0, total reward: 13.0 eps: 1.0 avg reward (last 100): 13.0
episode: 100, total reward: 128.0 eps: 0.09950371902099892 avg reward (last 100): 111.70297029702971
episode: 200, total reward: 181.0 eps: 0.07053456158585983 avg reward (last 100): 171.15841584158414
episode: 300, total reward: 199.0 eps: 0.0576390417704235 avg reward (last 100): 167.23762376237624
episode: 400, total reward: 184.0 eps: 0.04993761694389223 avg reward (last 100): 184.801980198019
episode: 500, total reward: 199.0 eps: 0.04466767051608770703 avg reward (last 100): 186.46534653465346
episode: 600, total reward: 199.0 eps: 0.04079085082240021 avg reward (last 100): 181.5742574257426
episode: 700, total reward: 199.0 eps: 0.0377694787300249 avg reward (last 100): 173.21782178217822
episode: 800, total reward: 199.0 eps: 0.03533326266687867 avg reward (last 100): 194.04950495049505
episode: 900, total reward: 199.0 eps: 0.03331483023263848 avg reward (last 100): 178.47524752475246
avg reward for last 100 episodes: 195.7
total steps: 174300.0

Process finished with exit code 0
```

6. 下图是智能体在游戏中学习获得的总奖励和平均奖励。根据 Cart-Pole wiki 上的表述，奖励 200 意味着智能体在训练 1000 次后获胜了一次，而这里的智能体在训练 100 次时就达到了平均奖励 195.7，这是非常不错的：

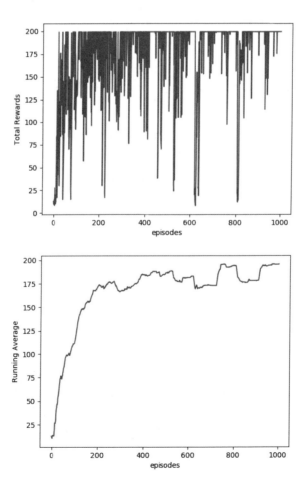

更多内容

相同的逻辑可以用于在其他 OpenAI 环境中构建的智能体上，然而，对于像 Breakout 和 Pac-Man 这样的 Atari 游戏来说，观测空间并不是一个只有四个数字的数组。相反，它非常大（210 × 160 = 33600 像素，3 个 RGB 通道），没有一定形式的量化，可能的状态是无限的，用简单的神经网络不会产生好结果，在深度 Q learning 章节中使用 CNN 来解决这个问题。

拓展阅读

虽然 Q learning 中有大量的网页链接，但有一些有用的链接如下：

- https://en.wikipedia.org/wiki/Q_learning
- http://mnemstudio.org/path-finding-q-learning-tutorial.htm
- http://artint.info/html/ArtInt_265.html
- https://medium.com/emergent-future/simple-reinforcement-learning-with-tensorflow-part-0-q-learning-with-tables-and-neural-networks-d195264329d0

9.5 用 DQN 玩 Atari 游戏

深度 Q 网络（DQN）是将 Q learning 和**卷积神经网络**（CNN）结合在一起，由 Mnih 等人在 2013 年首次提出（https://arxiv.org/pdf/1312.5602.pdf）。CNN 由于能够提取空间信息，能够从原始像素数据中学习得到控制策略。第 4 章已经介绍了卷积神经网络，所以本节不再介绍基础知识。

> 本节内容基于原始的 DQN 论文，DeepMind 使用深度强化学习玩转 Atari，这篇论文中提到了一种称为**经验回放**（experience replay）的概念，随机抽样前一个游戏动作（状态、动作奖励、下一个状态）。

准备工作

正如上一节提到的那样，对于像 Pac-Man 或 Breakout 这样的 Atari 游戏，需要预处理观测状态空间，它由 33600 个像素（RGB 的 3 个通道）组成。这些像素中的每一个都可以取 0 ~ 255 之间的任意值。`preprocess` 函数需要能够量化可能的像素值，同时减小观测状态空间。

这里利用 Scipy 的 `imresize` 函数来下采样图像。函数 `preprocess` 会在将图像输入到 DQN 之前，对图像进行预处理：

```
def preprocess(img):
    img_temp = img[31:195]   # Choose the important area of the image
    img_temp = img_temp.mean(axis=2)   # Convert to Grayscale#
    # Downsample image using nearest neighbour interpolation
    img_temp = imresize(img_temp, size=(IM_SIZE, IM_SIZE),
interp='nearest')
    return img_temp
```

`IM_SIZE` 是一个全局参数,这里设置为 80。该函数具有描述每个步骤的注释。下面是预处理前后的观测空间:

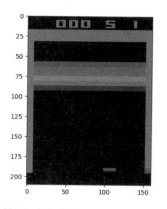 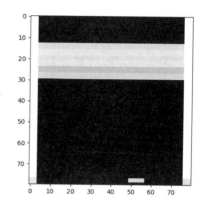

另外需要注意的是当前观测空间并没有给出完整的游戏画面,例如,只看上图不能确定下面的板子是向左还是向右。因此,要完全理解游戏的当前状态,需要考虑动作和观测的序列。本节考虑四个动作和观测序列来确定当前情况并训练智能体。`state_update` 函数用来将当前观测状态附加到以前的状态,从而产生状态序列。

```
def update_state(state, obs):
    obs_small = preprocess(obs)
    return np.append(state[1:], np.expand_dims(obs_small, 0), axis=0)
```

最后,为了训练稳定性,使用 `target_network`(目标网络)的概念,这是 DQN 的副本,但并不如同 DQN 一样更新。这里使用目标网络为 DQN 网络生成目标价值函数,在每一步中正常更新 DQN,同时在规律性的时间间隔之后更新 `target_network`(与 DQN 相同)。由于所有更新都在 TensorFlow 会话中进行,因此需要使用名称作用域来区分 `target_network` 和 DQN 网络。

具体做法

1. 导入必要的模块。使用 `sys` 模块的 `stdout.flush()` 来刷新标准输出(此例中是计算机屏幕)中的数据。`random` 模块用于从经验回放缓存(存储过去经验的缓存)中获得随机样本。`datatime` 模块用于记录训练花费的时间:

```
import gym
import sys
```

```python
import random
import numpy as np
import tensorflow as tf
import matplotlib.pyplot as plt
from datetime import datetime
from scipy.misc import imresize
```

2.定义训练的超参数,可以尝试改变它们,定义了经验回放缓存的最小和最大尺寸,以及目标网络更新的次数。

```python
MAX_EXPERIENCES = 500000
MIN_EXPERIENCES = 50000
TARGET_UPDATE_PERIOD = 10000
IM_SIZE = 80
K = 4
```

3.定义DQN类,构造器使用`tf.contrib.layers.conv2d`函数构建CNN网络,定义损失和训练操作。

```python
class DQN:
    def __init__(self, K, scope, save_path= 'models/atari.ckpt'):

        self.K = K
        self.scope = scope
        self.save_path = save_path

        with tf.variable_scope(scope):

            # inputs and targets
            self.X = tf.placeholder(tf.float32, shape=(None, 4, IM_SIZE, IM_SIZE), name='X')
            # tensorflow convolution needs the order to be:
            # (num_samples, height, width, "color")
            # so we need to tranpose later
            self.G = tf.placeholder(tf.float32, shape=(None,), name='G')
            self.actions = tf.placeholder(tf.int32, shape=(None,), name='actions')

            # calculate output and cost
            # convolutional layers
            Z = self.X / 255.0
            Z = tf.transpose(Z, [0, 2, 3, 1])
            cnn1 = tf.contrib.layers.conv2d(Z, 32, 8, 4, activation_fn=tf.nn.relu)
            cnn2 = tf.contrib.layers.conv2d(cnn1, 64, 4, 2, activation_fn=tf.nn.relu)
            cnn3 = tf.contrib.layers.conv2d(cnn2, 64, 3, 1, activation_fn=tf.nn.relu)

            # fully connected layers
            fc0 = tf.contrib.layers.flatten(cnn3)
            fc1 = tf.contrib.layers.fully_connected(fc0, 512)

            # final output layer
            self.predict_op = tf.contrib.layers.fully_connected(fc1, K)
```

```
            selected_action_values = tf.reduce_sum(self.predict_op
    * tf.one_hot(self.actions, K),
                 reduction_indices=[1]
            )

            self.cost = tf.reduce_mean(tf.square(self.G -
selected_action_values))
            self.train_op = tf.train.RMSPropOptimizer(0.00025,
0.99, 0.0, 1e-6).minimize(self.cost)
```

4. 类中用 `set_session()` 函数建立会话，用 `predict()` 预测动作值函数，用 `update()` 更新网络，在 `sample_action()` 函数中用 Epsilon 贪婪算法选择动作：

```
    def set_session(self, session):
        self.session = session

    def predict(self, states):
        return self.session.run(self.predict_op, feed_dict={self.X: states})
    def update(self, states, actions, targets):
        c, _ = self.session.run(
            [self.cost, self.train_op],
            feed_dict={
                self.X: states,
                self.G: targets,
                self.actions: actions
            }
        )
        return c

    def sample_action(self, x, eps):
        """Implements epsilon greedy algorithm"""
        if np.random.random() < eps:
            return np.random.choice(self.K)
        else:
            return np.argmax(self.predict([x])[0])
```

5. 另外还定义了加载和保存网络的方法，因为训练需要消耗大量时间：

```
def load(self):
    self.saver = tf.train.Saver(tf.global_variables())
    load_was_success = True
    try:
        save_dir = '/'.join(self.save_path.split('/')[:-1])
        ckpt = tf.train.get_checkpoint_state(save_dir)
        load_path = ckpt.model_checkpoint_path
        self.saver.restore(self.session, load_path)
    except:
        print("no saved model to load. starting new session")
        load_was_success = False
    else:
        print("loaded model: {}".format(load_path))
        saver = tf.train.Saver(tf.global_variables())
        episode_number = int(load_path.split('-')[-1])


def save(self, n):
```

```
        self.saver.save(self.session, self.save_path, global_step=n)
        print("SAVED MODEL #{}".format(n))
```

6.定义将主DQN网络的参数复制到目标网络的方法如下：

```
def copy_from(self, other):
    mine = [t for t in tf.trainable_variables() if
t.name.startswith(self.scope)]
    mine = sorted(mine, key=lambda v: v.name)
    others = [t for t in tf.trainable_variables() if
t.name.startswith(other.scope)]
    others = sorted(others, key=lambda v: v.name)

    ops = []
    for p, q in zip(mine, others):
        actual = self.session.run(q)
        op = p.assign(actual)
        ops.append(op)

    self.session.run(ops)
```

7.定义函数 `learn()`，预测价值函数并更新原始的DQN网络：

```
def learn(model, target_model, experience_replay_buffer, gamma,
batch_size):
    # Sample experiences
    samples = random.sample(experience_replay_buffer, batch_size)
    states, actions, rewards, next_states, dones = map(np.array,
zip(*samples))

    # Calculate targets
    next_Qs = target_model.predict(next_states)
    next_Q = np.amax(next_Qs, axis=1)
    targets = rewards + np.invert(dones).astype(np.float32) * gamma
* next_Q

    # Update model
    loss = model.update(states, actions, targets)
    return loss
```

8.现在已经在主代码中定义了所有要素，下面构建和训练一个DQN网络来玩Atari的游戏。代码中有详细的注释，这主要是之前Q learning代码的一个扩展，增加了经验回放缓存，所以不难理解：

```
if __name__ == '__main__':
    # hyperparameters
    gamma = 0.99
    batch_sz = 32
    num_episodes = 500
    total_t = 0
    experience_replay_buffer = []
    episode_rewards = np.zeros(num_episodes)
    last_100_avgs = []

    # epsilon for Epsilon Greedy Algorithm
    epsilon = 1.0
    epsilon_min = 0.1
```

```python
    epsilon_change = (epsilon - epsilon_min) / 500000

    # Create Atari Environment
    env = gym.envs.make("Breakout-v0")


    # Create original and target  Networks
    model = DQN(K=K, gamma=gamma, scope="model")
    target_model = DQN(K=K, gamma=gamma, scope="target_model")

    with tf.Session() as sess:
        model.set_session(sess)
        target_model.set_session(sess)
        sess.run(tf.global_variables_initializer())
        model.load()

        print("Filling experience replay buffer...")
        obs = env.reset()
        obs_small = preprocess(obs)
        state = np.stack([obs_small] * 4, axis=0)

        # Fill experience replay buffer
        for i in range(MIN_EXPERIENCES):

            action = np.random.randint(0,K)
            obs, reward, done, _ = env.step(action)
            next_state = update_state(state, obs)

            experience_replay_buffer.append((state, action, reward,
next_state, done))

            if done:
                obs = env.reset()
                obs_small = preprocess(obs)
                state = np.stack([obs_small] * 4, axis=0)

            else:
                state = next_state

        # Play a number of episodes and learn
        for i in range(num_episodes):
            t0 = datetime.now()

            # Reset the environment
            obs = env.reset()
            obs_small = preprocess(obs)
            state = np.stack([obs_small] * 4, axis=0)
            assert (state.shape == (4, 80, 80))
            loss = None

            total_time_training = 0
            num_steps_in_episode = 0
            episode_reward = 0

            done = False
            while not done:
```

```python
                # Update target network
                if total_t % TARGET_UPDATE_PERIOD == 0:
                    target_model.copy_from(model)
                    print("Copied model parameters to target network. total_t = %s, period = %s" % (
                        total_t, TARGET_UPDATE_PERIOD))

                # Take action
                action = model.sample_action(state, epsilon)
                obs, reward, done, _ = env.step(action)
                obs_small = preprocess(obs)
                next_state = np.append(state[1:], np.expand_dims(obs_small, 0), axis=0)

                episode_reward += reward

                # Remove oldest experience if replay buffer is full
                if len(experience_replay_buffer) == MAX_EXPERIENCES:
                    experience_replay_buffer.pop(0)
                # Save the recent experience
                experience_replay_buffer.append((state, action, reward, next_state, done))

                # Train the model and keep measure of time
                t0_2 = datetime.now()
                loss = learn(model, target_model, experience_replay_buffer, gamma, batch_sz)
                dt = datetime.now() - t0_2

                total_time_training += dt.total_seconds()
                num_steps_in_episode += 1

                state = next_state
                total_t += 1

                epsilon = max(epsilon - epsilon_change, epsilon_min)

            duration = datetime.now() - t0

            episode_rewards[i] = episode_reward
            time_per_step = total_time_training / num_steps_in_episode

            last_100_avg = episode_rewards[max(0, i - 100):i + 1].mean()
            last_100_avgs.append(last_100_avg)
            print("Episode:", i,"Duration:", duration, "Num steps:", num_steps_in_episode,
                  "Reward:", episode_reward, "Training time per step:", "%.3f" % time_per_step,
                  "Avg Reward (Last 100):", "%.3f" % last_100_avg,"Epsilon:", "%.3f" % epsilon)

            if i % 50 == 0:
                model.save(i)
```

```
        sys.stdout.flush()

#Plots
plt.plot(last_100_avgs)
plt.xlabel('episodes')
plt.ylabel('Average Rewards')
plt.show()
env.close()
```

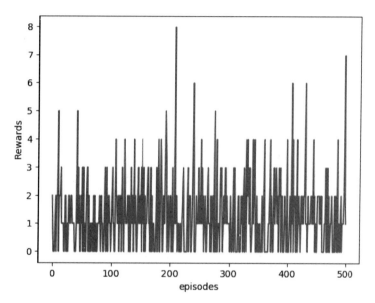

从上图可以看出，随着训练，智能体获得越来越高的奖励，下图是每 100 次运行的平均奖励，更清晰地展示了奖励的提高。

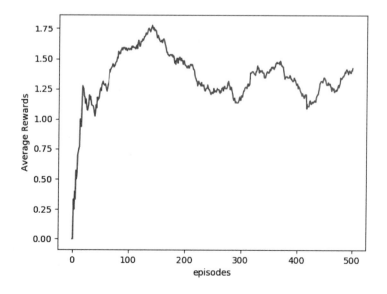

这只是在前 500 次运行后的训练结果。要想获得更好的结果，需要训练更多次，大约 1 万次。

更多内容

训练智能体需要运行很多次游戏，消耗大量的时间和内存。OpenAI Gym 提供了一个封装，将游戏保存为一个视频，因此，无须 render 函数，你可以使用这个封装来保存视频并在以后查看智能体是如何学习的。AI 工程师和爱好者可以上传这些视频来展示他们的结果。要使用的话，首先要导入 wrappers，然后创建环境并调用 wrappers。默认情况下，它会存储 1,8,27,64 等次的视频，1000 次训练以后将每 1000 次的视频（次数是整数立方）默认保存在一个文件夹中。为此添加的代码如下所示：

```
import gym
from gym import wrappers
env = gym.make('Breakout-v0')
env = wrappers.Monitor(env, '/save-path')
```

如果你想在下次训练中使用相同的文件夹，可以在 Monitor 中传入参数 `force = True`。

拓展阅读

- Mnih, Volodymyr, and others, Playing Atari with deep reinforcement learning, arXiv preprint arXiv:1312.5602 (2013) (https://arxiv.org/pdf/1312.5602.pdf)
- Mnih, Volodymyr, et al. Human-level control through deep reinforcement learning, Nature 518.7540 (2015): 529-533
- A cool implementation of DQN to play Atari: https://github.com/devsisters/DQN-tensorflow

9.6 用策略梯度网络玩 Pong 游戏

到目前为止，策略梯度是最常用的 RL 算法之一。研究表明，经过适当的调整，它们的性能要优于 DQN，同时内存和计算消耗又不会过高。与 Q learning 不同，策略梯度使用参数化策略，可以在无须价值函数的情况下选择动作。在策略梯度中有一个性能指标 $\eta(\theta_p)$，目标是最大化性能，同时根据梯度上升算法更新神经网络的权重。然而，TensorFlow 没有 `maximum` 优化器，因此，需要使用指标梯度负值 $-\nabla\eta(\theta_p)$ 的最小化来替代。

准备工作

Pong 是一个双人游戏，游戏的玩法是将球弹给对方，智能体可以上下移动球拍，也可以不操作。OpenAI 环境中的一个玩家是熟悉游戏规则的 AI 玩家，这里的目标是使用策略梯度来训练第二个智能体，使其成为每个玩过的游戏的专家。代码中只运行了 500 次游戏，并在指定检查点保存智能体状态，这样在下一次运行时加载上一个检查点就可以了。为此，声明一个保存器（saver），然后调用 TensorFlow 的 `saver.save` 方法保存当前网络状态（检查点），最后从上次保存的检查点加载网络，下一节中定义的 `PolicyNetwork` 类用下面的方法完成这个工作。

```
def load(self):
    self.saver = tf.train.Saver(tf.global_variables())
    load_was_success = True  # yes, I'm being optimistic
    try:
        save_dir = '/'.join(self.save_path.split('/')[:-1])
        ckpt = tf.train.get_checkpoint_state(save_dir)
        load_path = ckpt.model_checkpoint_path
        self.saver.restore(self.session, load_path)
    except:
        print("no saved model to load. starting new session")
        load_was_success = False
    else:
        print("loaded model: {}".format(load_path))
        saver = tf.train.Saver(tf.global_variables())
        episode_number = int(load_path.split('-')[-1])
```

使用以下代码每 50 次运行保存一次模型：

```
def save(self):
    self.saver.save(self.session, self.save_path, global_step=n)
    print("SAVED MODEL #{}".format(n))
```

具体做法

1. 本节代码基于 Andrej Karpathy 的博客（`http://karpathy.github.io/2016/05/31/rl/`），部分代码改编自 Sam Greydanus 的代码（`https://gist.github.com/karpathy/a4166c7fe253700972fcbc77e4ea32c5`）。

2. 导入模块：

```
import numpy as np
import gym
import matplotlib.pyplot as plt
import tensorflow as tf
```

3. 定义 `PolicyNetwork` 类。在类构建过程中初始化模型超参数，定义输入状态 `self.tf_x`、预测动作 `self.tf_y`、相应奖励 `self.tf_epr` 的占位符，还定义了网络权重和预测动作价值、训练和更新的操作。同时初始化了一个交互式的 TensorFlow 会话：

```
class PolicyNetwork(object):
    def __init__(self, N_SIZE, h=200, gamma=0.99, eta=1e-3,
decay=0.99, save_path = 'models2/pong.ckpt' ):

        self.gamma = gamma
```

```python
        self.save_path = save_path
        # Placeholders for passing state....
        self.tf_x = tf.placeholder(dtype=tf.float32, shape=[None, N_SIZE * N_SIZE], name="tf_x")
        self.tf_y = tf.placeholder(dtype=tf.float32, shape=[None, n_actions], name="tf_y")
        self.tf_epr = tf.placeholder(dtype=tf.float32, shape=[None, 1], name="tf_epr")

        # Weights
        xavier_l1 = tf.truncated_normal_initializer(mean=0, stddev=1. / N_SIZE, dtype=tf.float32)
        self.W1 = tf.get_variable("W1", [N_SIZE * N_SIZE, h], initializer=xavier_l1)
        xavier_l2 = tf.truncated_normal_initializer(mean=0, stddev=1. / np.sqrt(h), dtype=tf.float32)
        self.W2 = tf.get_variable("W2", [h, n_actions], initializer=xavier_l2)

        # Build Computation
        # tf reward processing (need tf_discounted_epr for policy gradient wizardry)
        tf_discounted_epr = self.tf_discount_rewards(self.tf_epr)
        tf_mean, tf_variance = tf.nn.moments(tf_discounted_epr, [0], shift=None, name="reward_moments")
        tf_discounted_epr -= tf_mean
        tf_discounted_epr /= tf.sqrt(tf_variance + 1e-6)

        # Define Optimizer, compute and apply gradients
        self.tf_aprob = self.tf_policy_forward(self.tf_x)
        loss = tf.nn.l2_loss(self.tf_y - self.tf_aprob)
        optimizer = tf.train.RMSPropOptimizer(eta, decay=decay)
        tf_grads = optimizer.compute_gradients(loss, var_list=tf.trainable_variables(), grad_loss=tf_discounted_epr)
        self.train_op = optimizer.apply_gradients(tf_grads)

        # Initialize Variables
        init = tf.global_variables_initializer()
        self.session = tf.InteractiveSession()
        self.session.run(init)
        self.load()
```

4. 定义计算折扣奖励的方法,这确保智能体不仅考虑到当前的奖励,也考虑到未来的奖励。折扣奖励用 $R_t = \sum \gamma^k r_{t+k}$ 计算,t 是时间,求和中 $k \in [0, \infty]$,折扣因子 γ 取值为 $0 \sim 1$,代码中设置 gamma = 0.99:

```python
def tf_discount_rewards(self, tf_r):  # tf_r ~ [game_steps,1]
    discount_f = lambda a, v: a * self.gamma + v;
    tf_r_reverse = tf.scan(discount_f, tf.reverse(tf_r, [0]))
    tf_discounted_r = tf.reverse(tf_r_reverse, [0])
    return tf_discounted_r
```

5. 定义 `tf_policy_forward` 方法计算在给定输入观测状态下向上移动球拍的概率,这里使用两层神经网络来实现。网络以预处理过的游戏状态图像为输入,生成移动球拍的概率。在 TensorFlow 中,由于网络计算图仅在 TensorFlow 会话中运算,因此这里定义另一个方法 predict_UP 来计算概率：

```
def tf_policy_forward(self, x): #x ~ [1,D]
    h = tf.matmul(x, self.W1)
    h = tf.nn.relu(h)
    logp = tf.matmul(h, self.W2)
    p = tf.nn.softmax(logp)
    return p

def predict_UP(self,x):
    feed = {self.tf_x: np.reshape(x, (1, -1))}
    aprob = self.session.run(self.tf_aprob, feed);
    return aprob
```

6. PolicyNetwork 智能体使用 update 方法更新权重：

```
def update(self, feed):
    return self.session.run(self.train_op, feed)
```

7. 定义一个辅助函数来预处理观测状态空间：

```
# downsampling
def preprocess(I):
    """ prepro 210x160x3 uint8 frame into 6400 (80x80) 1D float vector """
    I = I[35:195] # crop
    I = I[::2,::2,0] # downsample by factor of 2
    I[I == 144] = 0  # erase background (background type 1)
    I[I == 109] = 0  # erase background (background type 2)
    I[I != 0] = 1    # everything else (paddles, ball) just set to 1
    return I.astype(np.float).ravel()
```

8. 创建一个游戏环境定义保存（状态、动作、奖励、状态）的数组，并让智能体多次学习游戏（中断或连续只取决于你的计算资源）。这里要注意的是，智能体并不是每步都学习，相反，智能体使用每一次游戏的（状态、动作、奖励、状态）集合来修正策略。这非常消耗内存：

```
if __name__ == '__main__':
    # Create Game Environment
    env_name = "Pong-v0"
    env = gym.make(env_name)
    env = wrappers.Monitor(env, '/tmp/pong', force=True)
    n_actions = env.action_space.n # Number of possible actions
    # Initializing Game and State(t-1), action, reward, state(t)
    xs, rs, ys = [], [], []
    obs = env.reset()
    prev_x = None

    running_reward = None
    running_rewards = []
    reward_sum = 0
    n = 0
    done = False
    n_size = 80
```

```python
    num_episodes = 500

    #Create Agent
    agent = PolicyNetwork(n_size)

    # training loop
    while not done and n< num_episodes:
        # Preprocess the observation
        cur_x = preprocess(obs)
        x = cur_x - prev_x if prev_x is not None else np.zeros(n_size*n_size)
        prev_x = cur_x

        #Predict the action
        aprob = agent.predict_UP(x) ; aprob = aprob[0,:]
        action = np.random.choice(n_actions, p=aprob)
        #print(action)
        label = np.zeros_like(aprob) ; label[action] = 1

        # Step the environment and get new measurements
        obs, reward, done, info = env.step(action)
        env.render()
        reward_sum += reward

        # record game history
        xs.append(x) ; ys.append(label) ; rs.append(reward)

        if done:
            # update running reward
            running_reward = reward_sum if running_reward is None else running_reward * 0.99 + reward_sum * 0.01
            running_rewards.append(running_reward)
            feed = {agent.tf_x: np.vstack(xs), agent.tf_epr: np.vstack(rs), agent.tf_y: np.vstack(ys)}
            agent.update(feed)
            # print progress console
            if n % 10 == 0:
                print ('ep {}: reward: {}, mean reward: {:3f}'.format(n, reward_sum, running_reward))
            else:
                print ('\tep {}: reward: {}'.format(n, reward_sum))

            # Start next episode and save model
            xs, rs, ys = [], [], []
            obs = env.reset()
            n += 1 # the Next Episode

            reward_sum = 0
            if n % 50 == 0:
                agent.save()
            done = False

    plt.plot(running_rewards)
    plt.xlabel('episodes')
    plt.ylabel('Running Averge')
    plt.show()
    env.close()
```

下图显示了智能体学习前 500 次游戏的平均奖励：

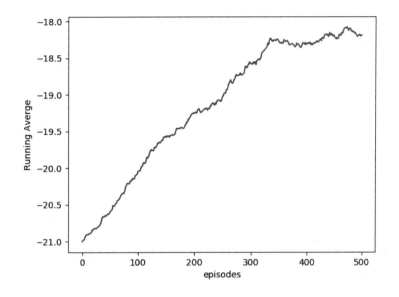

解读分析

权重使用 Xavier 方式进行初始化，权重既不能太大也不能太小，不会阻碍网络的学习。在这种方式下权重赋值服从均值为 0 和方差为 $2/(n_{in}+n_{out})$ 的分布，其中 n_{in} 和 n_{out} 分别是输入和输出的数量。要了解有关 Xavier 初始化的更多信息，请参阅 Glorot 和 Bengio 在 2009 年的论文。

更多内容

任何第一次看到智能体学习玩游戏的人都会惊叹于它如此像人类。起初的举动很笨拙，慢慢地智能体知道如何移动，尽管速度很慢，而且经常接不住球。然而，随着学习次数的增加，智能体成了专家。

但这与人类又非常不一样，我们玩游戏时可以轻松地使用其他类似场景中的知识。RL 智能体却做不到这一点——即使是简单的改变，比如改变环境空间的大小，也会将其能力打回到 0。研究者正在研究的迁移学习，可以帮助智能体将一个环境空间中学习到的知识应用到另一个环境空间，有朝一日可能会为真正的通用人工智能铺平道路。

AlphaGo Zero

前期 DeepMind 发表了一篇关于 AlphaGo Zero 的文章，是 AlphaGo 的最新改进版。根据他们发表的结果，AlphaGo Zero 更加强大，是史上最强的围棋玩家。AlphaGo 最初一无所知，它只使用棋盘状态和自我对弈的游戏来训练神经网络并预测正确的落子。

AlphaGo Zero 使用深度神经网络，将棋盘状态（当前和历史）作为输入，并输出落子概率和一个评估值，因此，这个神经网络结合了策略网络和价值网络的功能。不像以前的 AlphaGo 版本（使用监督学习进行训练），该网络训练时自己对弈。在每个位置，都会基于神经网络进行蒙特卡罗树搜索（MCTS）。神经网络通过自我强化学习算法进行训练，该算法使用 MCTS 来执行每个动作。

最初，神经网络的权重随机初始化。在每个迭代步骤中都自己对弈进行多次游戏。每一步，使用前一步迭代的神经网络对可能的策略执行 MCTS 搜索，然后通过对搜索概率进行采样来落子。重复以上步骤直到游戏结束。其中游戏每一步都存储下游戏状态、采取的策略和奖励。与此同时，从上一步迭代时自我对弈过程中产生的数据中均匀采样来训练神经网络。调整神经网络的权重以最小化预测值与自我对弈胜者之间的误差，并最大化神经网络落子概率与搜索概率的相似度。

在四个 TPU 的单机上仅训练了 3 天，AlphaGo Zero 就以 100 比 0 击败了 AlphaGo。AlphaGo Zero 完全基于 RL，细节参考 2017 年 10 月的 Nature 论文 "*Mastering the game of Go without human knowledge*"。

拓展阅读

- https://arxiv.org/pdf/1602.01783.pdf
- http://ufal.mff.cuni.cz/~straka/courses/npfl114/2016/sutton-bookdraft2016sep.pdf
- http://karpathy.github.io/2016/05/31/rl/
- Xavier Glorot and Yoshua Bengio, Understanding the difficulty of training deep feedforward neural networks, Proceedings of the Thirteenth International Conference on Artificial Intelligence and Statistics, 2010, http://proceedings.mlr.press/v9/glorot10a/glorot10a.pdf

CHAPTER 10

第 10 章

移动端计算

本章讨论如何在移动设备上使用深度学习,并给出一系列相关的设计模块案例。

10.1 引言

本节将介绍一些用于移动端深度学习的实例。它的学习情况与桌面学习或云端深度学习完全不同(在这两种学习中,GPU 和电力通常是可利用的)。实际上,在移动设备上,节省电量和减少 GPU 的使用是非常重要的。无论如何,深度学习在很多情况下都是非常有用的。现在回顾一下:

- **图像识别**:现代手机拥有强大的摄像头,用户渴望尝试对图像和图片进行处理。通常情况下,理解图片中的内容也是很重要的,正如第 4 章中讨论的那样,有多个预设的训练模型适用于处理此类问题。使用图像识别模型的一个好的示例见链接 https://github.com/TensorFlow/models/tree/master/official/resnet。
- **目标定位**:移动目标识别非常重要,是视频和图像处理所必需的。例如,想象一下,如果要在图像中将多个人识别出来,那么照相机将会使用多个焦点。相关的示例见链接 https://github.com/TensorFlow/models/tree/master/research/object_detection。
- **光学字符识别**:手写字符识别是许多工作(如文本分类和文本推荐)的基础,深度学习可以为执行这些工作提供基本的帮助。在第 4 章中已经看到了几个 MNIST 识别的例子。有关 MNIST 的信息也可参考:https://github.com/TensorFlow/models/tree/master/official/mnist。
- **语音识别**:语音识别是访问现代手机的通用交互接口。在此,深度学习被用于识别声音和口头指令。在最近几年里,关于这方面的进步是令人印象深刻的。
- **翻译**:处理多种语言是现代多元文化世界的一部分。手机在跨语言即时翻译方面正在变得越来越准确,这在前些年几乎是不可想象的,深度学习对于打破这些障碍提供了一定的帮助。在第 6 章中，有一些关于机器翻译的例子。
- **手势识别**：手机开始使用手势作为接收命令的接口，有相应的模型。
- **压缩**：压缩是手机的一个关键方面。可以想象，在通过网络发送图像或视频之前，压缩空间是非常有好处的。类似地，在进行本地设备存储前进行压缩也非常方便。在所有这些情况下，深度学习都可以提供帮助。用于压缩的一个 RNN 模型可参考 `https://github.com/TensorFlow/models/tree/master/research/compression`。

TensorFlow、移动端和云端

正如所讨论的，手机通常没有 GPU，而且节省电量是很重要的。因此，很多代价高的计算需要借助于云端以减小其代价。当然，它需要根据多种因素折中，如在移动设备上执行深度学习模型的代价、将数据传递到云端的代价、传递的电量代价以及云计算的代价。没有单一的解决方案，最佳的策略取决于你的具体情况。

10.2 安装适用于 macOS 和 Android 的 TensorFlow mobile

在下面的案例中将学习如何为移动环境安装 TensorFlow。设定的环境是 macOS，以 Android 系统为例，其他配置将在随后的案例中进行描述。

准备工作

使用 Android Studio，采用 Google 的 Android 操作系统中的官方**集成开发环境**（IDE）。

具体做法

按照如下步骤为 macOS 和 Android 安装 TensorFlow mobile：

1. 安装 Android Studio：`https://developer.android.com/studio/install.html`。
2. 创建一个新的项目名称 `AndroidExampleTensorflow`，如下图所示：

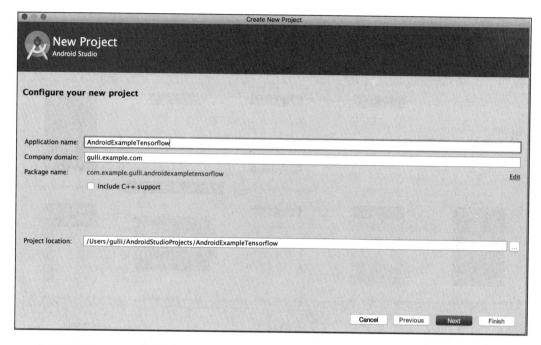

在 Android Studio 中创建 TensorFlow mobile 应用程序的示例,第一步选择 Phone and Tablet 选项,如下图所示:

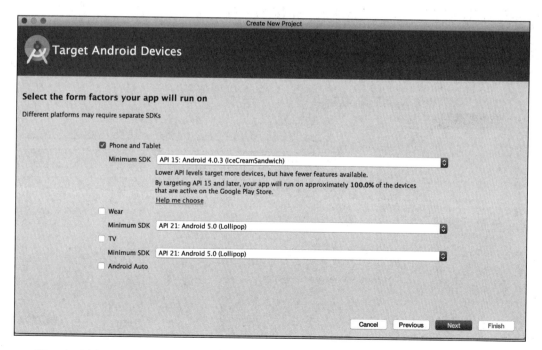

第二步选择 Empty Activity,如下图所示。

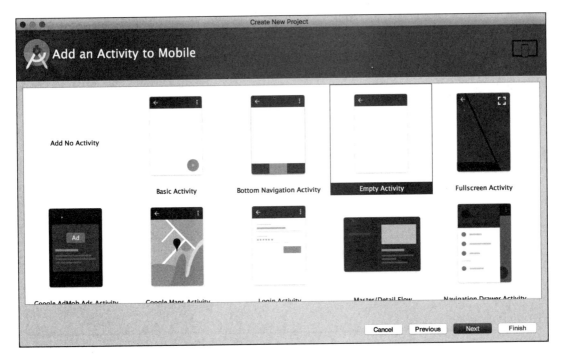

第三步自定义 MainActivity，如下图所示：

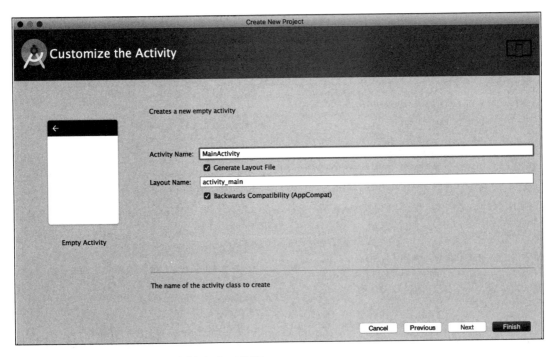

3. 在 `build.gradle` 中插入以下代码。

```
// added for automatically connect to TensorFlow via maven
repositories {
jcenter()
maven {
url 'https://google.bintray.com/TensorFlow'
}
}
dependencies {
compile fileTree(dir: 'libs', include: ['*.jar'])
androidTestCompile('com.android.support.test.espresso:espresso-
core:2.2.2', {
exclude group: 'com.android.support', module: 'support-annotations'
})
compile 'com.android.support:appcompat-v7:26.+'
compile 'com.android.support.constraint:constraint-layout:1.0.2'
// added for automatically compile TensorFlow
compile 'org.TensorFlow:TensorFlow-android:+'
testCompile 'junit:junit:4.12'
}
```

下图显示了插入的代码：

4. 运行该项目并获得如下结果。

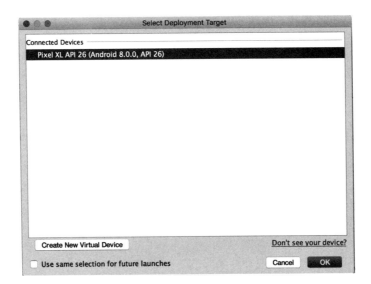

下面是一个 Android Studio 编译的示例，显示了连接的设备：

以上是在 Android Studio 中创建的 TensorFlow mobile 应用程序的一个示例，这是一个简单的 `Hello World` 应用程序。

解读分析

使用 Android Studio 安装 Android TensorFlow 非常简单，只需要将一些配置代码添加到应用的 `build.gradle` 文件中，Android Studio 就将替你完成所有工作。

更多内容

如果从源码安装 TensorFlow，那么就需要安装 Bazel 和 TensorFlow。Bazel 是一个快速、可伸缩、多语言、可扩展的构建系统。Google 在内部使用构建工具 Blaze，并以 Bazel 的形式发布了 Blaze 工具的开源部分。

以下网页将指导你完成整个过程：`https://github.com/TensorFlow/TensorFlow/tree/master/TensorFlow/examples/android/`

如果在 macOS 上进行操作，整个过程会非常简单：

1. 按照 `https://docs.bazel.build/versions/master/install.html` 上的说明安装 Bazel，对于 macOS 使用自制软件：

```
/usr/bin/ruby -e "$(curl -fsSL \
https://raw.githubusercontent.com/Homebrew/install/master/install)"
brew install bazel
bazel version
brew upgrade bazel
```

2. 从 GitHub 克隆 TensorFlow 发行版：

```
git clone https://github.com/TensorFlow/TensorFlow.git
```

10.3 玩转 TensorFlow 和 Android 的示例

在这个案例中，将考虑在 TensorFlow 发行版中提供的标准 Android 示例，并将它们安装在移动设备上。

准备工作

TensorFlow mobile Android app 可通过 GitHub 地址：`https://github.com/TensorFlow/TensorFlow/tree/master/TensorFlow/examples/android` 获得。2017 年 10 月，该网页包含以下示例：

- **TF 分类**：使用 Google 的 Inception 模型实时分类相机的帧，在相机图像上直接覆盖显示结果。
- **TF 检测**：对使用 TensorFlow 目标检测 API 进行训练的 SSD-Mobilenet 模型进行演

示。现代卷积目标探测需要对速度和精度进行折中,能够在相机预览中实时定位和跟踪目标(来自 80 个类别)。
- **TF 风格化**:使用基于论文"A Learned Representation For Artistic Style"的模型将相机预览图像重新设置为具有多种不同艺术风格的图像。
- **TF 语音**:运行音频训练教程中的一个简单的语音识别模型。听一小部分单词,识别时在 UI 中突出显示。

具体做法

1.安装软件包的最好方法是使用预设的 APK。单击浏览器进入 `https://ci.TensorFlow.org/view/Nightly/job/nightly-android/`,并下载 `TensorFlow_demo.apk`,如下图所示:

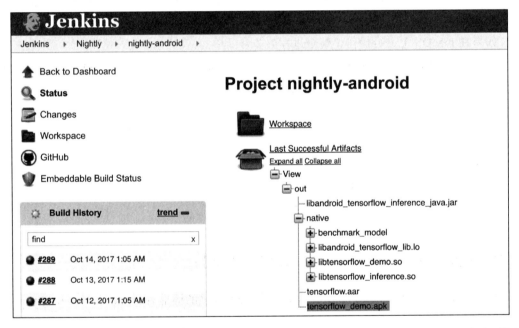

2.在设备上安装应用程序。在接下来的示例中,将使用 Android Studio 中提供的 Pixel XL 模拟设备。它是一个直接从 Android Studio 内部模拟的终端设备。使用命令 `adb devices` 列出所有连接的设备。在本例中,对于 Pixel XL,有一个相应的模拟器,可以用于安装 `TensorFlow_demo apk`。

```
adb devices
List of devices attached
emulator-5554   device
adb install -r TensorFlow_demo.apk
```

安装完成后,模拟器将有一组新的 TensorFlow 应用程序供使用,如下图所示。

3. 运行你喜欢的应用程序，例如，以下图像是 TF 风格化的示例，用于通过迁移学习将相机预览图像重新设置为多种不同艺术风格的图像：

下图是 TF 语音的一个例子（记住需要激活模拟器的麦克风）：

解读分析

如果你使用每夜版构建 demo 和用于在设备上安装 APK 的 `adb` 工具，为 Android 安装 TensorFlow 示例会非常容易。

10.4　安装适用于 macOS 和 iPhone 的 TensorFlow mobile

在这个案例中将学习如何在移动环境中安装 TensorFlow。此处的环境系统采用 macOS，主要是为 iOS 和 iPhone 开发的。

准备工作

使用 Xcode 开发环境和预安装 TensorFlow 的 CocoaPods。假设 Xcode 已经安装在你的环境中。如果没有，请从 `https://developer.apple.com/xcode/` 下载。

具体做法

1. 使用以下命令安装 CocoaPods。

```
sudo gem install cocoapods
pod setup
Setting up CocoaPods master repo
```

```
$ /usr/local/git/current/bin/git clone
https://github.com/CocoaPods/Specs.git master --progress
Cloning into 'master'...
remote: Counting objects: 1602077, done.
remote: Compressing objects: 100% (243/243), done.
remote: Total 1602077 (delta 125), reused 172 (delta 74), pack-
reused 1601747
Receiving objects: 100% (1602077/1602077), 432.12 MiB | 1.83 MiB/s,
done.
Resolving deltas: 100% (849517/849517), done.
Checking out files: 100% (188907/188907), done.
```

2. 用 CocoaPods 安装 TensorFlow 发行版：

```
cd TensorFlow/TensorFlow/examples/ios/benchmark
pod install
Analyzing dependencies
Downloading dependencies
Installing TensorFlow-experimental (1.1.1)
Generating Pods project
Integrating client project
[!] Please close any current Xcode sessions and use
`tf_benchmark_example.xcworkspace` for this project from now on.
Sending stats
Pod installation complete! There is 1 dependency from the Podfile
and 1 total pod installed.
```

3. 从 Inception v1 中下载一些示例数据。将标签和图形文件提取到 `simple` 文件夹和 `camera` 文件夹内的数据文件夹中：

```
mkdir -p ~/graphs
 curl -o ~/graphs/inception5h.zip \
https://storage.googleapis.com/download.TensorFlow.org/models/incep
tion5h.zip \
 && unzip ~/graphs/inception5h.zip -d ~/graphs/inception5h
 cp ~/graphs/inception5h/* TensorFlow/examples/ios/benchmark/data/
 cp ~/graphs/inception5h/* TensorFlow/examples/ios/camera/data/
 cp ~/graphs/inception5h/* TensorFlow/examples/ios/simple/data/
```

4. 下载用作测试的图像并将其复制到基准测试目录中：

```
   https://upload.wikimedia.org/wikipedia/commons/5/55/Grace_Hopper.jpg cp grace_
hopper.jpg ../../benchmark/data/
```

一幅 Grace Hopper 的图像

5. 打开之前使用的示例项目。下面的命令将打开已经提供了 TensorFlow 的 Xcode，然后运行编译，如下面的代码和图所示：

```
open tf_benchmark_example.xcworkspace
```

6. 现在看看 iPhone 模拟器的结果。使用 Inception v1 分类器，步骤 4 的图像被识别为一个军装（`military uniform`）的图像：

iPhone 应用 TensorFlow 计算的一个示例

解读分析

Xcode 和 CocoaPods 被用作编译 TensorFlow 应用程序，用于在不同的 Inception 分类器中分类图像，将结果用 iPhone 模拟器可视化。

拓展阅读

可以直接从应用程序中使用 TensorFlow。更多信息请参考：

`https://github.com/TensorFlow/TensorFlow/blob/master/TensorFlow/examples/ios/README.md`。

10.5 为移动设备优化 TensorFlow 计算图

在这个案例中将考虑采用不同的方式来优化运行在移动设备上的 TensorFlow 代码。从减少模型的大小到量化等方面进行相关的分析。

准备工作

使用 Bazel 来构建 TensorFlow 的不同组件，因此，第一步是确保已经安装了 Bazel 和 TensorFlow。

具体做法

1. 从网址 `https://developer.android.com/studio/install.html` 安装 Android Studio。

2. 按照网址 `https://docs.bazel.build/versions/master/install.html` 的说明安装 Bazel。对于 macOS 使用 Homebrew：

```
/usr/bin/ruby -e "$(curl -fsSL \
https://raw.githubusercontent.com/Homebrew/install/master/install)"
brew install bazel
bazel version
brew upgrade bazel
```

3. 从 GitHub 克隆 TensorFlow 发行版：

```
git clone https://github.com/TensorFlow/TensorFlow.git
```

4. 建立总结计算图本身的计算图变换器：

```
cd ~/TensorFlow/
bazel build TensorFlow/tools/graph_transforms:summarize_graph
[2,326 / 2,531] Compiling
TensorFlow/core/kernels/cwise_op_greater.cc
```

```
INFO: From Linking
TensorFlow/tools/graph_transforms/summarize_graph:
clang: warning: argument unused during compilation: '-pthread' [-
Wunused-command-line-argument]
Target //TensorFlow/tools/graph_transforms:summarize_graph up-to-
date:
bazel-bin/TensorFlow/tools/graph_transforms/summarize_graph
INFO: Elapsed time: 1521.260s, Critical Path: 103.87s
```

5. 下载一个 TensorFlow 计算图作为示例。本例中将使用 Inception v1 TensorFlow 计算图：

```
mkdir -p ~/graphs
 curl -o ~/graphs/inception5h.zip \
https://storage.googleapis.com/download.TensorFlow.org/models/incep
tion5h.zip \
 && unzip ~/graphs/inception5h.zip -d ~/graphs/inception5h
```

6. 总结 Inception 计算图，并注意常量参数的数量：1346 万。每个参数都存储 32 位的浮点数，这非常占用资源：

```
bazel-bin/TensorFlow/tools/graph_transforms/summarize_graph --
in_graph=/Users/gulli/graphs/TensorFlow_inception_graph.pb
Found 1 possible inputs: (name=input, type=float(1), shape=[])
No variables spotted.
Found 3 possible outputs: (name=output, op=Identity) (name=output1,
op=Identity) (name=output2, op=Identity)
Found 13462015 (13.46M) const parameters, 0 (0) variable
parameters, and 0 control_edges
370 nodes assigned to device '/cpu:0'Op types used: 142 Const, 64
BiasAdd, 61 Relu, 59 Conv2D, 13 MaxPool, 9 Concat, 5 Reshape, 5
MatMul, 3 Softmax, 3 Identity, 3 AvgPool, 2 LRN, 1 Placeholder
To use with TensorFlow/tools/benchmark:benchmark_model try these
arguments:
bazel run TensorFlow/tools/benchmark:benchmark_model -- --
graph=/Users/gulli/graphs/TensorFlow_inception_graph.pb --
show_flops --input_layer=input --input_layer_type=float --
input_layer_shape= --output_layer=output,output1,output2
```

7. 编译用于将常值运算量化为 8 位的工具：

```
bazel build TensorFlow/tools/graph_transforms:transform_graph
INFO: From Linking
TensorFlow/tools/graph_transforms/transform_graph:
clang: warning: argument unused during compilation: '-pthread' [-
Wunused-command-line-argument]
Target //TensorFlow/tools/graph_transforms:transform_graph up-to-
date:
bazel-bin/TensorFlow/tools/graph_transforms/transform_graph
INFO: Elapsed time: 294.421s, Critical Path: 28.83s
```

8. 运行用于量化 Inception v1 计算图的工具：

```
bazel-bin/TensorFlow/tools/graph_transforms/transform_graph --
in_graph=/Users/gulli/graphs/inception5h/TensorFlow_inception_graph
.pb --out_graph=/tmp/TensorFlow_inception_quantized.pb --
inputs='Mul:0' --outputs='softmax:0' --
```

```
transforms='quantize_weights'
2017-10-15 18:56:01.192498: I
TensorFlow/tools/graph_transforms/transform_graph.cc:264] Applying
quantize_weights
```

9. 比较这两个模型：

```
ls -lah
/Users/gulli/graphs/inception5h/TensorFlow_inception_graph.pb
-rw-r----- 1 gulli 5001 51M Nov 19 2015
/Users/gulli/graphs/inception5h/TensorFlow_inception_graph.pb
ls -lah /tmp/TensorFlow_inception_quantized.pb
-rw-r--r-- 1 gulli wheel 13M Oct 15 18:56
/tmp/TensorFlow_inception_quantized.pb
```

解读分析

量化有助于将常值运算从 32 位减小到 8 位，以减少模型的大小。一般来说，该模型性能不会显著下降，但是，必须要在一个个案例中加以核实。

10.6 为移动设备分析 TensorFlow 计算图

在本案例中考虑采用不同的方式来优化运行在移动设备上的 TensorFlow 代码。从减少模型的大小到量化等方面对其进行相关分析。

准备工作

使用 Bazel 来构建 TensorFlow 的不同组件。因此，第一步是确保已经安装了 Bazel 和 TensorFlow。

具体做法

1. 从网址 https://developer.android.com/studio/install.html 安装 Android Studio。

2. 按照网址 https://docs.bazel.build/versions/master/install.html 的说明安装 Bazel。对于 macOS 使用 Homebrew：

```
/usr/bin/ruby -e "$(curl -fsSL \
https://raw.githubusercontent.com/Homebrew/install/master/install)"
brew install bazel
bazel version
brew upgrade bazel
```

3. 从 GitHub 克隆 TensorFlow 发行版：

```
git clone https://github.com/TensorFlow/TensorFlow.git
```

4. 构建图形变换器，用于剖析计算图本身：

```
cd ~/TensorFlow/
bazel build -c opt TensorFlow/tools/benchmark:benchmark_model
INFO: Found 1 target...
Target //TensorFlow/tools/benchmark:benchmark_model up-to-date:
bazel-bin/TensorFlow/tools/benchmark/benchmark_model
INFO: Elapsed time: 0.493s, Critical Path: 0.01s
```

5. 通过在桌面上运行以下命令来对模型进行基准测试：

```
bazel-bin/TensorFlow/tools/benchmark/benchmark_model --
graph=/Users/gulli/graphs/TensorFlow_inception_graph.pb --
show_run_order=false --show_time=false --show_memory=false --
show_summary=true --show_flops=true
Graph: [/Users/gulli/graphs/TensorFlow_inception_graph.pb]
Input layers: [input:0]
Input shapes: [1,224,224,3]
Input types: [float]
Output layers: [output:0]
Num runs: [1000]
Inter-inference delay (seconds): [-1.0]
Inter-benchmark delay (seconds): [-1.0]
Num threads: [-1]
Benchmark name: []
Output prefix: []
Show sizes: [0]
Warmup runs: [2]
Loading TensorFlow.
Got config, 0 devices
Running benchmark for max 2 iterations, max -1 seconds without
detailed stat logging, with -1s sleep between inferences
count=2 first=279182 curr=41827 min=41827 max=279182 avg=160504
std=118677
Running benchmark for max 1000 iterations, max 10 seconds without
detailed stat logging, with -1s sleep between inferences
count=259 first=39945 curr=44189 min=36539 max=51743 avg=38651.1
std=1886
Running benchmark for max 1000 iterations, max 10 seconds with
detailed stat logging, with -1s sleep between inferences
count=241 first=40794 curr=39178 min=37634 max=153345 avg=41644.8
std=8092
Average inference timings in us: Warmup: 160504, no stats: 38651,
with stats: 41644
```

```
Number of nodes executed: 141
============================ Summary by node type ============================
        [Node type]  [count]    [avg ms]    [avg %]    [cdf %]    [mem KB]  [times called]
             Conv2D       22      45.729     77.520%    77.520%   10077.888              22
                LRN        2       4.495      7.620%    85.140%    3211.264               2
            MaxPool        6       3.119      5.287%    90.427%    3562.496               6
            BiasAdd       24       2.269      3.846%    94.274%       0.000              24
               Relu       23       1.071      1.816%    96.089%       0.000              23
             MatMul        2       0.851      1.443%    97.532%       8.128               2
             Concat        3       0.726      1.231%    98.763%    2706.368               3
            AvgPool        1       0.514      0.871%    99.634%      32.512               1
              Const       51       0.162      0.275%    99.908%       0.000              51
            Softmax        1       0.027      0.046%    99.954%       0.000               1
               _Arg        1       0.008      0.014%    99.968%       0.000               1
               NoOp        1       0.008      0.014%    99.981%       0.000               1
            Reshape        2       0.005      0.008%    99.990%       0.000               2
            _Retval        1       0.004      0.007%    99.997%       0.000               1
           Identity        1       0.002      0.003%   100.000%       0.000               1

Timings (microseconds): count=241 first=57594 curr=57181 min=50932 max=265096 avg=59057.9 std=14447
Memory (bytes): count=241 curr=19598656(all same)
141 nodes observed
```

6. 在目标 Android 设备上运行 64 位 ARM 处理器，并运行以下命令来对模型进行基准测试。请注意，以下命令会将 Inception 计算图推送至设备并运行一个脚本，在该脚本中可以执行基准测试：

```
bazel build -c opt --config=android_arm64 \
TensorFlow/tools/benchmark:benchmark_model
adb push bazel-bin/TensorFlow/tools/benchmark/benchmark_model \
/data/local/tmp
adb push /tmp/TensorFlow_inception_graph.pb /data/local/tmp/
adb push ~gulli/graphs/inception5h/TensorFlow_inception_graph.pb
/data/local/tmp/
/Users/gulli/graphs/inception5h/TensorFlow_inception_graph.pb: 1
file pushed. 83.2 MB/s (53884595 bytes in 0.618s)
adb shell
generic_x86:/ $
/data/local/tmp/benchmark_model --
graph=/data/local/tmp/TensorFlow_inception_graph.pb --
show_run_order=false --show_time=false --show_memory=false --
show_summary=true
```

解读分析

正如预期的那样，这个模型在 Conv2D 节点上花费了很多时间，大约占用了台式机 77.5% 的平均时间。如果你在移动设备上运行这个程序，那么花时间去执行神经网络中的每一层并确保它们处于控制中是非常重要的。另一方面要考虑的是内存占用量，本例中使用台式机执行，约占用 10 MB 内存。

10.7 为移动设备转换 TensorFlow 计算图

在这个案例中将学习如何转换 TensorFlow 计算图，所有的纯训练节点都将被删除，这将减少计算图的大小，使其更适合移动设备。

什么是计算图转换工具？ 参考 https://github.com/tensorflow/tensorflow/blob/master/tensorflow/tools/graph_transforms/README.md，当完成了一个模型的训练并且想要在生产环境中进行部署时，你通常会想修改它使之可以更好地运行在最终的环境中。例如，如果你的目标是手机，你可能希望通过量化权重或者优化批量标准化或其他纯训练功能来减少文件大小，计算图转换工具提供了一套用于修改计算图的工具，以及支持自编写修改程序的框架。

准备工作

使用 Bazel 来构建 TensorFlow 的不同组件。因此，第一步是确保已经安装了 Bazel 和 TensorFlow。

具体做法

1. 从网址 `https://developer.android.com/studio/install.html` 安装 Android Studio。

2. 按照网址 `https://docs.bazel.build/versions/master/install.html` 的说明安装 Bazel。对于 macOS 使用 Homebrew：

```
/usr/bin/ruby -e "$(curl -fsSL \
https://raw.githubusercontent.com/Homebrew/install/master/install)"
brew install bazel
bazel version
brew upgrade bazel
```

3. 从 GitHub 克隆 TensorFlow 发行版：

```
git clone https://github.com/TensorFlow/TensorFlow.git
```

4. 建立总结计算图本身的计算图变换器：

```
bazel run TensorFlow/tools/graph_transforms:summarize_graph -- --
in_graph=/Users/gulli/graphs/inception5h/TensorFlow_inception_graph
.pb
WARNING: /Users/gulli/TensorFlow/TensorFlow/core/BUILD:1783:1: in
includes attribute of cc_library rule
//TensorFlow/core:framework_headers_lib:
'../../external/nsync/public' resolves to 'external/nsync/public'
not below the relative path of its package 'TensorFlow/core'. This
will be an error in the future. Since this rule was created by the
macro 'cc_header_only_library', the error might have been caused by
the macro implementation in
/Users/gulli/TensorFlow/TensorFlow/TensorFlow.bzl:1054:30.
INFO: Found 1 target...
Target //TensorFlow/tools/graph_transforms:summarize_graph up-to-
date:
bazel-bin/TensorFlow/tools/graph_transforms/summarize_graph
INFO: Elapsed time: 0.395s, Critical Path: 0.01s
INFO: Running command line: bazel-
bin/TensorFlow/tools/graph_transforms/summarize_graph '--
in_graph=/Users/gulli/graphs/inception5h/TensorFlow_inception_graph
.pb'
Found 1 possible inputs: (name=input, type=float(1), shape=[])
No variables spotted.
Found 3 possible outputs: (name=output, op=Identity) (name=output1,
op=Identity) (name=output2, op=Identity)
Found 13462015 (13.46M) const parameters, 0 (0) variable
parameters, and 0 control_edges
370 nodes assigned to device '/cpu:0'Op types used: 142 Const, 64
BiasAdd, 61 Relu, 59 Conv2D, 13 MaxPool, 9 Concat, 5 Reshape, 5
MatMul, 3 Softmax, 3 Identity, 3 AvgPool, 2 LRN, 1 Placeholder
To use with TensorFlow/tools/benchmark:benchmark_model try these
arguments:
bazel run TensorFlow/tools/benchmark:benchmark_model -- --
graph=/Users/gulli/graphs/inception5h/TensorFlow_inception_graph.pb
--show_flops --input_layer=input --input_layer_type=float --
input_layer_shape= --output_layer=output,output1,output2
```

5. 删除只用于训练的所有节点，当计算图在移动设备上进行推理时，不需要这些节点：

```
bazel run TensorFlow/tools/graph_transforms:transform_graph -- --
in_graph=/Users/gulli/graphs/inception5h/TensorFlow_inception_graph
.pb --out_graph=/tmp/optimized_inception_graph.pb --
transforms="strip_unused_nodes fold_constants(ignore_errors=true)
fold_batch_norms fold_old_batch_norms"

WARNING: /Users/gulli/TensorFlow/TensorFlow/core/BUILD:1783:1: in
includes attribute of cc_library rule
//TensorFlow/core:framework_headers_lib:
'../../external/nsync/public' resolves to 'external/nsync/public'
not below the relative path of its package 'TensorFlow/core'. This
will be an error in the future. Since this rule was created by the
macro 'cc_header_only_library', the error might have been caused by
the macro implementation in
/Users/gulli/TensorFlow/TensorFlow/TensorFlow.bzl:1054:30.
INFO: Found 1 target...
Target //TensorFlow/tools/graph_transforms:transform_graph up-to-
date:
bazel-bin/TensorFlow/tools/graph_transforms/transform_graph
INFO: Elapsed time: 0.578s, Critical Path: 0.01s
INFO: Running command line: bazel-
bin/TensorFlow/tools/graph_transforms/transform_graph '--
in_graph=/Users/gulli/graphs/inception5h/TensorFlow_inception_graph
.pb' '--out_graph=/tmp/optimized_inception_graph.pb' '--
transforms=strip_unused_nodes fold_constants(ignore_errors=true)
fold_batch_norms fold_old_batch_norms'
2017-10-15 22:26:59.357129: I
TensorFlow/tools/graph_transforms/transform_graph.cc:264] Applying
strip_unused_nodes
2017-10-15 22:26:59.367997: I
TensorFlow/tools/graph_transforms/transform_graph.cc:264] Applying
fold_constants
2017-10-15 22:26:59.387800: I
TensorFlow/core/platform/cpu_feature_guard.cc:137] Your CPU
supports instructions that this TensorFlow binary was not compiled
to use: SSE4.2 AVX AVX2 FMA
2017-10-15 22:26:59.388676: E
TensorFlow/tools/graph_transforms/transform_graph.cc:279]
fold_constants: Ignoring error Must specify at least one target to
fetch or execute.
2017-10-15 22:26:59.388695: I
TensorFlow/tools/graph_transforms/transform_graph.cc:264] Applying
fold_batch_norms
2017-10-15 22:26:59.388721: I
TensorFlow/tools/graph_transforms/transform_graph.cc:264] Applying
fold_old_batch_norms
```

解读分析

为了创建一个可以加载到设备上的简化模型，已经使用计算图变换工具应用的 **strip_unused_nodes** 规则删除了所有不需要的节点。此操作移除了用于学习的所有节点，保留了用于推理的完整节点。

CHAPTER 11

第 11 章

生成式模型和 CapsNet

11.1 引言

在本章中将讨论如何将**生成对抗网络**（GAN）应用于深度学习的某个领域。其核心方法是在训练生成器的同时，也对鉴别器进行训练，以达到改进后者的目的。同样的方法可以应用于图像处理的不同领域中。另外也会讨论变分自动编码机。

Yann LeCun（深度学习创始人之一）提出的 GAN 已经被认为是过去的 10 年里在 ML、领域中最有吸引力的想法（https://www.quora.com/What-are-some-recent-and-potentially-upcoming-breakthroughs-in-deep-learning）。GAN 可以学习如何再现看似真实的合成数据。例如，计算机可以学习如何绘画和创造逼真的图像。这个想法最初是由 Ian Goodfellow 提出的，他曾就职于蒙特利尔大学的 Google Brain 团队，现就职于 OpenAI 团队（https://openai.com/）。

什么是 GAN

GAN 的关键过程很容易理解，就好像制作赝品的过程一样，赝品制作是仿造未被原创者授权艺术品的过程，原创者通常是更有名的艺术家。GAN 同时训练两个神经网络。

生成器 $G(Z)$ 是用来生成赝品的模块，而**鉴别器** $D(Y)$ 是可以根据对真实艺术品和副本的观察来判断赝品的真实性的模块。鉴别器 $D(Y)$ 取一个输入 Y（例如一个图像）并发起投票来判断该输入的真实程度。一般来说，数值越接近于 0 表示输入越真实，而数值越接近 1 表示输入越虚假。生成器 $G(Z)$ 从一个随机噪声 Z 中生成一个输入并训练自己骗过鉴别器 D，使之认为其生成的输入都是真实的。因此训练鉴别器 $D(Y)$ 的目标是使鉴别器 $D(Y)$ 最大化来自真实数据分布的图像，并最小化不是来自真实数据分布的图像。所以生成器 G 和鉴别器 D 在玩一个对立的游戏：名为对抗训练。需要注意的是，以交替方式训练生成器 G 和鉴别器 D，其中每个目标都表示为通过梯度下降优化的损失函数。生成器模块学会如何使输入越来越逼真，而鉴别器模块学会如何越来越准确地识别虚假输入。

鉴别器网络（通常是标准的卷积神经网络）用来界定输入图像是真实的还是生成的。一个重要的新想法是通过倒置鉴别器和发生器来调整生成器的参数，使得生成器可以学习如何在各种情况下骗过鉴别器。最终，生成器将学会如何生成与真实图像无法区分的图像：

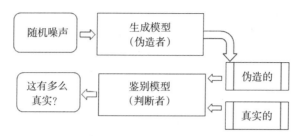

生成器（伪造者）-鉴别器（判断者）模型示例图，鉴别器接收伪造的和真实的图像输入

当然，GAN 在两人比赛中找到了平衡点。为了有效地学习，如果一个选手在一轮更新中成功收敛，则相同的更新必须使另一个选手也能收敛。试想一下！如果伪造者每次都学会如何欺骗法官，那么伪造者自己就没有什么可学的了。

有时两人最终达到一个平衡，但并不能确保一定能够达到平衡，两位选手可以继续对弈很长时间。下图提供了双方的一个例子：

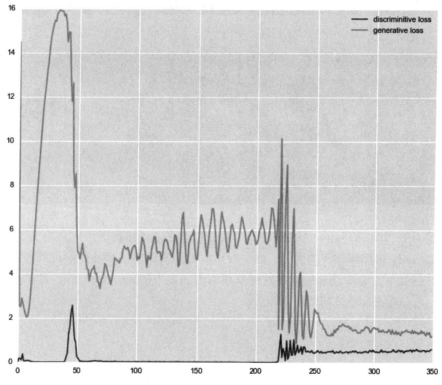

生成器和鉴别器的收敛性示意图

一些很酷的 GAN 应用

之前已经建立了生成器学习如何伪造数据，这意味着它学会了如何生成新的合成数据，这些数据是由网络和人所创建的，并且看起来很逼真。在详细介绍 GAN 代码之前，我想分享一篇论文的结果（代码可通过 `https://github.com/hanzhanggit/StackGAN` 获取），其中 GAN 被用来从一个文本描述开始合成伪造图像。试验结果令人印象深刻。第一列是测试集中的实际图像，其他所有列都是根据 StackGAN 的 Stage-I 和 Stage-II 中的相同文本描述生成的图像。YouTube 上还有更多的例子（`https://www.youtube.com/watch?V=SuRyL5vhCIMfeature = youtu.be`）：

现在来看看 GAN 如何学习伪造 MNIST 数据集。在这种情况下，它是用于发生器和鉴别器网络的 GAN 和 ConvNet 的组合。在一开始时，生成器无法生成任何可以理解的东西，但经过几次迭代后，合成伪造的数字就越来越清晰。在下图中，面板是通过增加训练次数来整理的，可以看到面板之间的质量改进。

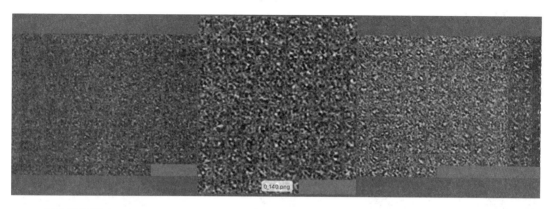

改进的图像如下:

可以在下面的图像中看到进一步的改进:

GAN 的最酷用途之一是在生成器的矢量 Z 中对脸进行算术运算。换句话说,如果留在合成伪造图像的空间中,可以看到像这样的东西:[微笑的女人] − [平静的女人] + [平静的男人] = [微笑的男人],或者像这样:[戴眼镜的男人] − [没有戴眼镜的男人] + [没有戴眼镜的女人] = [戴眼镜的女人]。下图来自：*Unsupervised Representation Learning with Deep Convolutional Generative Adversarial Networks*，Alec Radford，Luke Metz，Soumith Chintala，2016，https://arxiv.org/abs/1511.06434。

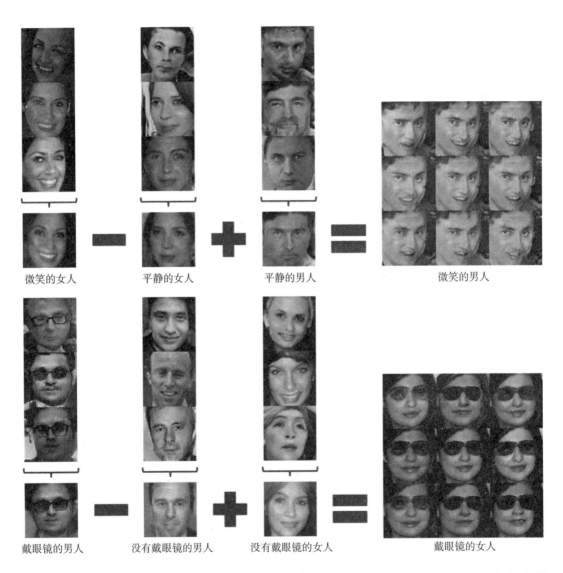

GAN 其他的酷例可访问 https://github.com/Newmu/dcgan_code。本文中的所有图像都是由神经网络生成的。它们都不是真实的。阅读全文可访问：http://arxiv.org/abs/1511.06434。

卧室：经过 5 次训练后生成的卧室如下图所示。

第 11 章 生成式模型和 CapsNet 283

生成的卧室示意图

专辑封面：这些图像不是真实的，而是由 GAN 生成的。专辑封面看起来像是真实的：

生成的专辑封面示意图

11.2 学习使用简单 GAN 虚构 MNIST 图像

更好地理解 GAN 的一篇很好的论文为："Generative Adversarial Networks"（2014），作者为 Ian J.Goodfellow、Jean Pouget-Abadie、Mehdi Mirza、Bing Xu、David Warde-Farley、Sherjil Ozair、Aaron Courville 和 Yoshua Bengio。在本文方法中，将学习如何使用由生成器–鉴别器架构组织的完全连接层网络来伪造 MNIST 手写的数字。

准备工作

相关的可用代码请参阅：https://github.com/TengdaHan/GaN-TensorFlow。

具体做法

1. 从 GitHub 克隆代码：

```
git clone https://github.com/TengdaHan/GAN-TensorFlow
```

2. 参照 *Xavier Glorot* 和 *Yoshua Bengio* 写的论文 "*Understanding the difficulty of training deep feedforward neural networks*"（2009），定义一个 Xavier 初始化器。文章网页链接为 `http://citeseerx.ist.psu.edu/viewdoc/download?doi=10.1.1.207.2059&rep=rep1&t ype = pdf`。初始化器被证明可以使 GAN 更好地收敛。

```
def xavier_init(size):
    in_dim = size[0]
    xavier_stddev = 1. / tf.sqrt(in_dim / 2.)
    return xavier_stddev
```

3. 定义输入 *X* 的生成器。首先定义一个维数为 `[100，K=128]` 的矩阵 *W1*，按照正态分布进行初始化。需要注意的是，100 是生成器使用的初始化噪音 *Z* 的任意值。接着，定义维度为 `[K=256]` 的偏置 *B1*。同样，定义一个维数为 `[K=128，L=784]` 的矩阵 *W2* 和一个维数为 `[L=784]` 的偏置 *B2*。通过使用步骤 1 中定义的 `xavier_init` 来初始化两个矩阵 *W1* 和 *W2*，而使用 `tf.constant_initializer()` 来初始化 *B1* 和 *B2*。之后，计算 *X* * *W1*，加上偏置 *B1*，并代入 ReLU 激活函数得到 *fc1*。然后将这个密集层与通过将矩阵 *fc1* 与 *W2* 相乘并加上偏置 B2 而产生的下一个密集层连接。将得到的结果代入 Sigmoid 函数。这些步骤用来定义用于生成器的两层神经网络：

```
def generator(X):
  with tf.variable_scope('generator'):
    K = 128
    L = 784
    W1 = tf.get_variable('G_W1', [100, K],
  initializer=tf.random_normal_initializer(stddev=xavier_init([100, K])))
    B1 = tf.get_variable('G_B1', [K],
initializer=tf.constant_initializer())
    W2 = tf.get_variable('G_W2', [K, L],
      initializer=tf.random_normal_initializer(stddev=xavier_init([K, L])))
```

```
    B2 = tf.get_variable('G_B2', [L],
initializer=tf.constant_initializer())
    # summary
    tf.summary.histogram('weight1', W1)
    tf.summary.histogram('weight2', W2)
    tf.summary.histogram('biases1', B1)
    tf.summary.histogram('biases2', B2)
    fc1 = tf.nn.relu((tf.matmul(X, W1) + B1))
    fc2 = tf.matmul(fc1, W2) + B2
    prob = tf.nn.sigmoid(fc2)
    return prob
```

4. 定义输入 *X* 的鉴别器。原则上，这与生成器非常相似。主要区别在于，如果参数重用为真，那么调用 `scope.reuse_variables()` 来触发重用。接着定义两个密集层。第一层使用维数为 [J=784，K=128] 的矩阵 *W1* 和维数为 [K=128] 的偏置 *B1*，并且它基于由 *W1* 生成的 *X* 的标准乘积。这个结果与偏置 *B1* 求和并代入一个 ReLU 激活函数来得到结果 *fc1*。第二层使用维数为 [K=128，L=1] 的矩阵 *W2* 与维数为 [L=1] 的偏置 *B2*，它基于由 *W2* 生成的 *fc1* 的标准乘积。这个结果与 *B2* 求和并代入一个 Sigmoid 函数：

```
def discriminator(X, reuse=False):
  with tf.variable_scope('discriminator'):
    if reuse:
      tf.get_variable_scope().reuse_variables()
    J = 784
    K = 128
    L = 1
    W1 = tf.get_variable('D_W1', [J, K],
      initializer=tf.random_normal_initializer(stddev=xavier_init([J, K])))
    B1 = tf.get_variable('D_B1', [K],
initializer=tf.constant_initializer())
    W2 = tf.get_variable('D_W2', [K, L],
initializer=tf.random_normal_initializer(stddev=xavier_init([K, L])))
    B2 = tf.get_variable('D_B2', [L],
initializer=tf.constant_initializer())
    # summary
    tf.summary.histogram('weight1', W1)
    tf.summary.histogram('weight2', W2)
    tf.summary.histogram('biases1', B1)
    tf.summary.histogram('biases2', B2)
    fc1 = tf.nn.relu((tf.matmul(X, W1) + B1))
    logits = tf.matmul(fc1, W2) + B2
    prob = tf.nn.sigmoid(logits)
    return prob, logits
```

5. 定义一些有用的附加函数。首先，导入一堆标准模块：

```
import tensorflow as tf
import numpy as np
import matplotlib.pyplot as plt
import matplotlib.gridspec as gridspec
import os
import argparse
```

6. 接着从 MNIST 数据集中读取数据并定义一个绘制样本的辅助函数。

```python
def read_data():
    from tensorflow.examples.tutorials.mnist import input_data
    mnist = input_data.read_data_sets("../MNIST_data/", one_hot=True)
    return mnist

def plot(samples):
    fig = plt.figure(figsize=(8, 8))
    gs = gridspec.GridSpec(8, 8)
    gs.update(wspace=0.05, hspace=0.05)
    for i, sample in enumerate(samples):
        ax = plt.subplot(gs[i])
        plt.axis('off')
        ax.set_xticklabels([])
        ax.set_yticklabels([])
        ax.set_aspect('equal')
        plt.imshow(sample.reshape(28, 28), cmap='Greys_r')
    return fig
```

7. 定义训练函数。首先，读取 MNIST 数据，然后为一个标准的 MNIST 手写字符定义一个形状为 28×28 的单通道矩阵 X。接着定义大小为 100 的噪声矢量 z ——这是在高质量 GAN 论文中采用的常见选择。下一步是在 z 上调用生成器并将结果赋值给 G。之后，将 X 代入鉴别器而不再重复使用。接着把伪造/虚假的 G 结果代入鉴别器，并重新使用经过训练形成的权重。其中一个重要的方面是如何选择鉴别器的损失函数，它是两个交叉熵的总和：一个熵用于真实字符，其中所有真实的 MNIST 字符都有设定为 1 的标签，而另一个熵用于伪造字符，其中所有伪造的字符都有设定为 1 的标签。鉴别器和生成器在 10 万步迭代中交替运行。每 500 步，从经过训练形成的分布中抽取一个样本，以展示生成器目前为止学到的内容。它定义了一个新的迭代，结果将在下一节中给出。实现上述描述的代码片段如下：

```python
def train(logdir, batch_size):
    from model_fc import discriminator, generator
    mnist = read_data()
    with tf.variable_scope('placeholder'):
        # Raw image
        X = tf.placeholder(tf.float32, [None, 784])
        tf.summary.image('raw image', tf.reshape(X, [-1, 28, 28, 1]), 3)
        # Noise
        z = tf.placeholder(tf.float32, [None, 100])  # noise
        tf.summary.histogram('Noise', z)

    with tf.variable_scope('GAN'):
        G = generator(z)
        D_real, D_real_logits = discriminator(X, reuse=False)
        D_fake, D_fake_logits = discriminator(G, reuse=True)
        tf.summary.image('generated image', tf.reshape(G, [-1, 28, 28, 1]), 3)
    with tf.variable_scope('Prediction'):
        tf.summary.histogram('real', D_real)
        tf.summary.histogram('fake', D_fake)

    with tf.variable_scope('D_loss'):
        d_loss_real = tf.reduce_mean(
            tf.nn.sigmoid_cross_entropy_with_logits(
```

```python
      logits=D_real_logits, labels=tf.ones_like(D_real_logits)))
    d_loss_fake = tf.reduce_mean(
      tf.nn.sigmoid_cross_entropy_with_logits(
        logits=D_fake_logits, labels=tf.zeros_like(D_fake_logits)))
    d_loss = d_loss_real + d_loss_fake

  tf.summary.scalar('d_loss_real', d_loss_real)
  tf.summary.scalar('d_loss_fake', d_loss_fake)
  tf.summary.scalar('d_loss', d_loss)

  with tf.name_scope('G_loss'):
    g_loss = tf.reduce_mean(tf.nn.sigmoid_cross_entropy_with_logits
    (logits=D_fake_logits, labels=tf.ones_like(D_fake_logits)))
    tf.summary.scalar('g_loss', g_loss)
  tvar = tf.trainable_variables()
  dvar = [var for var in tvar if 'discriminator' in var.name]
  gvar = [var for var in tvar if 'generator' in var.name]

  with tf.name_scope('train'):
    d_train_step = tf.train.AdamOptimizer().minimize(d_loss, var_list=dvar)
    g_train_step = tf.train.AdamOptimizer().minimize(g_loss, var_list=gvar)

  sess = tf.Session()
  init = tf.global_variables_initializer()
  sess.run(init)
  merged_summary = tf.summary.merge_all()
  writer = tf.summary.FileWriter('tmp/mnist/'+logdir)
  writer.add_graph(sess.graph)
  num_img = 0
  if not os.path.exists('output/'):
    os.makedirs('output/')

  for i in range(100000):
    batch_X, _ = mnist.train.next_batch(batch_size)
    batch_noise = np.random.uniform(-1., 1., [batch_size, 100])
    if i % 500 == 0:
      samples = sess.run(G, feed_dict={z: np.random.uniform(-1., 1., [64, 100])})
      fig = plot(samples)
      plt.savefig('output/%s.png' % str(num_img).zfill(3), bbox_inches='tight')
      num_img += 1
      plt.close(fig)

    _, d_loss_print = sess.run([d_train_step, d_loss],
    feed_dict={X: batch_X, z: batch_noise})
    _, g_loss_print = sess.run([g_train_step, g_loss],
    feed_dict={z: batch_noise})
    if i % 100 == 0:
      s = sess.run(merged_summary, feed_dict={X: batch_X, z: batch_noise})
      writer.add_summary(s, i)
      print('epoch:%d g_loss:%f d_loss:%f' % (i, g_loss_print, d_loss_print))

if __name__ == '__main__':
  parser = argparse.ArgumentParser(description='Train vanila GAN using
```

```
fully-connected layers networks')
    parser.add_argument('--logdir', type=str, default='1', help='logdir for
Tensorboard, give a string')
    parser.add_argument('--batch_size', type=int, default=64, help='batch
size: give a int')
    args = parser.parse_args()
    train(logdir=args.logdir, batch_size=args.batch_size)
```

解读分析

在每个迭代中,生成器进行了大量的预测(它生成伪造的 MNIST 图像),鉴别器试图学习如何生成伪造的图像,该图像由预测与真实的 MNIST 图像混合产生。经过 32 次迭代后,生成器学会了伪造这组手写数字。没有人编写程序使机器能够写作,但机器已经学会了如何编写与人类写的无法区分的数字。值得注意的是,训练 GAN 可能非常困难,因为需要找到两名选手之间的平衡点。如果对这个话题感兴趣,建议看一下从业者收集的一系列技巧(https://github.com/soumith/ganhacks)。

下面看一下不同迭代时的一些实际例子,以了解机器如何学会改进写作:

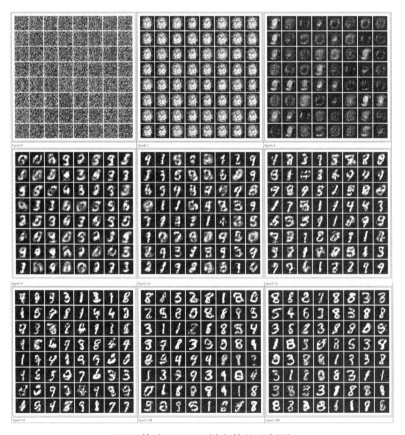

GAN 伪造 MNIST 样字符的示例图

11.3　学习使用 DCGAN 虚构 MNIST 图像

在本方法中将使用一个简单的 GAN，它使用 CNN 来学习如何伪造 MNIST 图像并生成不属于原始数据集的新图像。这个想法是将 CNN 与 GAN 一起使用来提高处理图像数据集的能力。值得注意的是，之前的方法是使用具有完全连接网络的 GAN，而在这里将专注于 CNN。

准备工作

本方法的可行代码可访问网页：https://github.com/TengdaHan/GAN-TensorFlow。

具体做法

1. 从 GitHub 中克隆代码：

```
git clone https://github.com/TengdaHan/GAN-TensorFlow
```

2. 参照 Xavier Glorot 和 Yoshua Bengio 的论文"*Understanding the difficulty of training deep feedforward neural networks*"（2009），定义一个 Xavier 初始化器。初始化器被证明可以使 GAN 更好地收敛：

```
def xavier_init(size):
    in_dim = size[0]
    xavier_stddev = 1. / tf.sqrt(in_dim / 2.)
    # return tf.random_normal(shape=size, stddev=xavier_stddev)
    return xavier_stddev
```

3. 为给定输入 *x*、权重 *w*、偏置 *b* 和给定步幅定义一个卷积运算。代码使用标准的 `tf.nn.conv2d(...)` 模块。需要注意的是，将使用在第 4 章中定义的 `'SAME'` 进行填充：

```
def conv(x, w, b, stride, name):
    with tf.variable_scope('conv'):
        tf.summary.histogram('weight', w)
        tf.summary.histogram('biases', b)
        return tf.nn.conv2d(x,
            filter=w,
            strides=[1, stride, stride, 1],
            padding='SAME',
            name=name) + b
```

4. 为给定的输入 *x*、权重 *w*、偏置 *b* 和给定的步幅定义一个去卷积运算。代码使用标准的 `tf.nn.conv2d_transpose(...)` 模块。再一次使用 `'SAME'` 进行填充。

```
def deconv(x, w, b, shape, stride, name):
    with tf.variable_scope('deconv'):
        tf.summary.histogram('weight', w)
        tf.summary.histogram('biases', b)
        return tf.nn.conv2d_transpose(x,
            filter=w,
            output_shape=shape,
            strides=[1, stride, stride, 1],
```

```
            padding='SAME',
            name=name) + b
```

5. 定义一个标准的 `LeakyReLU`，它对于 GAN 来说是一个非常有效的激活函数：

```
def lrelu(x, alpha=0.2):
  with tf.variable_scope('leakyReLU'):
    return tf.maximum(x, alpha * x)
```

6. 定义生成器。首先定义一个输入大小为 100 的全连接层（Z 为生成器选取的初始噪声，其大小是任意的）。全连接层由维数为 `[100,7*7*256]`、按照正态分布初始化的矩阵 **W1** 和维数为 `[7*7*256]` 的偏置 *B1* 组成。该图层使用 ReLU 作为激活函数。有了全连接层之后，生成器将进行两个解卷积运算，deconv1 和 deconv2，这两个运算的步幅均为 2。第一次 deconv1 运算完成后，其结果被批量标准化。值得注意的是，第二个解卷积运算有 40% 的概率在运行之前被退出。最后一个阶段是一个用作非线性激活的 Sigmoid 函数，如下代码片段所示：

```
def generator(X, batch_size=64):
  with tf.variable_scope('generator'):
    K = 256
    L = 128
    M = 64
    W1 = tf.get_variable('G_W1', [100, 7*7*K], initializer=tf.random_normal_initializer(stddev=0.1))
    B1 = tf.get_variable('G_B1', [7*7*K], initializer=tf.constant_initializer())
    W2 = tf.get_variable('G_W2', [4, 4, M, K], initializer=tf.random_normal_initializer(stddev=0.1))
    B2 = tf.get_variable('G_B2', [M], initializer=tf.constant_initializer())
    W3 = tf.get_variable('G_W3', [4, 4, 1, M], initializer=tf.random_normal_initializer(stddev=0.1))
    B3 = tf.get_variable('G_B3', [1], initializer=tf.constant_initializer())
    X = lrelu(tf.matmul(X, W1) + B1)
    X = tf.reshape(X, [batch_size, 7, 7, K])
    deconv1 = deconv(X, W2, B2, shape=[batch_size, 14, 14, M], stride=2, name='deconv1')
    bn1 = tf.contrib.layers.batch_norm(deconv1)
    deconv2 = deconv(tf.nn.dropout(lrelu(bn1), 0.4), W3, B3, shape=[batch_size, 28, 28, 1], stride=2, name='deconv2')
    XX = tf.reshape(deconv2, [-1, 28*28], 'reshape')
    return tf.nn.sigmoid(XX)
```

7. 定义鉴别器。如前方法所述，如果参数重用是真实的，那么就调用 `scope.reuse_variables()` 来触发重用。鉴别器使用两个卷积层。第一个卷积层接着批量标准化，第二个卷积层在运行前有 40% 的概率被退出，之后才接着批量标准化步骤。在这之后，是一个基于激活函数 ReLU 的密集层，而后是另一个基于 Sigmoid 的激活函数的密集层：

```
def discriminator(X, reuse=False):
  with tf.variable_scope('discriminator'):
    if reuse:
```

```
        tf.get_variable_scope().reuse_variables()
    K = 64
    M = 128
    N = 256
    W1 = tf.get_variable('D_W1', [4, 4, 1, K],
initializer=tf.random_normal_initializer(stddev=0.1))
    B1 = tf.get_variable('D_B1', [K],
initializer=tf.constant_initializer())
    W2 = tf.get_variable('D_W2', [4, 4, K, M],
initializer=tf.random_normal_initializer(stddev=0.1))
    B2 = tf.get_variable('D_B2', [M],
initializer=tf.constant_initializer())
    W3 = tf.get_variable('D_W3', [7*7*M, N],
initializer=tf.random_normal_initializer(stddev=0.1))
    B3 = tf.get_variable('D_B3', [N],
initializer=tf.constant_initializer())
    W4 = tf.get_variable('D_W4', [N, 1],
initializer=tf.random_normal_initializer(stddev=0.1))
    B4 = tf.get_variable('D_B4', [1],
initializer=tf.constant_initializer())
    X = tf.reshape(X, [-1, 28, 28, 1], 'reshape')
    conv1 = conv(X, W1, B1, stride=2, name='conv1')
    bn1 = tf.contrib.layers.batch_norm(conv1)
    conv2 = conv(tf.nn.dropout(lrelu(bn1), 0.4), W2, B2, stride=2,
name='conv2')
    bn2 = tf.contrib.layers.batch_norm(conv2)
    flat = tf.reshape(tf.nn.dropout(lrelu(bn2), 0.4), [-1, 7*7*M],
name='flat')
    dense = lrelu(tf.matmul(flat, W3) + B3)
    logits = tf.matmul(dense, W4) + B4
    prob = tf.nn.sigmoid(logits)
    return prob, logits
```

8. 从 MNIST 数据集读取数据，并定义一个绘制样本的辅助函数：

```
import numpy as np
import matplotlib.pyplot as plt
import matplotlib.gridspec as gridspec
import os
import argparse

def read_data():
  from tensorflow.examples.tutorials.mnist import input_data
  mnist = input_data.read_data_sets("../MNIST_data/", one_hot=True)
  return mnist

def plot(samples):
  fig = plt.figure(figsize=(8, 8))
  gs = gridspec.GridSpec(8, 8)
  gs.update(wspace=0.05, hspace=0.05)
  for i, sample in enumerate(samples):
    ax = plt.subplot(gs[i])
    plt.axis('off')
    ax.set_xticklabels([])
    ax.set_yticklabels([])
    ax.set_aspect('equal')
    plt.imshow(sample.reshape(28, 28), cmap='Greys_r')
  return fig
```

9. 定义训练函数。首先，读取 MNIST 数据，然后为一个标准的 MNIST 手写字符定义一个形状为 28×28 的单通道矩阵 **X**。接着定义大小为 100 的噪声矢量 z ——这是在高质量 GAN 论文中采用的常见选择。下一步是在 z 上调用生成器并将结果赋值给 G。之后，将 **X** 代入鉴别器而不再重复使用。接着把伪造／虚假的 G 结果代入鉴别器，并重新使用经训练形成的权重。其中一个重要的方面是如何选择鉴别器的损失函数，它是两个交叉熵的总和：一个熵用于真实字符，其中所有真实的 MNIST 字符都有设定为 1 的标签，而另一个熵用于伪造字符，其中所有伪造的字符都有设定为 1 的标签。鉴别器和生成器在 10 万步迭代中交替运行。每 500 步，从经过训练形成的分布中抽取一个样本，以展示生成器到目前为止学到的内容。它定义了一个新的迭代，结果将在下一节中给出。训练函数的代码片段如下所示：

```
def train(logdir, batch_size):
  from model_conv import discriminator, generator
  mnist = read_data()
  with tf.variable_scope('placeholder'):
    # Raw image
    X = tf.placeholder(tf.float32, [None, 784])
    tf.summary.image('raw image', tf.reshape(X, [-1, 28, 28, 1]), 3)
    # Noise
    z = tf.placeholder(tf.float32, [None, 100]) # noise
    tf.summary.histogram('Noise', z)
  with tf.variable_scope('GAN'):
    G = generator(z, batch_size)
    D_real, D_real_logits = discriminator(X, reuse=False)
    D_fake, D_fake_logits = discriminator(G, reuse=True)
    tf.summary.image('generated image', tf.reshape(G, [-1, 28, 28, 1]), 3)

  with tf.variable_scope('Prediction'):
    tf.summary.histogram('real', D_real)
    tf.summary.histogram('fake', D_fake)

  with tf.variable_scope('D_loss'):
    d_loss_real = tf.reduce_mean(
      tf.nn.sigmoid_cross_entropy_with_logits(
    logits=D_real_logits, labels=tf.ones_like(D_real_logits)))

    d_loss_fake = tf.reduce_mean(
    tf.nn.sigmoid_cross_entropy_with_logits(
    logits=D_fake_logits, labels=tf.zeros_like(D_fake_logits)))
    d_loss = d_loss_real + d_loss_fake
    tf.summary.scalar('d_loss_real', d_loss_real)
    tf.summary.scalar('d_loss_fake', d_loss_fake)
    tf.summary.scalar('d_loss', d_loss)

  with tf.name_scope('G_loss'):
    g_loss =   tf.reduce_mean(tf.nn.sigmoid_cross_entropy_with_logits
(logits=D_fake_logits, labels=tf.ones_like(D_fake_logits)))
    tf.summary.scalar('g_loss', g_loss)
    tvar = tf.trainable_variables()
    dvar = [var for var in tvar if 'discriminator' in var.name]
    gvar = [var for var in tvar if 'generator' in var.name]
```

```python
    with tf.name_scope('train'):
      d_train_step = tf.train.AdamOptimizer().minimize(d_loss, var_list=dvar)
      g_train_step = tf.train.AdamOptimizer().minimize(g_loss, var_list=gvar)
  sess = tf.Session()
  init = tf.global_variables_initializer()
  sess.run(init)
  merged_summary = tf.summary.merge_all()
  writer = tf.summary.FileWriter('tmp/'+'gan_conv_'+logdir)
  writer.add_graph(sess.graph)
  num_img = 0

  if not os.path.exists('output/'):
    os.makedirs('output/')
  for i in range(100000):
    batch_X, _ = mnist.train.next_batch(batch_size)
    batch_noise = np.random.uniform(-1., 1., [batch_size, 100])
    if i % 500 == 0:
      samples = sess.run(G, feed_dict={z: np.random.uniform(-1., 1., [64, 100])})
      fig = plot(samples)
      plt.savefig('output/%s.png' % str(num_img).zfill(3), bbox_inches='tight')
      num_img += 1
      plt.close(fig)

    _, d_loss_print = sess.run([d_train_step, d_loss], feed_dict={X: batch_X, z: batch_noise})
    _, g_loss_print = sess.run([g_train_step, g_loss], feed_dict={z: batch_noise})

    if i % 100 == 0:
      s = sess.run(merged_summary, feed_dict={X: batch_X, z: batch_noise})
      writer.add_summary(s, i)
      print('epoch:%d g_loss:%f d_loss:%f' % (i, g_loss_print, d_loss_print))

  if __name__ == '__main__':
    parser = argparse.ArgumentParser(description='Train vanila GAN using convolutional networks')
    parser.add_argument('--logdir', type=str, default='1', help='logdir for Tensorboard, give a string')
    parser.add_argument('--batch_size', type=int, default=64, help='batch size: give a int')
    args = parser.parse_args()
    train(logdir=args.logdir, batch_size=args.batch_size)
```

解读分析

将 CNN 与 GAN 一起使用可以提高学习的速度。通过对比不同迭代的一些实际例子，了解机器如何学会改进写作的过程。例如，将使用上述方法进行四次迭代后的结果与使用之前的方法进行四次迭代后的结果进行比较。能看到差别吗？

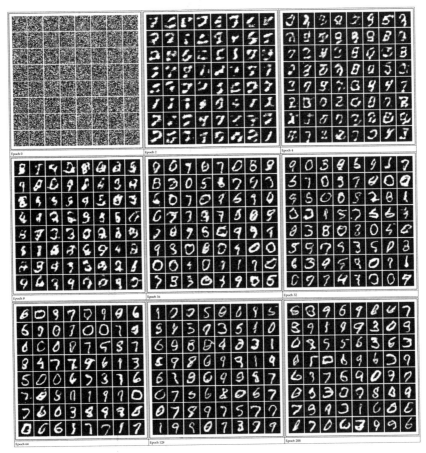

DCGAN 伪造 MNIST 样字符的示例图

11.4 学习使用 DCGAN 虚构名人面孔和其他数据集

用于伪造 MNIST 图像的想法同样可以应用于其他图像领域。在本方法中，将学习如何使用软件包在不同的数据集上训练 DCGAN 模型。参阅该软件包可访问网址：`https://github.com/carpedm20/DCGAN-tensorflow`。本方法基于论文 "Unsupervised Representation Learning with Deep Convolutional Generative Adversarial Networks" 由 Alec Radford、Luke Metz、Soumith Chintal 于 2015 年编写。摘要引用如下：

近年来，卷积网络（CNN）的监督学习在计算机视觉中得到了广泛的应用。相比之下，无监督的 CNN 学习受到的关注较少。本节希望能够帮助弥合有监督学习和无监督学习的可用 CNN 之间的差距。引入了一类称为深度卷积生成对抗网络（DCGAN）的 CNN，它具有一定的架构约束，表明它们是无监督学习的有力候选者。通过训练各种图像数据集，展示了令人信服的证据，深度卷积对抗双方学习了从对象部分到生成器和鉴别器中场景的表示层。另外，将经过训练生成的特征应用于新任务——将它作为一般图像特征进行应用。

请注意，生成器具有下图所示的体系结构：

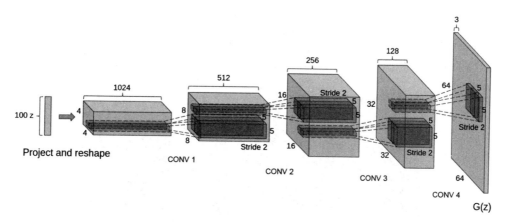

值得注意的是，在封装方面，为了避免 D(鉴别器) 网络的快速收敛，原始纸张有变化，G（生成器）网络在每个 D 网络更新周期中被更新两次。

准备工作

本方法的可行代码可访问网页：https://github.com/carpedm20/DCGAN-tensorflow。

具体做法

1. 从 GitHub 中克隆代码：

```
git clone https://github.com/carpedm20/DCGAN-tensorflow
```

2. 下载以下数据集：

```
python download.py mnist celebA
```

3. 要使用下载的数据集来训练模型，需要使用以下内容：

```
python main.py --dataset celebA --input_height=108 --train --crop
```

4. 要使用现有模型进行测试，需要使用以下命令：

```
python main.py --dataset celebA --input_height=108 --crop
```

5. 还可以通过执行以下操作来使用自己定义的数据集：

```
$ mkdir data/DATASET_NAME
... add images to data/DATASET_NAME ...
$ python main.py --dataset DATASET_NAME --train
```

```
$ python main.py --dataset DATASET_NAME
$ # example
$ python main.py --dataset=eyes --input_fname_pattern="*_cropped.png" --train
```

解读分析

生成器学习如何生成名人的伪造图像，而鉴别器学会如何从真实的图像中识别伪造的图像。在两个网络的每次迭代都争取改善和减少损失。前五个迭代的报告如下：

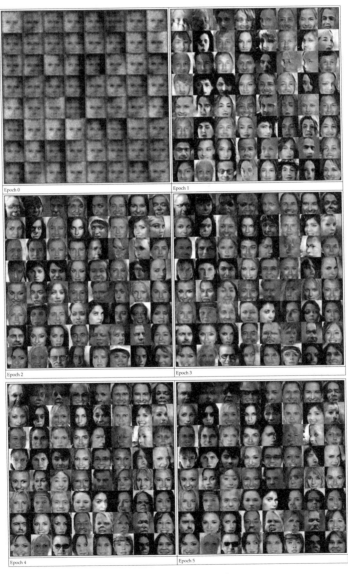

DCGAN 伪造名人的示例图

解读分析

内容感知填充是摄影师用来填充不需要或缺少的图像部分的工具。Raymond A. Yeh、Chen Chen、Teck Yian Lim、Alexander G. Schwing、Mark Hasegawa-Johnson 和 Minh N 在 2016 年发表的论文"*Semantic Image Inpainting with Perceptual and Contextual Losses*"中使用 DCGAN 进行图像生成，它学会了如何填充图像的所需部分。

11.5 实现变分自动编码机

变分自动编码机（VAE）组合了神经网络和贝叶斯推理这两种最好的方法，是最酷的神经网络，已经成为无监督学习的流行方法之一。它是一个扭曲的自动编码机。同自动编码机的传统编码器和解码器网络（见第 8 章）一起，它们具有附加的随机层。

编码器网络之后的随机层使用高斯分布对数据进行采样，而解码器网络之后的随机层使用伯努利分布对数据进行采样。

与 GAN 一样，变分自动编码机根据它们所接受的分布来生成图像和数字。VAE 允许设置潜在的复杂先验，从而学习强大的潜在表征。

下图描述了一个 VAE。编码器 $q_\phi(z|x)$ 网络近似于真实，但后验分布 $p(z|x)$ 很难处理，其中 x 是 VAE 的输入，z 是潜在表示。解码器网络 $p_\theta(x|z)$ 将 d 维潜在变量（也称为潜在空间）作为其输入并生成与 $P(x)$ 相同分布的新图像。从 $z|x \sim N(\mu_{z|x}, \Sigma z|x)$ 采样得到潜在表示 z，而解码器网络的输出从 $x|z \sim N(\mu_{x|z}, \Sigma x|z)$ 采样得到 x | z：

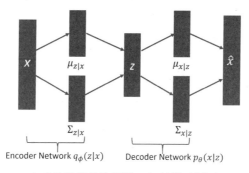

自动编码机的编码器 – 解码器示例图

准备工作

现在已经有了 VAE 的基本结构，问题在于如何对它们进行训练，因为训练数据和后验密度的最大可能性是难以处理的。通过最大化 log 数据可能性的下限来训练网络。因此，损失项由两部分组成：通过采样从解码器网络获得的生成损失，以及被称为潜在损失的 KL 发散项。

发生损耗确保了由解码器生成的图像和用于训练网络的图像是相同的，并且潜在损失

能够确保后验分布 $q_\phi(z|x)$ 接近于先验分布 $p_\theta(z)$。由于编码器使用高斯分布进行采样,所以潜在损耗测量潜在变量与单位高斯匹配的匹配程度。

一旦 VAE 接受训练,只能使用解码器网络来生成新的图像。

具体做法

代码基于 Kingma 和 Welling 撰写的论文 "Autoencoding Variational Bayes"(`https://arxiv.org/pdf/1312.6114.pdf`),并通过 GitHub 进行了调整:`https://jmetzen.github.io/2015-11-27/vae.html`。

1. 导入必要的模块。本方法中,需要调用 Numpy、Matplolib 和 TensorFlow 函数:

```
import numpy as np
import tensorflow as tf
import matplotlib.pyplot as plt
%matplotlib inline
```

2. 定义 VariationalAutoencoder 类。采用 __init__ 类方法来定义超参数,如学习率、批量大小、用于输入的占位符、编码器及解码器网络的权重和偏置变量。它还根据 VAE 的网络体系结构建立计算图。在本方法中使用 Xavier 初始化器初始化权重。与使用自己定义的方法进行 Xavier 初始化不同,本方法使用 `tf.contrib.layers.xavier_initializer()` 来进行初始化。最后,定义损失(生成和潜在)及优化器操作:

```
class VariationalAutoencoder(object):
    def __init__(self, network_architecture,  transfer_fct=tf.nn.softplus,
learning_rate=0.001, batch_size=100):
        self.network_architecture = network_architecture
        self.transfer_fct = transfer_fct
        self.learning_rate = learning_rate
        self.batch_size = batch_size
        # Place holder for the input
        self.x = tf.placeholder(tf.float32, [None,
network_architecture["n_input"]])
        # Define weights and biases
        network_weights =
self._initialize_weights(**self.network_architecture)
        # Create autoencoder network
        # Use Encoder Network to determine mean and
        # (log) variance of Gaussian distribution in latent
        # space
        self.z_mean, self.z_log_sigma_sq = \
            self._encoder_network(network_weights["weights_encoder"],
    network_weights["biases_encoder"])
        # Draw one sample z from Gaussian distribution
        n_z = self.network_architecture["n_z"]
        eps = tf.random_normal((self.batch_size, n_z), 0, 1,
dtype=tf.float32)
        # z = mu + sigma*epsilon
        self.z =
tf.add(self.z_mean,tf.multiply(tf.sqrt(tf.exp(self.z_log_sigma_sq)), eps))
        # Use Decoder network to determine mean of
        # Bernoulli distribution of reconstructed input
        self.x_reconstr_mean = \
```

```python
            self._decoder_network(network_weights["weights_decoder"],
    network_weights["biases_decoder"])
        # Define loss function based variational upper-bound and
        # corresponding optimizer
        # define generation loss
        generation_loss = \
   -tf.reduce_sum(self.x * tf.log(1e-10 + self.x_reconstr_mean)
 + (1-self.x) * tf.log(1e-10 + 1 - self.x_reconstr_mean), 1)
        latent_loss = -0.5 * tf.reduce_sum(1 + self.z_log_sigma_sq
 - tf.square(self.z_mean)- tf.exp(self.z_log_sigma_sq), 1)
        self.cost = tf.reduce_mean(generation_loss + latent_loss)      #
average over batch
        # Define the optimizer
        self.optimizer = \
 tf.train.AdamOptimizer(learning_rate=self.learning_rate).minimize(self.cost
)
        # Initializing the tensor flow variables
        init = tf.global_variables_initializer()
  # Launch the session
        self.sess = tf.InteractiveSession()
        self.sess.run(init)
def _initialize_weights(self, n_hidden_recog_1, n_hidden_recog_2,
n_hidden_gener_1, n_hidden_gener_2,
n_input, n_z):
    initializer = tf.contrib.layers.xavier_initializer()
    all_weights = dict()
    all_weights['weights_encoder'] = {
    'h1': tf.Variable(initializer(shape=(n_input, n_hidden_recog_1))),
    'h2': tf.Variable(initializer(shape=(n_hidden_recog_1,
n_hidden_recog_2))),
    'out_mean': tf.Variable(initializer(shape=(n_hidden_recog_2, n_z))),
    'out_log_sigma': tf.Variable(initializer(shape=(n_hidden_recog_2,
n_z)))}
    all_weights['biases_encoder'] = {
    'b1': tf.Variable(tf.zeros([n_hidden_recog_1], dtype=tf.float32)),
    'b2': tf.Variable(tf.zeros([n_hidden_recog_2], dtype=tf.float32)),
    'out_mean': tf.Variable(tf.zeros([n_z], dtype=tf.float32)),
    'out_log_sigma': tf.Variable(tf.zeros([n_z], dtype=tf.float32))}

    all_weights['weights_decoder'] = {
    'h1': tf.Variable(initializer(shape=(n_z, n_hidden_gener_1))),
    'h2': tf.Variable(initializer(shape=(n_hidden_gener_1,
n_hidden_gener_2))),
    'out_mean': tf.Variable(initializer(shape=(n_hidden_gener_2, n_input))),
    'out_log_sigma': tf.Variable(initializer(shape=(n_hidden_gener_2,
n_input)))}

    all_weights['biases_decoder'] = {
    'b1': tf.Variable(tf.zeros([n_hidden_gener_1],    dtype=tf.float32)),
    'b2': tf.Variable(tf.zeros([n_hidden_gener_2],
dtype=tf.float32)),'out_mean': tf.Variable(tf.zeros([n_input],
dtype=tf.float32)),
    'out_log_sigma': tf.Variable(tf.zeros([n_input], dtype=tf.float32))}
    return all_weights
```

3. 创建网络编码器和网络解码器。网络编码器的第一层接收输入并生成输入的递减式潜在表示；第二层将输入映射到高斯分布。网络学习这些转变：

```python
def _encoder_network(self, weights, biases):
    # Generate probabilistic encoder (recognition network), which
    # maps inputs onto a normal distribution in latent space.
    # The transformation is parametrized and can be learned.
    layer_1 = self.transfer_fct(tf.add(tf.matmul(self.x,    weights['h1']),
biases['b1']))
    layer_2 = self.transfer_fct(tf.add(tf.matmul(layer_1,   weights['h2']),
biases['b2']))
    z_mean = tf.add(tf.matmul(layer_2, weights['out_mean']),
biases['out_mean'])
    z_log_sigma_sq = \
tf.add(tf.matmul(layer_2, weights['out_log_sigma']),
biases['out_log_sigma'])
    return (z_mean, z_log_sigma_sq)

def _decoder_network(self, weights, biases):
    # Generate probabilistic decoder (decoder network), which
    # maps points in latent space onto a Bernoulli distribution in data
space.
    # The transformation is parametrized and can be learned.
    layer_1 = self.transfer_fct(tf.add(tf.matmul(self.z, weights['h1']),
biases['b1']))
    layer_2 = self.transfer_fct(tf.add(tf.matmul(layer_1, weights['h2']),
biases['b2']))
    x_reconstr_mean = \
tf.nn.sigmoid(tf.add(tf.matmul(layer_2, weights['out_mean']),
    biases['out_mean']))
    return x_reconstr_mean
```

4.VariationalAutoencoder 类还包含一些帮助函数来生成和重建数据，并适应 VAE：

```python
def fit(self, X):
    opt, cost = self.sess.run((self.optimizer, self.cost),
    feed_dict={self.x: X})
    return cost

def generate(self, z_mu=None):
""" Generate data by sampling from latent space.
If z_mu is not None, data for this point in latent space is
generated. Otherwise, z_mu is drawn from prior in latent
space.
"""
    if z_mu is None:
        z_mu = np.random.normal(size=self.network_architecture["n_z"])
# Note: This maps to mean of distribution, we could alternatively
# sample from Gaussian distribution
    return self.sess.run(self.x_reconstr_mean,
        feed_dict={self.z: z_mu})

def reconstruct(self, X):
""" Use VAE to reconstruct given data. """
    return self.sess.run(self.x_reconstr_mean,
        feed_dict={self.x: X})
```

5. 一旦 VAE 类完成，定义一个函数序列，它使用 VAE 类对象并通过给定的数据进行训练。

```
def train(network_architecture, learning_rate=0.001,
batch_size=100, training_epochs=10, display_step=5):
  vae = VariationalAutoencoder(network_architecture,
  learning_rate=learning_rate,
  batch_size=batch_size)
  # Training cycle
  for epoch in range(training_epochs):
    avg_cost = 0.
    total_batch = int(n_samples / batch_size)
    # Loop over all batches
    for i in range(total_batch):
      batch_xs, _ = mnist.train.next_batch(batch_size)
      # Fit training using batch data
      cost = vae.fit(batch_xs)
      # Compute average loss
      avg_cost += cost / n_samples * batch_size
      # Display logs per epoch step
    if epoch % display_step == 0:
      print("Epoch:", '%04d' % (epoch+1),
          "cost=", "{:.9f}".format(avg_cost))
  return vae
```

6. 使用 VAE 类和序列函数。采用 MNIST 数据集：

```
# Load MNIST data in a format suited for tensorflow.
# The script input_data is available under this URL:
#https://raw.githubusercontent.com/tensorflow/tensorflow/master/tensorflow/
examples/tutorials/mnist/input_data.py

from tensorflow.examples.tutorials.mnist import input_data
mnist = input_data.read_data_sets('MNIST_data', one_hot=True)
n_samples = mnist.train.num_examples
```

7. 定义网络架构，并在 MNIST 数据集上进行 VAE 的训练。在这种情况下，为了简单保留了潜在维度 2。

```
network_architecture = \
dict(n_hidden_recog_1=500, # 1st layer encoder neurons
n_hidden_recog_2=500, # 2nd layer encoder neurons
n_hidden_gener_1=500, # 1st layer decoder neurons
n_hidden_gener_2=500, # 2nd layer decoder neurons
n_input=784, # MNIST data input (img shape: 28*28)
n_z=2) # dimensionality of latent space
vae = train(network_architecture, training_epochs=75)
```

8. 看一下 VAE 是否重构了输入。输出表明那些数字确实被重构了，而且由于使用了二维的潜在空间，所以图像显得模糊了：

```
x_sample = mnist.test.next_batch(100)[0]
x_reconstruct = vae.reconstruct(x_sample)
plt.figure(figsize=(8, 12))
for i in range(5):
  plt.subplot(5, 2, 2*i + 1)
```

```
plt.imshow(x_sample[i].reshape(28, 28),  vmin=0, vmax=1, cmap="gray")
plt.title("Test input")
plt.colorbar()
plt.subplot(5, 2, 2*i + 2)
plt.imshow(x_reconstruct[i].reshape(28, 28), vmin=0, vmax=1, cmap="gray")
plt.title("Reconstruction")
plt.colorbar()
plt.tight_layout()
```

下图是上述代码的输出：

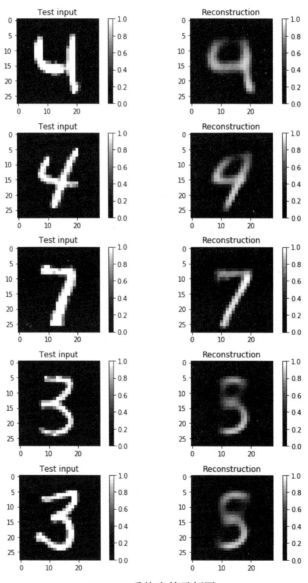

MNIST 重构字符示例图

9. 以下是使用经过训练的 VAE 生成的手写数字样本:

```
nx = ny = 20
x_values = np.linspace(-3, 3, nx)
y_values = np.linspace(-3, 3, ny)
canvas = np.empty((28*ny, 28*nx))
for i, yi in enumerate(x_values):
    for j, xi in enumerate(y_values):
        z_mu = np.array([[xi, yi]]*vae.batch_size)
        x_mean = vae.generate(z_mu)
        canvas[(nx-i-1)*28:(nx-i)*28, j*28:(j+1)*28] = x_mean[0].reshape(28, 28)
plt.figure(figsize=(8, 10))
Xi, Yi = np.meshgrid(x_values, y_values)
plt.imshow(canvas, origin="upper", cmap="gray")
plt.tight_layout()
```

以下是由自动编码机生成的 MNIST 样字符的范围:

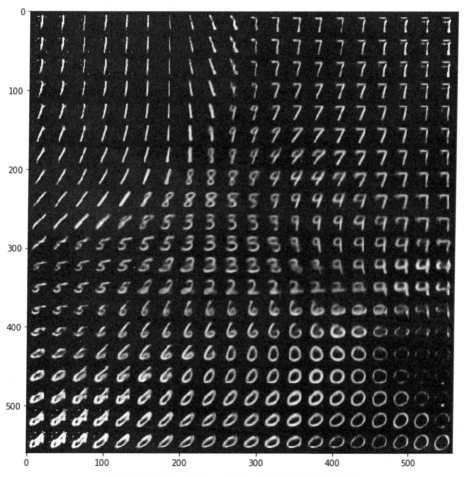

由自动编码机生成的一系列 MNIST 样字符示意图

解读分析

VAE 学习重构，同时产生新的图像。生成的图像依赖于潜在空间。生成的图像与训练它们的数据集具有相同的分布。

通过在 VariationalAutoencoder 类中定义一个变换函数来查看潜在空间中的数据：

```
def transform(self, X):
    """Transform data by mapping it into the latent space."""
    # Note: This maps to mean of distribution, we could alternatively sample from Gaussian distribution
    return self.sess.run(self.z_mean, feed_dict={self.x: X})
```

使用变换函数的 MNIST 数据集的潜在表示如下：

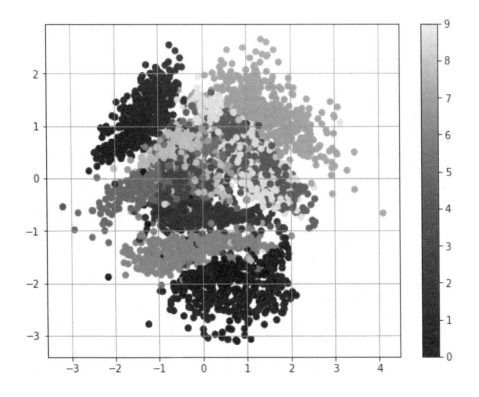

更多内容

生成的 VAE 图像取决于潜在的空间维度。使潜在空间尺寸减小的模糊在增加。5-d、10-d 和 20-d 潜在维度的重构图像分别如下。

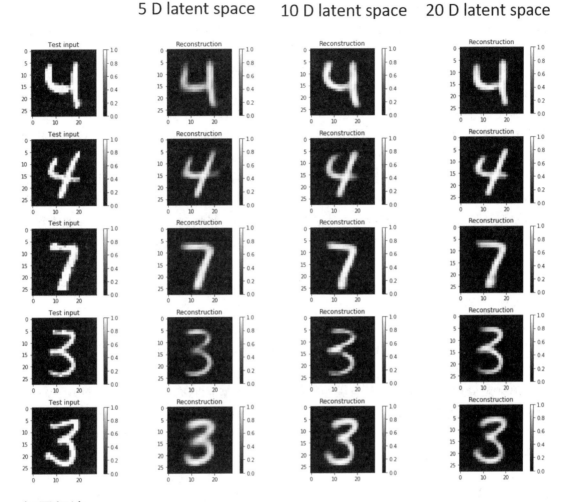

拓展阅读

- Kingma 和 Welling 的论文是这个领域的开创性论文。他们进行了完整的想法构建过程以及数学表述。对于任何对 VAE 感兴趣的人,这是必读的,参见 `https://arxiv.org/pdf/1312.6114.pdf`。
- 另一篇有趣的论文是 Carl Doersch 的变分编码机教程,参见 `https://arxiv.org/pdf/1606.05908.pdf`。
- GitHub 链接包含了 VAE 的另一个应用,来自 Kingma 和 Welling 论文的图像复制,参见 `https://github.com/hwalsuklee/tensorflow-mnist-VAE`。

11.6 学习使用胶囊网络击败 MNIST 前期的最新成果

胶囊(capsule)网络(或 CapsNet)是一个非常新颖的创新型深度学习网络。该技术于 2017 年 10 月底在 Sara Sabour、Nicholas Frost 和 Geoffrey Hinton（https://arxiv.org/abs/1710.09829）发表的一篇名为"Dynamic Routing Between Capsule"的开创性论文中被提出。Hinton 是深度学习的先驱之一，因此，整个深度学习社区很高兴看到 capsule 带来的进展。事实上，令人印象深刻的是 CapsNet 已经在 MNIST 分类中击败了最好的 CNN！

那么 CNN 有什么问题呢？ 在 CNN 中，每一层都以渐进级别的粒度识别图像。正如在多个方法中描述的那样，第一层很可能会识别直线或简单的曲线和边，而后面的层将开始识别更复杂的形状，如矩形和更复杂的形式，如人脸。

当前用于 CNN 的一个关键操作是池化。池化旨在建立位置不变性，并且通常在每个 CNN 层之后使用以使任何问题在计算上易于处理。但是，池化引入了一个重要的问题，因为它迫使我们失去所有的位置数据。这是很不利的。想象一张脸：包含两只眼睛、一张嘴和一个鼻子，而重要的是这些部位之间有空间关系（嘴巴在鼻子下方，而鼻子通常在眼睛的下面）。

的确如此，Hinton 说：

卷积神经网络中使用的池化操作是一个很大的错误，即使它运行得很好，也将是一场数据灾难。

从技术上讲，需要的不是位置不变性，而是等值性。等值是表明想要了解图像中的旋转或比例变化的一个特定术语，希望相应地调整网络。这样，图像中不同分量的空间定位就不会丢失。

那么胶囊网络有什么创新点呢？ 据作者说，大脑有被称为 capsule 的模块，每个 capsule 专门处理特定类型的信息。capsule 对于理解位置的概念、尺寸的概念、定向的概念、变形的概念、纹理等尤其有效。除此之外，作者推断大脑具有特别有效的机制，能将每条信息动态传输到 capsule。这被认为最适合用来处理特定类型的信息。

所以，CNN 和 CapsNet 之间的主要区别在于，CNN 不断添加用于创建深层网络的图层，而 CapsNet 将神经层嵌入另一个。一个 capsule 是一组神经元，它在网络中引入更多的结构，它产生一个矢量来表示图像中存在的一个实体。特别是，Hinton 使用活动矢量的长度来表示实体存在的概率和表示实例化参数的取向。当多个预测一致时，更高级别的 capsule 变得活跃。对于每个可能的父代，capsule 产生额外的预测矢量。

现在又有了一个创新：使用 capsule 的动态路由，不再使用原始的池化思想。较低级别的 capsule 更倾向于将其输出发送到较高级别的 capsule，而活动矢量与来自较低级别 capsule 的预测具有较高的标量乘积。具有最大标量预测矢量积的父代增加了 capsule 键。所有其他父代减少了它们的 capsule 键。

换句话说，这个想法是如果一个更高层次的 capsule 与更低层次的 capsule 达成了协议，那么它会要求发送更多的这种类型的信息。如果没有达成协议，它会要求发送更少的这种类型的信息。这种使用协议方法的动态路由优于当前的机制，如最大池化。根据 Hinton 所述，路由最终是解析图像的一种方式。确实，最大池化只关注了最大值，而动态路由通过低层和高层之间的协议有选择性地传递信息。

第三个区别是引入了一个新的非线性激活函数。CapsNet 并没有像在 CNN 中一样为每个图层添加一个压缩函数，而是将一个压缩函数添加到一个嵌套的图层集合中。非线性激活函数如下所示，它被称为压缩函数（等式 1）：

$$V_j = \frac{\|s_j\|^2}{1+\|s_j\|^2} \frac{s_j}{\|s_j\|} \qquad (1)$$

其中，v_j 是 capsule j 的矢量输出，s_j 是它总的输入。

在 Hinton 的开创性论文中提出的挤压函数

此外，Hinton 等人也表明：一种有差别训练的多层 capsule 系统在 MNIST 上实现了最新的性能，并且在识别高度重叠的数字方面比卷积网络好得多。

文章"Dynamic Routing Between Capsules"展示了简单的 CapsNet 架构：

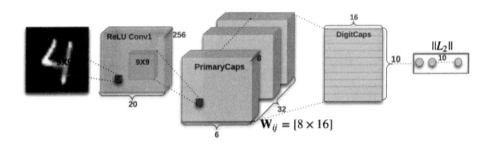

一个简单的 CapsNet 架构图

该架构非常浅显易懂，只有两个卷积层和一个全连接层。Conv1 有 256 个 9×9 卷积核，它的步长为 1，由 ReLU 激活。该层的作用是将像素强度转换为局部特征检测器的活动，然后将其作为 PrimaryCapsules 的输入。PrimaryCapsules 是一个 32 通道的卷积 capsule 层；每个主 capsule 包含 8 个卷积单元，带有一个 9×9 内核，步幅为 2。总之，PrimaryCapsules 有 [32, 6, 6] capsule 输出（每个输出是一个 8D 矢量），[6,6] 网格中的每个 capsule 共享彼此的权重。最后一层（DigitCaps）在每个数字类中有一个 16D capsule，这些 capsule 中的每一个都接收来自下面所有其他 capsule 的输入。路由只产生在两个连续的 capsule 层（如 PrimaryCapsules 和 DigitCaps）之间。

准备工作

相关代码可访问网页：https://github.com/debarko/CapsNet-Tensorflow，更多相关代码可访问：https://github.com/naturomics/CapsNet-Tensorflow.git。

具体做法

1. 根据 Apache 许可证从 GitHub 克隆代码：

```
git clone https://github.com/naturomics/CapsNet-Tensorflow.git
 $ cd CapsNet-Tensorflow
```

2. 下载 MNIST 并创建适当的结构：

```
mkdir -p data/mnist
wget -c -P data/mnist \\
http://yann.lecun.com/exdb/mnist/{train-images-idx3-ubyte.gz,train-labels-idx1-ubyte.gz,t10k-images-idx3-ubyte.gz,t10k-labels-idx1-ubyte.gz}
gunzip data/mnist/*.gz
```

3. 开始训练过程：

```
python main.py
```

4. 看一下用于定义 capsule 的代码。每个 capsule 以一个 4D 张量作为输入并返回一个 4D 张量。可以将 capsule 定义为一个完全连接的网络（DigiCap）或一个卷积型网络（Primary capsule）。需要注意的是，Primary 是 ConvNet 的集合，之后非线性压缩函数被应用。Primary capsule 将通过动态路由与 DigiCap 进行通信：

```python
# capsLayer.py
#
import numpy as np
import tensorflow as tf
from config import cfg
epsilon = 1e-9
class CapsLayer(object):
''' Capsule layer.
Args:
input: A 4-D tensor.
num_outputs: the number of capsule in this layer.
vec_len: integer, the length of the output vector of a capsule.
layer_type: string, one of 'FC' or "CONV", the type of this layer,
fully connected or convolution, for the future expansion capability
with_routing: boolean, this capsule is routing with the
lower-level layer capsule.
Returns:
A 4-D tensor.
'''
def __init__(self, num_outputs, vec_len, with_routing=True,
layer_type='FC'):
  self.num_outputs = num_outputs
  self.vec_len = vec_len
  self.with_routing = with_routing
  self.layer_type = layer_type

def __call__(self, input, kernel_size=None, stride=None):
'''
The parameters 'kernel_size' and 'stride' will be used while 'layer_type'
equal 'CONV'
'''
  if self.layer_type == 'CONV':
```

```
        self.kernel_size = kernel_size
        self.stride = stride

        if not self.with_routing:
        # the PrimaryCaps layer, a convolutional layer
        # input: [batch_size, 20, 20, 256]
            assert input.get_shape() == [cfg.batch_size, 20, 20, 256]
            capsules = []
            for i in range(self.vec_len):
              # each capsule i: [batch_size, 6, 6, 32]
              with tf.variable_scope('ConvUnit_' + str(i)):
                caps_i = tf.contrib.layers.conv2d(input,    self.num_outputs,
self.kernel_size, self.stride,
padding="VALID")
                caps_i = tf.reshape(caps_i, shape=(cfg.batch_size, -1, 1, 1))
                capsules.append(caps_i)
            assert capsules[0].get_shape() == [cfg.batch_size, 1152, 1, 1]
# [batch_size, 1152, 8, 1]
            capsules = tf.concat(capsules, axis=2)
            capsules = squash(capsules)
            assert capsules.get_shape() == [cfg.batch_size, 1152, 8, 1]
            return(capsules)

    if self.layer_type == 'FC':
      if self.with_routing:
        # the DigitCaps layer, a fully connected layer
        # Reshape the input into [batch_size, 1152, 1, 8, 1]
        self.input = tf.reshape(input, shape=(cfg.batch_size, -1, 1,
input.shape[-2].value, 1))
        with tf.variable_scope('routing'):
          # b_IJ: [1, num_caps_l, num_caps_l_plus_1, 1, 1]
          b_IJ = tf.constant(np.zeros([1, input.shape[1].value,
self.num_outputs, 1, 1], dtype=np.float32))
          capsules = routing(self.input, b_IJ)
          capsules = tf.squeeze(capsules, axis=1)
      return(capsules)
```

5. 路由算法在论文 Dynamic Routing Between Capsules 中进行了描述，同时本文的相关部分与方程 2 和方程 3 的定义一起进行了解释。路由算法的目标是将信息从低层 capsule 传递到高层 capsule 并识别哪里有协议。通过简单地使用在上层中的每个 capsule j 的当前输出 v_j 和由 capsule i 做出的预测 $\hat{u}_{j|i}$ 之间的标量积来计算该协议：

> 对于除了第一层之外的所有 capsule，一个 capsule s_j 的总输入是来自下层 capsule 的所有 "预测矢量" $\hat{u}_{j|i}$ 的加权和，预测矢量由下层 capsule 的输出与权重矩阵 W_{ij} 的乘积获得。
>
> $$s_j = \sum_i c_{ij}\hat{u}_{j|i}, \qquad u_{j|i} = W_{ij}u_i \qquad (2)$$
>
> 其中，c_{ij} 为取决于迭代动态路由过程的耦合系数。capsules i 与所有上层的 capsule 的耦合系数之和为 1，它取决于路由 softmax。路由 softmax 的初始分对数 b_{ij} 是 capsule i 与 capsule j 的先验概率的对数。

应该与capsule j 耦合：

$$c_{ij} = \frac{\exp(b_{ij})}{\sum_k \exp(b_{ik})} \quad (3)$$

先验概率的对数可以同时被分别训练成所有其他的权重。它们取决于位置和两个capsule的类型而不是取决于当前的输入图像。初始的耦合系数通过衡量每个capsule j 在上一层级的当前输出 v_j 和由 capsule i 做出的预测 $\hat{u}_{j|i}$ 之间的协议而被循环优化。

在卷积 capsule 层中，capsule 的每个单元都是一个卷积单元。因此，每个capsule 都会输出一系列矢量而不是只有一个矢量输出。

协议就是一个简单的数量积 $a_{ij}=v_j*\hat{u}_{j|i}$。在计算所有使 capsule i 与更高层 capsule 相连的耦合系数的新取值之前，该协议被训练为最大似然的对数并代入初始分对数 b_{ij}。

Procedure 1 路由算法

1: **procedure** ROUTING($\hat{u}_{j|i}, r, l$)
2:     for all capsule i in layer l and capsule j in layer $(l+1)$: $b_{ij} \leftarrow 0$.
3:     **for** r iterations **do**
4:         for all capsule i in layer l: $c_i \leftarrow \text{softmax}(b_i)$ ▷ softmax computes Eq. 3
5:         for all capsule j in layer $(l+1)$: $s_j \leftarrow \sum_i c_{ij} \hat{u}_{j|i}$
6:         for all capsule j in layer $(l+1)$: $v_j \leftarrow \text{squash}(s_j)$ ▷ squash computes Eq. 1
7:         for all capsule i in layer l and capsule j in layer $(l+1)$: $b_{ij} \leftarrow b_{ij} + \hat{u}_{j|i} \cdot v_j$
    return v_j

以下方法在前面的图像中实现了步骤 1 中描述的步骤。值得注意的是，输入是来自第 1 层中的 1152 个 capsule 的 4D 张量。输出是在 $l+1$ 层中 capsule j 的矢量输出 "v_j" 的形状张量 [`batch_size, 1, length(v_j) = 16, 1`]

```
def routing(input, b_IJ):
    ''' The routing algorithm.
    Args:
    input: A Tensor with [batch_size, num_caps_l=1152, 1, length(u_i)=8, 1]
    shape, num_caps_l meaning the number of capsule in the layer l.
    Returns:
    A Tensor of shape [batch_size, num_caps_l_plus_1, length(v_j)=16, 1]
    representing the vector output `v_j` in the layer l+1
    Notes:
    u_i represents the vector output of capsule i in the layer l, and
    v_j the vector output of capsule j in the layer l+1.
    '''
    # W: [num_caps_j, num_caps_i, len_u_i, len_v_j]
    W = tf.get_variable('Weight', shape=(1, 1152, 10, 8, 16), dtype=tf.float32,
        initializer=tf.random_normal_initializer(stddev=cfg.stddev))
    # Eq.2, calc u_hat
    # do tiling for input and W before matmul
    # input => [batch_size, 1152, 10, 8, 1]
    # W => [batch_size, 1152, 10, 8, 16]
    input = tf.tile(input, [1, 1, 10, 1, 1])
    W = tf.tile(W, [cfg.batch_size, 1, 1, 1, 1])
```

```
    assert input.get_shape() == [cfg.batch_size, 1152, 10, 8, 1]
    # in last 2 dims:
    # [8, 16].T x [8, 1] => [16, 1] => [batch_size, 1152, 10, 16, 1]
    u_hat = tf.matmul(W, input, transpose_a=True)
    assert u_hat.get_shape() == [cfg.batch_size, 1152, 10, 16, 1]
    # line 3,for r iterations do
    for r_iter in range(cfg.iter_routing):
        with tf.variable_scope('iter_' + str(r_iter)):
            # line 4:
            # => [1, 1152, 10, 1, 1]
            c_IJ = tf.nn.softmax(b_IJ, dim=2)
            c_IJ = tf.tile(c_IJ, [cfg.batch_size, 1, 1, 1, 1])
            assert c_IJ.get_shape() == [cfg.batch_size, 1152, 10, 1, 1]
            # line 5:
            # weighting u_hat with c_IJ, element-wise in the last two dims
            # => [batch_size, 1152, 10, 16, 1]
            s_J = tf.multiply(c_IJ, u_hat)
            # then sum in the second dim, resulting in [batch_size, 1, 10, 16, 1]
            s_J = tf.reduce_sum(s_J, axis=1, keep_dims=True)
            assert s_J.get_shape() == [cfg.batch_size, 1, 10, 16, 16
            # line 6:
            # squash using Eq.1,
            v_J = squash(s_J)
            assert v_J.get_shape() == [cfg.batch_size, 1, 10, 16, 1]
            # line 7:
            # reshape & tile v_j from [batch_size ,1, 10, 16, 1] to [batch_size, 10, 1152, 16, 1]
            # then matmul in the last tow dim: [16, 1].T x [16, 1] => [1, 1], reduce mean in the
            # batch_size dim, resulting in [1, 1152, 10, 1, 1]
            v_J_tiled = tf.tile(v_J, [1, 1152, 1, 1, 1])
            u_produce_v = tf.matmul(u_hat, v_J_tiled, transpose_a=True)
            assert u_produce_v.get_shape() == [cfg.batch_size, 1152, 10, 1, 1]
            b_IJ += tf.reduce_sum(u_produce_v, axis=0, keep_dims=True)
    return(v_J)
```

6. 回顾一下非线性激活压缩函数。它的输入是一个形状为 `[batch_size, num_caps, vec_len, 1]` 的4D矢量，输出是一个与矢量形状相同但在第三维和第四维上压缩的4D张量。给定一个矢量输入，目标是计算公式1中表示的值，如下所示：

$$V_j = \frac{\|s_j\|^2}{1+\|s_j\|^2} \frac{s_j}{\|s_j\|} \quad (1)$$

其中，v_j 是 capsule j 的矢量输出，s_j 是总的输入。

```
def squash(vector):
'''Squashing function corresponding to Eq. 1
Args:
vector: A 5-D tensor with shape [batch_size, 1, num_caps, vec_len, 1],
Returns:
A 5-D tensor with the same shape as vector but squashed in 4rd and 5th
dimensions.
'''
```

```
    vec_squared_norm = tf.reduce_sum(tf.square(vector), -2, keep_dims=True)
    scalar_factor = vec_squared_norm / (1 + vec_squared_norm) /
tf.sqrt(vec_squared_norm + epsilon)
    vec_squashed = scalar_factor * vector # element-wise
return(vec_squashed)
```

7. 在前面的步骤中，已经定义了 capsule 是什么，capsule 之间的动态路由算法以及非线性压缩函数。现在我们可以定义合适的 CapsNet。损失函数是为训练而建立的，选择 Adam 优化器。`build_arch(...)` 方法定义了下图所示的 CapsNet:

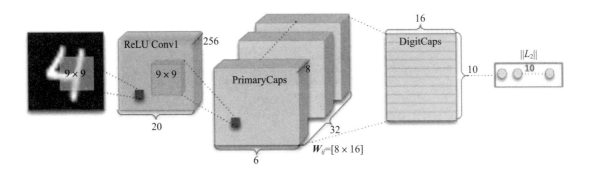

值得注意的是，本文描述了一种重构技术作为正则化方法。来自论文：

使用额外的重构损失来鼓励数字 capsule 对输入数字的实例化参数进行编码。在训练期间，掩盖了除了正确数字 capsule 的活动矢量之外的所有数据。

接着使用这个活动向量来重构。

数字 capsule 的输出被反馈到由三个完全连接层构成的解码器中，解码器对如下图所述的像素强度进行建模。最小化逻辑单元的输出与像素强度之间的差的平方和。把这个重构损失降低了 0.0005，这样它就不会在训练期间占主导地位。下面给出的 `build_arch(..)` 方法也被用于创建解码器：

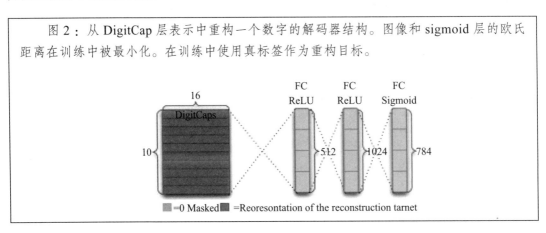

图 2：从 DigitCap 层表示中重构一个数字的解码器结构。图像和 sigmoid 层的欧氏距离在训练中被最小化。在训练中使用真标签作为重构目标。

```python
#capsNet.py
#
import tensorflow as tf
from config import cfg
from utils import get_batch_data
from capsLayer import CapsLayer
epsilon = 1e-9

class CapsNet(object):
    def __init__(self, is_training=True):
        self.graph = tf.Graph()
        with self.graph.as_default():
            if is_training:
                self.X, self.labels = get_batch_data()
                self.Y = tf.one_hot(self.labels, depth=10, axis=1, dtype=tf.float32)
                self.build_arch()
                self.loss()
                self._summary()

                # t_vars = tf.trainable_variables()
                self.global_step = tf.Variable(0, name='global_step', trainable=False)
                self.optimizer = tf.train.AdamOptimizer()
                self.train_op =    self.optimizer.minimize(self.total_loss, global_step=self.global_step) # var_list=t_vars)

            elif cfg.mask_with_y:
                self.X = tf.placeholder(tf.float32,
                    shape=(cfg.batch_size, 28, 28, 1))
                self.Y = tf.placeholder(tf.float32, shape=(cfg.batch_size, 10, 1))
                self.build_arch()
            else:
                self.X = tf.placeholder(tf.float32,
                    shape=(cfg.batch_size, 28, 28, 1))
                self.build_arch()

            tf.logging.info('Setting up the main structure')

    def build_arch(self):
      with tf.variable_scope('Conv1_layer'):
        # Conv1, [batch_size, 20, 20, 256]
        conv1 = tf.contrib.layers.conv2d(self.X, num_outputs=256,
            kernel_size=9, stride=1,
            padding='VALID')
        assert conv1.get_shape() == [cfg.batch_size, 20, 20, 256]# Primary Capsules layer, return [batch_size, 1152, 8, 1]

      with tf.variable_scope('PrimaryCaps_layer'):
        primaryCaps = CapsLayer(num_outputs=32, vec_len=8, with_routing=False, layer_type='CONV')
        caps1 = primaryCaps(conv1, kernel_size=9, stride=2)
        assert caps1.get_shape() == [cfg.batch_size, 1152, 8, 1]

      # DigitCaps layer, return [batch_size, 10, 16, 1]
      with tf.variable_scope('DigitCaps_layer'):
        digitCaps = CapsLayer(num_outputs=10, vec_len=16,    with_routing=True, layer_type='FC')
```

```python
    self.caps2 = digitCaps(caps1)

# Decoder structure in Fig. 2
# 1. Do masking, how:
with tf.variable_scope('Masking'):
    # a). calc ||v_c||, then do softmax(||v_c||)
    # [batch_size, 10, 16, 1] => [batch_size, 10, 1, 1]
    self.v_length = tf.sqrt(tf.reduce_sum(tf.square(self.caps2),
axis=2, keep_dims=True) + epsilon)
    self.softmax_v = tf.nn.softmax(self.v_length, dim=1)
    assert self.softmax_v.get_shape() == [cfg.batch_size, 10, 1, 1]

    # b). pick out the index of max softmax val of the 10 caps
    # [batch_size, 10, 1, 1] => [batch_size] (index)
    self.argmax_idx = tf.to_int32(tf.argmax(self.softmax_v, axis=1))
    assert self.argmax_idx.get_shape() == [cfg.batch_size, 1, 1]
    self.argmax_idx = tf.reshape(self.argmax_idx, shape=(cfg.batch_size, ))

    # Method 1.
    if not cfg.mask_with_y:
        # c). indexing
        # It's not easy to understand the indexing process with   argmax_idx
        # as we are 3-dim animal
        masked_v = []
        for batch_size in range(cfg.batch_size):
            v = self.caps2[batch_size][self.argmax_idx[batch_size], :]
            masked_v.append(tf.reshape(v, shape=(1, 1, 16, 1)))
            self.masked_v = tf.concat(masked_v, axis=0)
            assert self.masked_v.get_shape() == [cfg.batch_size, 1, 16, 1]

    # Method 2. masking with true label, default mode
    else:
        self.masked_v = tf.matmul(tf.squeeze(self.caps2), tf.reshape(self.Y,
(-1, 10, 1)), transpose_a=True)
        self.v_length = tf.sqrt(tf.reduce_sum(tf.square(self.caps2), axis=2,
keep_dims=True) + epsilon)

# 2. Reconstruct the MNIST images with 3 FC layers
# [batch_size, 1, 16, 1] => [batch_size, 16] => [batch_size, 512]
with tf.variable_scope('Decoder'):
    vector_j = tf.reshape(self.masked_v, shape=(cfg.batch_size, -1))
    fc1 = tf.contrib.layers.fully_connected(vector_j, num_outputs=512)
    assert fc1.get_shape() == [cfg.batch_size, 512]
    fc2 = tf.contrib.layers.fully_connected(fc1, num_outputs=1024)
    assert fc2.get_shape() == [cfg.batch_size, 1024]
    self.decoded = tf.contrib.layers.fully_connected(fc2, num_outputs=784,
activation_fn=tf.sigmoid)
```

8. 边际损失函数是本文定义中的另一个重要部分。在以下文章的片段引用（等式 4）中进行了解释，并在 `loss(..)` 方法中执行，该方法包括三项损失：差额损失、重构损失和总损失：

> 使用实例化向量的长度来表示一个 capsule 的实体存在概率，因此希望高等级 capsule 的数字类 k 有一个长的实例化向量，当且仅当该数字在图像中存在时可实现。为了满足多数字，对于每个数字 capsule k，使用一个分离差额损失 L_k:

$$L_c = T_c \max\left(0, m^+ - \|v_c\|\right)^2 + \lambda(1-T_c)\max\left(0, \|v_c\| - m^-\right)^2 \qquad (4)$$

其中，当且仅当类别 c 的一个数字存在，m⁺=0.9 并且 m⁻=0.1 时，T_c=1。缺席数字类的损失的 λ 下权重阻止初始训练减小所有数字 capsule 的活动向量的长度。选取 λ=0.5。总损失为所有数字 capsule 的简单求和。

```python
def loss(self):
    # 1. The margin loss
    # [batch_size, 10, 1, 1]
    # max_l = max(0, m_plus-||v_c||)^2
    max_l = tf.square(tf.maximum(0., cfg.m_plus - self.v_length))
    # max_r = max(0, ||v_c||-m_minus)^2
    max_r = tf.square(tf.maximum(0., self.v_length - cfg.m_minus))
    assert max_l.get_shape() == [cfg.batch_size, 10, 1, 1]
    # reshape: [batch_size, 10, 1, 1] => [batch_size, 10]
    max_l = tf.reshape(max_l, shape=(cfg.batch_size, -1))
    max_r = tf.reshape(max_r, shape=(cfg.batch_size, -1))
    # calc T_c: [batch_size, 10]
    T_c = self.Y
    # [batch_size, 10], element-wise multiply
    L_c = T_c * max_l + cfg.lambda_val * (1 - T_c) * max_r

    self.margin_loss = tf.reduce_mean(tf.reduce_sum(L_c, axis=1))
    # 2. The reconstruction loss
    orgin = tf.reshape(self.X, shape=(cfg.batch_size, -1))
    squared = tf.square(self.decoded - orgin)
    self.reconstruction_err = tf.reduce_mean(squared)

    # 3. Total loss
    # The paper uses sum of squared error as reconstruction  error, but we
    # have used reduce_mean in `# 2 The reconstruction loss` to calculate
    # mean squared error. In order to keep in line with the paper,the
    # regularization scale should be 0.0005*784=0.392
    self.total_loss = self.margin_loss + cfg.regularization_scale * self.reconstruction_err
```

9. 此外，很容易定义一个 _summary（...）方法来报告损失和精确度：

```python
#Summary
def _summary(self):
    train_summary = []
    train_summary.append(tf.summary.scalar('train/margin_loss', self.margin_loss))train_summary.append(tf.summary.scalar('train/reconstruction_loss', self.reconstruction_err))
    train_summary.append(tf.summary.scalar('train/total_loss', self.total_loss))
    recon_img = tf.reshape(self.decoded, shape=(cfg.batch_size, 28, 28, 1))
    train_summary.append(tf.summary.image('reconstruction_img', recon_img))
    correct_prediction = tf.equal(tf.to_int32(self.labels), self.argmax_idx)
    self.batch_accuracy = tf.reduce_sum(tf.cast(correct_prediction, tf.float32))
    self.test_acc = tf.placeholder_with_default(tf.constant(0.), shape=[])
    test_summary = []
```

```
test_summary.append(tf.summary.scalar('test/accuracy', self.test_acc))
self.train_summary = tf.summary.merge(train_summary)
self.test_summary = tf.summary.merge(test_summary)
```

解读分析

CapsNet 与最先进的深度学习网络是不同的，它使用浅层网络，让 capsule 层嵌套在其他层中，而不需要增加更多的层和让网络深层次化。每个 capsule 被用于检测图像中的特定实体，一个动态的路由机制将检测到的实体发送给父层。对于 CNN 来说，要从不同角度识别一个物体，你必须从很多不同的方面来考虑成千上万的图像。Hinton 认为，这些层次中的冗余将会使 capsule 网络能够从多个角度和在不同情况下识别物体，相比 CNN，使用的数据更少。

检验一下 Tensorboard 所示的网络：

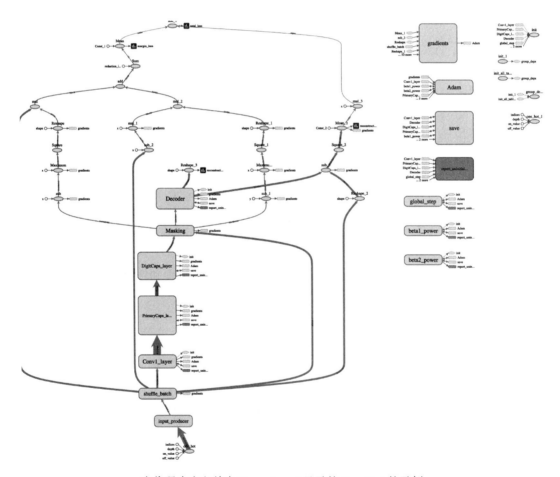

在代码中定义并由 Tensorboard 显示的 CapsNet 的示例

令人印象深刻的结果如下图所示,这些图表取自公开发表的论文。CapsNet 在以前只能在深度网络中实现三层网络上的低测试误差(0.25%)。基线是具有 3 个卷积层的 CNN 具有 256,256-128 的通道。每个基线有 5×5 个内核,步幅为 1。最终的卷积层之后是两个大小为 328 和 192 的完全连接层。最后的全连接层连接到具有交叉熵损失的 10 类 softmax 层:

图 3:3 个路由迭代的 CapsNet 的样本 MNIST 测试重构。(l,p,r) 分别表示标签、预测和重构目标。最右边的两列表示一个失败案例的两个重构,它解释了模型是如何混淆在图像中的 5 和 3 的。其他的列都是由正确的分类获得的,它表明模型选取了细节而平滑了噪声。

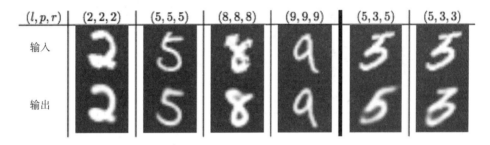

表 1:CapsNet 分类测试精度。MNIST 均值和标准差结果由 3 次试验获得。

方法	路由	是否重构	MINIST(%)	MultiMNIST(%)
Baseline	-	-	0.39	8
CapsNet	1	否	0.34 ± 0.032	-
CapsNet	1	是	0.29 ± 0.011	7
CapsNet	3	否	0.35 ± 0.036	-
CapsNet	3	是	0.25 ± 0.005	5

看看差额损失、重构损失和总损失的减少情况:

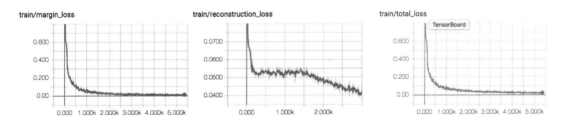

看一下精确度的提高情况。经过 500 次迭代达到 92%,在 3500 次迭代中达到 98.46%。

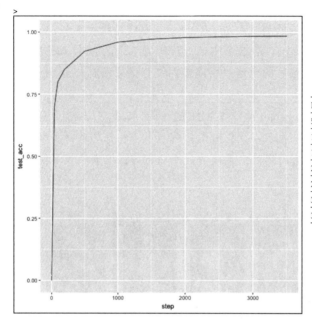

CapsNet 的精确度提高的例子

拓展阅读

如果要在其他数据集（如 CIFAR）或更一般的图像集合上实现同样出色的结果，CapsNet 在 MNIST 上应用得非常好，但是在理解方面还有很多的研究工作要做。如果你有兴趣了解更多，请参阅以下内容：

- 谷歌的人工智能向导揭示了神经网络的新局面：`https://www.wired.com/story/googles-ai-wizard-unveils-a-new-twist-on-neural-networks/`。
- 谷歌研究人员有一个新的选择替代传统的神经网络：`https://www.technologyreview.com/the-download/609297/google-researchers-have-a-new-alternative-to-traditional-neural-networks/`。
- Keras-CapsNet 是一个 Keras 应用，可参阅：`https://github.com/XifengGuo/CapsNet-Keras`。
- Geoffrey Hinton 谈及了卷积神经网络存在的问题，参见 `https://www.youtube.com/watch?v=rTawFwUvnLEfeature=youtu.be`。

CHAPTER 12

第 12 章

分布式 TensorFlow 和云深度学习

本章将讨论如何使用分布式 TensorFlow 和云深度学习。

12.1 引言

每次 TensorFlow 运算都被描述成计算图的形式，它允许结构和运算操作配置所具备的自由度能够被分配到各个分布式节点上。计算图可以分成多个子图，分配给服务器集群中的不同节点。

强烈推荐读者阅读论文"Large Scale Distributed Deep Networks"（Jeffrey Dean、Greg S. Corrado、Rajat Monga、Kai Chen、Matthieu Devin、Quoc V. Le、Mark Z. Mao、Marc'Aurelio Ranzato、Andrew Senior、Paul Tucker、Ke Yang 和 Andrew Y. Ng. NIPS, 2012, `https://research.google.com/archive/large_deep_networks_nips2012.html`）

本文的一个重要成果是证明了分布式**随机梯度下降算法**（SDG）可以运行，在该算法中，有多个节点在数据分片上并行工作，通过向参数服务器发送更新来异步独立更新梯度。论文摘要引用如下：

实验揭示了一些关于大规模非凸优化的令人惊喜的结果。首先，很少应用于非凸问题的异步 *SGD* 在训练深度网络方面效果很好，特别是在结合 *Adagrad* 自适应学习率时。

这篇论文本身的一个照片可以很好地解释这一点：

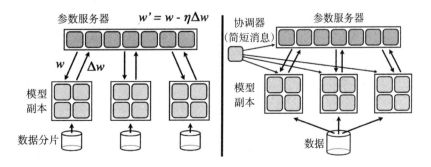

分布式梯度下降的示例（来自 https://research.google.com/archive/large_deep_networks_nips2012.html）

另一个值得阅读的文档是 TensorFlow 白皮书：《Large-Scale Machine Learning on Heterogeneous Distributed Systems》（Martín Abadi 等人，2015 年 11 月，http://download.tensorflow.org/paper/whitepaper2015.pdf）。

考虑其中包含的一些示例，可以在下面的图片中看见，左侧显示的是 TensorFlow 代码片段，右侧显示的是对应的图表：

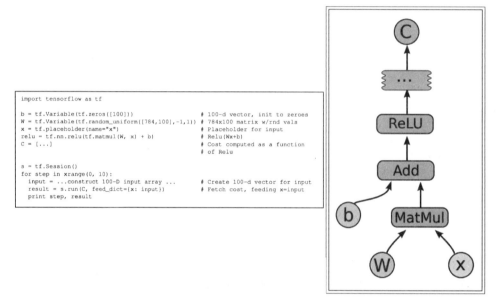

TensorFlow 计算图示例（来自 http://download.tensorflow.org/paper/whitepaper2015.pdf）

通过进行本地计算以及在必要时轻松添加计算图远程通信节点，计算图可以被切分覆盖多个节点，前面提到的论文中的图可以很好地解释这个问题：

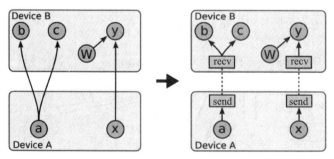

图 4：插入接收/发送节点前后的对比图

分布式 TensorFlow 计算图（来自 http://download.tensorflow.org/paper/whitepaper2015.pdf）

梯度下降和所有主要的优化器算法可以以集中方式（下图左侧）或分布式方式（下图右侧）进行计算，后者包括一个 master 进程，它与多个提供 GPU 和 CPU 的 worker 相连：

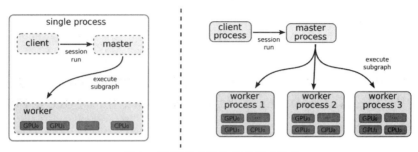

图 3：单机模式和分布式模式结构

单机模式和分布式模式结构示意图（引自 http://download.tensorflow.org/paper/whitepaper2015.pdf）

分布式计算可以是同步的（所有 worker 同时更新数据块上的梯度）或异步的（更新不会同时发生），后者通常允许更高的可扩展性，更大的图计算在最优解的收敛方面表现得更好。以下照片同样来自 TensorFlow 白皮书，如果想了解更多，我强烈鼓励有兴趣的读者看看这篇论文：

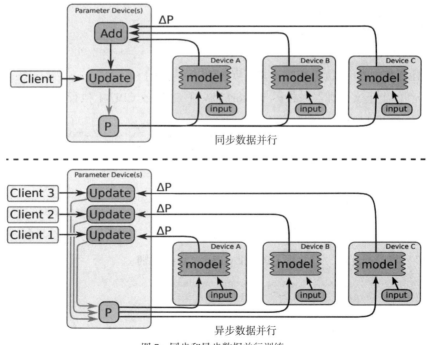

图 7：同步和异步数据并行训练。

同步和异步数据并行训练的示例

12.2 在 GPU 上使用 TensorFlow

在这个案例中将学习如何在 GPU 中使用 TensorFlow：所执行的运算操作是在 CPU 或 GPU 上进行简单的矩阵乘法。

准备工作

第一步是安装一个支持 GPU 的 TensorFlow 版本。官方的 TensorFlow 安装说明能够提供良好的开始：`https://www.tensorflow.org/`。请记住，对 GPU 的环境支持需要通过 CUDA 或 CuDNN。

具体做法

1. 导入几个模块。

```
import sys
import numpy as np
import tensorflow as tf
from datetime import datetime
```

2. 从命令行获得想要使用的处理器类型（GPU 或 CPU）。

```
device_name = sys.argv[1] # Choose device from cmd line. Options: gpu or cpu
shape = (int(sys.argv[2]), int(sys.argv[2]))
if device_name == "gpu":
  device_name = "/gpu:0"
else:
  device_name = "/cpu:0"
```

3. 在 GPU 或 CPU 上执行矩阵乘法。关键指令是使用 `tf.device(device_name)`，它创建一个新的上下文管理器来告诉 TensorFlow 在 GPU 或 CPU 上执行这些操作。

```
with tf.device(device_name):
  random_matrix = tf.random_uniform(shape=shape, minval=0, maxval=1)
  dot_operation = tf.matmul(random_matrix, tf.transpose(random_matrix))
  sum_operation = tf.reduce_sum(dot_operation)

startTime = datetime.now()
with tf.Session(config=tf.ConfigProto(log_device_placement=True)) as session:
  result = session.run(sum_operation)
  print(result)
```

4. 打印一些调试计时，以验证 CPU 和 GPU 之间的区别。

```
print("Shape:", shape, "Device:", device_name)
print("Time taken:", datetime.now() - startTime)
```

解读分析

此案例说明了如何将 TensorFlow 计算分配给 CPU 或 GPU。代码非常简单，并将作为下一个案例的基础。

12.3 玩转分布式 TensorFlow：多个 GPU 和一个 CPU

展示一个数据并行的例子，其中数据被切分到多个 GPU 上。

准备工作

这个案例灵感来自 Neil Tenenholtz 的一篇很好的博客文章，网址为 https://clindatsci.com/blog/2017/5/31/distributed-tensorflow。

具体做法

1. 考虑在单个 GPU 上运行矩阵乘法的这段代码。

```
# single GPU (baseline)
import tensorflow as tf
# place the initial data on the cpu
with tf.device('/cpu:0'):
  input_data = tf.Variable([[1., 2., 3.],
    [4., 5., 6.],
    [7., 8., 9.],
    [10., 11., 12.]])
b = tf.Variable([[1.], [1.], [2.]])

# compute the result on the 0th gpu
with tf.device('/gpu:0'):
  output = tf.matmul(input_data, b)
# create a session and run
with tf.Session() as sess:
  sess.run(tf.global_variables_initializer())
  print sess.run(output)
```

2. 通过图内拷贝模式中对代码进行了分割，如以下两个不同 GPU 的代码片段所示。请注意，CPU 充当 master 节点，分配计算图，并收集最终结果。

```
# in-graph replication
import tensorflow as tf
num_gpus = 2
# place the initial data on the cpu
with tf.device('/cpu:0'):
  input_data = tf.Variable([[1., 2., 3.],
    [4., 5., 6.],
    [7., 8., 9.],
    [10., 11., 12.]])
  b = tf.Variable([[1.], [1.], [2.]])

# split the data into chunks for each gpu
inputs = tf.split(input_data, num_gpus)
outputs = []
```

```
# loop over available gpus and pass input data
for i in range(num_gpus):
  with tf.device('/gpu:'+str(i)):
    outputs.append(tf.matmul(inputs[i], b))

# merge the results of the devices
with tf.device('/cpu:0'):
  output = tf.concat(outputs, axis=0)

# create a session and run
with tf.Session() as sess:
  sess.run(tf.global_variables_initializer())
  print sess.run(output)
```

解读分析

这是一个非常简单的案例，其中计算图由作为 master 的 CPU 分成两部分，并分配给作为 worker 的两个 GPU，计算结果由 CPU 收集。

12.4 玩转分布式 TensorFlow：多服务器

在这个案例中学习如何将 TensorFlow 计算分配到多个服务器中，其中需假设 worker 和参数服务器的代码是相同的，因此每个计算节点的作用就是传递命令行参数。

准备工作

这个案例的灵感来自 Neil Tenenholtz 的一篇很好的博客文章，可以在线获取，网址为：https://clindatsci.com/blog/2017/5/31/distributed-tensorflow。

具体做法

1. 考虑以下这段代码，所采用的集群架构包括一个在 192.168.1.1:1111 上运行的 master 和两个分别在 192.168.1.2:1111 和 192.168.1.3:1111 上运行的 worker。

```
import sys
import tensorflow as tf

# specify the cluster's architecture
cluster = tf.train.ClusterSpec({'ps': ['192.168.1.1:1111'],
    'worker': ['192.168.1.2:1111',
    '192.168.1.3:1111']
})
```

2. 请注意，代码被复制到多台机器上，因此知道当前执行节点的角色是很重要的，从命令行中能得到这些信息。机器可以是一个 worker 或一个参数服务器。

```
# parse command-line to specify machine
job_type = sys.argv[1]  # job type: "worker" or "ps"
task_idx = sys.argv[2]  # index job in the worker or ps list
# as defined in the ClusterSpec
```

3. 给定一个集群，运行训练服务，每个计算节点都能够有一个角色（worker 或者 ps）和一个 id。

```
# create TensorFlow Server. This is how the machines communicate.
server = tf.train.Server(cluster, job_name=job_type, task_index=task_idx)
```

4. 根据计算节点的角色，计算是不同的：

- 如果角色是参数服务器，则条件是加入服务。请注意，在这种情况下，没有要执行的代码，因为 worker 将不断推送更新，而参数服务器必须执行的唯一操作就是等待。
- 相反，worker 代码将在集群内的特定设备上执行。这部分代码与第一次构建模型然后进行本地训练的单个机器上执行的代码类似。请注意，TensoFlow 轻松地完成了所有的工作分配以及更新结果的收集，并且提供了非常方便的 `tf.train.replica_device_setter`，可自动将运算操作分配给设备。

```
# parameter server is updated by remote clients.
# will not proceed beyond this if statement.
if job_type == 'ps':
  server.join()
else:
  # workers only
  with tf.device(tf.train.replica_device_setter(
    worker_device='/job:worker/task:'+task_idx,
    cluster=cluster)):
# build your model here as if you only were using a single machine

with tf.Session(server.target):
  # train your model here
```

解读分析

在这个案例中，已经看到了如何创建一个具有多个计算节点的集群。节点既可以扮演参数服务器的角色，也可以扮演 worker 的角色。

在这两种情况下，执行的代码是相同的，但是根据从命令行收集的参数，代码的执行则是不同的。参数服务器只需要等待 worker 发送更新。请注意，`tf.train.replica_device_setter(..)` 的作用是自动将运算操作分配给可用设备，而 `tf.train.ClusterSpec(..)` 用于集群设置。

拓展阅读

一个 MNIST 分布式训练的示例可在线获得，网址为 https://github.com/ischlag/distributed-tensorflow-example/blob/master/example.py。

另外请注意，出于效率原因，你可以使用多个参数服务器。利用参数，服务器可以提供更好的网络利用率，并允许将模型扩展到更多的并行机器。感兴趣的读者可以参考 https://www.tensorflow.org/deploy/distributed。

12.5 训练分布式 TensorFlow MNIST 分类器

此案例用于以分布式方式训练完整的 MNIST 分类器。该案例受到下面博客文章的启发：`http://ischlag.github.io/2016/06/12/async-distributed-tensorflow/`，运行在 TensorFlow 1.2 上的代码可以在网址 `https://github.com/ischlag/distributed-tensorflow-example` 上找到。

准备工作

这个案例是基于上一个的，所以按顺序阅读可能会很方便。

具体做法

1. 导入一些标准模块并定义运行计算的 TensorFlow 集群，然后为指定任务启动服务。

```
import tensorflow as tf
import sys
import time
# cluster specification
parameter_servers = ["pc-01:2222"]
workers = [ "pc-02:2222",
"pc-03:2222",
"pc-04:2222"]
cluster = tf.train.ClusterSpec({"ps":parameter_servers, "worker":workers})
# input flags
tf.app.flags.DEFINE_string("job_name", "", "Either 'ps' or 'worker'")
tf.app.flags.DEFINE_integer("task_index", 0, "Index of task within the job")FLAGS = tf.app.flags.FLAGS
# start a server for a specific task
server = tf.train.Server(
  cluster,
  job_name=FLAGS.job_name,
  task_index=FLAGS.task_index)
```

2. 读取 MNIST 数据并定义用于训练的超参数。

```
# config
batch_size = 100
learning_rate = 0.0005
training_epochs = 20
logs_path = "/tmp/mnist/1"
# load mnist data set
from tensorflow.examples.tutorials.mnist import input_data
mnist = input_data.read_data_sets('MNIST_data', one_hot=True)
```

3. 检查角色是参数服务器还是 worker，如果是 worker 就定义一个简单的稠密神经网络，定义一个优化器以及用于评估分类器的度量（例如精确度）。

```
if FLAGS.job_name == "ps":
  server.join()
elif FLAGS.job_name == "worker":
# Between-graph replication
with tf.device(tf.train.replica_device_setter(
```

```
      worker_device="/job:worker/task:%d" % FLAGS.task_index,
      cluster=cluster)):
    # count the number of updates
      global_step = tf.get_variable( 'global_step', [], initializer =
tf.constant_initializer(0),
trainable = False)

    # input images
    with tf.name_scope('input'):
      # None -> batch size can be any size, 784 -> flattened mnist image
      x = tf.placeholder(tf.float32, shape=[None, 784], name="x-input")
      # target 10 output classes
      y_ = tf.placeholder(tf.float32, shape=[None, 10], name="y-input")

    # model parameters will change during training so we use tf.Variable
    tf.set_random_seed(1)
    with tf.name_scope("weights"):
      W1 = tf.Variable(tf.random_normal([784, 100]))
      W2 = tf.Variable(tf.random_normal([100, 10]))

    # bias
    with tf.name_scope("biases"):
      b1 = tf.Variable(tf.zeros([100]))
      b2 = tf.Variable(tf.zeros([10]))

    # implement model
    with tf.name_scope("softmax"):
      # y is our prediction
      z2 = tf.add(tf.matmul(x,W1),b1)
      a2 = tf.nn.sigmoid(z2)
      z3 = tf.add(tf.matmul(a2,W2),b2)
      y = tf.nn.softmax(z3)

    # specify cost function
    with tf.name_scope('cross_entropy'):
      # this is our cost
      cross_entropy = tf.reduce_mean(
-tf.reduce_sum(y_ * tf.log(y), reduction_indices=[1]))

    # specify optimizer
    with tf.name_scope('train'):
      # optimizer is an "operation" which we can execute in a session
      grad_op = tf.train.GradientDescentOptimizer(learning_rate)
      train_op = grad_op.minimize(cross_entropy, global_step=global_step)

    with tf.name_scope('Accuracy'):
      # accuracy
      correct_prediction = tf.equal(tf.argmax(y,1), tf.argmax(y_,1))
      accuracy = tf.reduce_mean(tf.cast(correct_prediction, tf.float32))

    # create a summary for our cost and accuracy
    tf.summary.scalar("cost", cross_entropy)
    tf.summary.scalar("accuracy", accuracy)
    # merge all summaries into a single "operation" which we can execute in a
session
    summary_op = tf.summary.merge_all()
    init_op = tf.global_variables_initializer()
    print("Variables initialized ...")
```

4.启动一个监督器作为分布式设置的主机,主机是管理集群其余部分的机器。会话由主机维护,关键指令是 `sv = tf.train.Supervisor(is_chief=(FLAGS.task_index == 0))`。另外,通过 `prepare_or_wait_for_session(server.target)`,监督器将等待模型投入使用。请注意,每个 worker 将处理不同的批量模型,然后将最终的模型提供给主机。

```
sv = tf.train.Supervisor(is_chief=(FLAGS.task_index == 0),
begin_time = time.time()
frequency = 100
with sv.prepare_or_wait_for_session(server.target) as sess:
  # create log writer object (this will log on every machine)
  writer = tf.summary.FileWriter(logs_path, graph=tf.get_default_graph())
  # perform training cycles
  start_time = time.time()
  for epoch in range(training_epochs):
    # number of batches in one epoch
    batch_count = int(mnist.train.num_examples/batch_size)
    count = 0
    for i in range(batch_count):
      batch_x, batch_y = mnist.train.next_batch(batch_size)
      # perform the operations we defined earlier on batch
      _, cost, summary, step = sess.run(
      [train_op, cross_entropy, summary_op, global_step],
      feed_dict={x: batch_x, y_: batch_y})
      writer.add_summary(summary, step)
      count += 1
      if count % frequency == 0 or i+1 == batch_count:
        elapsed_time = time.time() - start_time
        start_time = time.time()
        print("Step: %d," % (step+1),
          " Epoch: %2d," % (epoch+1), " Batch: %3d of %3d," % (i+1, batch_count),
          " Cost: %.4f," % cost,
          "AvgTime:%3.2fms" % float(elapsed_time*1000/frequency))
        count = 0
    print("Test-Accuracy: %2.2f" % sess.run(accuracy, feed_dict={x: mnist.test.images, y_: mnist.test.labels}))
    print("Total Time: %3.2fs" % float(time.time() - begin_time))
    print("Final Cost: %.4f" % cost)
  sv.stop()
  print("done")
```

解读分析

这个案例描述了一个分布式 MNIST 分类器的示例,在这个例子中,TensorFlow 允许定义一个三台机器的集群,一个用作参数服务器,另外两个用作独立批量训练数据的 worker。

12.6 基于 Docker 使用 TensorFlow Serving

在这个案例中将演示如何为 TensorFlow Serving 运行一个 Docker 容器,这是一组组件，用于导出经过训练的 TensorFlow 模型，并使用标准的 `tensorflow_model_server` 来为实现的模型服务。TensorFlow Serving 服务器发掘出新的导出模型并使用 gRPC 来提供服务。

准备工作

我们将使用 Docker 并假定你熟悉该系统，可参考 https://www.docker.com/ 了解 Docker 的相关知识并安装它。我们要做的是建立一个 TensorFlow Serving 版本。

具体做法

1. 从网址 https://github.com/tensorflow/serving/blob/master/tensorflow_serving/tools/docker/Dockerfile.devel 上下载 Dockerfile.devel。

2. 通过运行以下命令构建一个容器：

`docker build--pull-t$USER/tensorflow-serving-devel-f Dockerfile.devel`

3. 运行容器：

`docker run -it $USER/tensorflow-serving-devel`

4. 克隆 TensorFlow Serving、配置和测试服务器：

```
git clone --recurse-submodules https://github.com/tensorflow/serving
cd serving/tensorflow
./configure
cd ..
bazel test tensorflow_serving/...
```

5. 现在看一个保存模型的示例，以便服务器可以保存它。该例子是受一个示例启发得到的，构建 MNIST 训练器并为模型提供服务（https://github.com/tensorflow/serving/blob/master/tensorflow_serving/example/mnist_saved_model.py）。第一步是将构建器（builder）导入为 `saved_model_builder`，然后大部分的工作由 `SavedModelBuilder()` 来完成，它将被训练模型的 snapshot 保存到可靠的存储空间中。请注意，这里的 `export_path` 是 /tmp/mnist_model/：

```
from tensorflow.python.saved_model import builder as saved_model_builder
...
export_path_base = sys.argv[-1]
export_path = os.path.join(
  compat.as_bytes(export_path_base),
  compat.as_bytes(str(FLAGS.model_version)))
print 'Exporting trained model to', export_path
builder = saved_model_builder.SavedModelBuilder(export_path)
builder.add_meta_graph_and_variables(
  sess, [tag_constants.SERVING],
  signature_def_map={
    'predict_images':
    prediction_signature,
```

```
        signature_constants.DEFAULT_SERVING_SIGNATURE_DEF_KEY:
        classification_signature,
    },
    legacy_init_op=legacy_init_op)
builder.save()
```

6. 模型可以用一个简单的命令来执行:

```
tensorflow_model_server --port=9000 --model_name=mnist --
model_base_path=/tmp/mnist_model/
```

解读分析

2016 年 2 月谷歌发布 TensorFlow Serving（https://www.tensorflow.org/serving/），是针对机器学习模型的一个高性能服务系统，用于生产环境。2017 年 8 月，谷歌在生产中已经有超过 800 个项目使用了 TensorFlow Serving。

拓展阅读

TensorFlow Serving 是一款非常灵活的软件，在这个案例中只是学习了它潜在用途的表面而已。如果你有兴趣了解更多的高级功能，如大批量运行或动态加载模型，推荐参考 https://github.com/tensorflow/serving/blob/master/tensorflow_serving/g3doc/serving_advanced.md。

12.7 使用计算引擎在谷歌云平台上运行分布式 TensorFlow

在这个案例中将学习如何在谷歌云平台上使用 Google TensorFlow，回顾的例子是经典的 MNIST。

准备工作

在 https://cloud.google.com/ 中查看 GCP 的工作方式将是一件好事。请注意，GCP 提供 300 美元的免费积分以使用任何 GCP 产品。此外，在免费试用期间和之后，对符合条件的客户，某些产品可免费使用。（具体优惠请参阅 https://cloud.google.com/free/。）

具体做法

1. 从网络控制台 https://pantheon.google.com/cloud-resource-manager 上创建新的谷歌云项目。

第12章 分布式TensorFlow和云深度学习

单击"create project"时将显示以下界面：

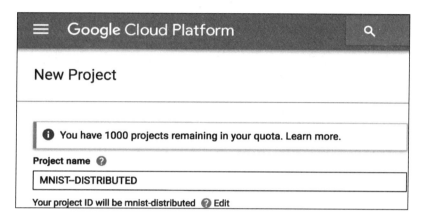

2. 通过在云控制台的左侧导航栏中选择相关的语音来为此项目启用计费，然后为项目启动Compute Engine（计算引擎）和Cloud Machine Learning APIs（云机器学习API）：

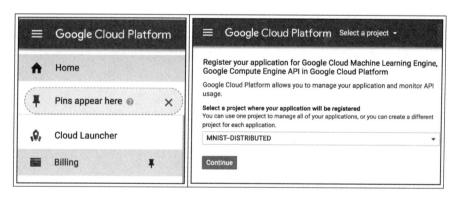

3. 登录到网络云端 https://pantheon.google.com/cloudshell/editor?

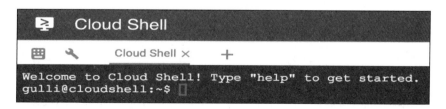

4. 在控制台上运行以下命令来配置要执行计算的区域，下载示例代码并创建用于运行代码的虚拟机，最后连接到机器上。

```
gcloud config set compute/zone us-east1-c
gcloud config set project [YOUR_PROJECT_ID]
git clone https://github.com/GoogleCloudPlatform/cloudml-dist-mnist-example
cd cloudml-dist-mnist-example
gcloud compute instances create template-instance \
 --image-project ubuntu-os-cloud \
 --image-family ubuntu-1604-lts \
 --boot-disk-size 10GB \
 --machine-type n1-standard-1
gcloud compute ssh template-instance
```

5. 登录到机器后，需要用以下命令安装 PIP 和 TensorFlow 来设置环境。

```
sudo apt-get update
sudo apt-get -y upgrade \
 && sudo apt-get install -y python-pip python-dev
sudo pip install tensorflow
sudo pip install --upgrade tensorflow
```

6. 有多个处理 MNIST 数据的 worker，所以最好的办法是创建一个在所有 worker 之间共享的存储桶，并复制这个桶中的 MNIST 数据。

```
BUCKET="mnist-$RANDOM-$RANDOM"
gsutil mb -c regional -l us-east1 gs://${BUCKET}
sudo ./scripts/create_records.py
gsutil cp /tmp/data/train.tfrecords gs://${BUCKET}/data/
gsutil cp /tmp/data/test.tfrecords gs://${BUCKET}/data/
```

7. 创建多个 worker（worker-0，worker-1），它们是初始模板——实例机器的克隆。不希望在机器关闭时删除磁盘，这就是为什么要有第一条命令。

```
gcloud compute instances set-disk-auto-delete template-instance \
 --disk template-instance --no-auto-delete
gcloud compute instances delete template-instance
gcloud compute images create template-image \
 --source-disk template-instance
gcloud compute instances create \
 master-0 worker-0 worker-1 ps-0 \
 --image template-image \
 --machine-type n1-standard-4 \
 --scopes=default,storage-rw
```

8. 最后一步是执行分布式训练的计算。

```
./scripts/start-training.sh gs://${BUCKET}
```

解读分析

Demo 脚本将代码移动到每个虚拟机并启动分布式计算，这两个 worker 基于一个共同的存储桶所共享的相同的 MNIST 数据，并行进行运算，计算结束后，脚本将打印训练模型的位置。

扩展阅读

如果不想管理 TensorFlow，那么可使用 Google 运行的托管版本，这是在下一个案例中所描述的 CloudML 服务。另外，如果你决定使用带有 CloudML 的 GPU，那么 `https://cloud.google.com/ml-engine/docs/using-gpus` 是一个很好的起点。

12.8 在谷歌 CloudML 上运行分布式 TensorFlow

CloudML 是由 Google 运行的 TensorFlow 托管版本，你可以不用自己运行 TensorFlow，只需轻松地使用 CloudML，并忘记与基础架构和可扩展性相关的所有问题。

准备工作

假设你已经创建了云平台项目，并为项目启动计费，并启用 Google 计算引擎和云机器学习 API，这些步骤与上一个案例描述的步骤相似。这个案例受到 MNIST 训练代码的启发：`https://cloud.google.com/ml-engine/docs/distributed-tensorflow-mnist-cloud-datalab`。

具体做法

继续在 Google CloudML 上运行分布式 TensorFlow：

1. 从 `https://github.com/GoogleCloudPlatform/cloudml-dist-mnist-example` 下载示例代码。

2. 然后下载数据并保存在 GCP 存储桶中：

```
PROJECT_ID=$(gcloud config list project --format "value(core.project)")
BUCKET="${PROJECT_ID}-ml"
gsutil mb -c regional -l us-central1 gs://${BUCKET}
./scripts/create_records.py
gsutil cp /tmp/data/train.tfrecords gs://${BUCKET}/data/
gsutil cp /tmp/data/test.tfrecords gs://${BUCKET}/data/
```

3. 提交训练工作非常简单：可以轻松地使用 CloudML 引擎调用训练步骤。在这个例子中，训练代码在 us-central1 区域运行 1000 次，输入数据来自存储桶，输出桶将被提交到一个不同的存储桶。

```
JOB_NAME="job_$(date +%Y%m%d_%H%M%S)"
gcloud ml-engine jobs submit training ${JOB_NAME} \
 --package-path trainer \
 --module-name trainer.task \
 --staging-bucket gs://${BUCKET} \
 --job-dir gs://${BUCKET}/${JOB_NAME} \
 --runtime-version 1.2 \
 --region us-central1 \
 --config config/config.yaml \
```

```
-- \
--data_dir gs://${BUCKET}/data \
--output_dir gs://${BUCKET}/${JOB_NAME} \
--train_steps 10000
```

4. 如果你愿意，可以通过访问 CloudML 控制台（`https://pantheon.google.com/mlengine/`）控制训练过程。

5. 一旦培训结束，可以直接从 CloudML 中提取模型。

```
MODEL_NAME=MNIST
gcloud ml-engine models create --regions us-central1 ${MODEL_NAME}
VERSION_NAME=v1
ORIGIN=$(gsutil ls gs://${BUCKET}/${JOB_NAME}/export/Servo | tail -1)
gcloud ml-engine versions create \
 --origin ${ORIGIN} \
 --model ${MODEL_NAME} \
${VERSION_NAME}
gcloud ml-engine versions set-default --model ${MODEL_NAME} ${VERSION_NAME}
```

6. 一旦模型在线提供，就可以获取服务并进行预测，request.json 是通过使用从 MNIST 读取数据的脚本 `make_request.py` 创建的，该脚本执行独热编码，然后使用格式良好的 json 模式编写特征。

```
gcloud ml-engine predict --model ${MODEL_NAME} --json-instances
request.json
```

解读分析

CloudML 是使用 Google TensorFlow 托管版本的便捷解决方案，它不直接关注基础架构和运算，而是关注开发机器学习模型。

拓展阅读

CloudML 的一个非常酷的功能是通过并行运行多个试验来自动调整模型中包含的超参数，这为超参数提供了优化值，从而最大限度地提高了模型的预测精度。如果有兴趣了解更多信息，请参考 `https://cloud.google.com/ml-engine/docs/hyperparameter-tuning-overview`。

12.9　在 Microsoft Azure 上运行分布式 TensorFlow

Microsoft Azure 提供了一个名为 Batch AI 的服务，它允许在 Azure 虚拟机群集上运行机器学习模型。

准备工作

第一步，需要一个 Azure 账户，如果你还没有账户，可以免费创建一个账户，网址为

https://azure.microsoft.com/en-us/services/batch-ai/。Azure 为新用户提供 30 天 200 美元的积分。这个案例将按照 Azure 提供的示例，使用分布式 TensorFlow 在两个 GPU 上运行 MNIST，相关的代码发布在 GitHub 上，网址为：https://github.com/Azure/batch-shipyard/tree/master/recipes/TensorFlow-Distributed。

具体做法

1. 安装 Azure CLI。不同操作系统平台上的安装细节请参考：https://docs.microsoft.com/en-us/cli/azure/install-azure-cli?view=azure-cli-latest。

2. 在创建集群之前，需要使用命令 `az login` 登录 Azure。它会生成一个口令和网址，并验证你的使用凭证。在这个网址上按照步骤依次操作，系统会要求关掉页面并验证你的凭证，az 证书将被验证。

3. 配置默认位置，创建和配置资源组。

```
az group create --name myResourceGroup --location eastus
az configure --defaults group=myResourceGroup
az configure --defaults location=eastus
```

4. 使用 <az storage account create> 命令创建存储，并根据操作系统设置环境变量，有关环境变量及其值的详细信息可从网址 https://docs.microsoft.com/en-us/azure/batch-ai/quickstart-cli 获取。

5. 下载并提取预处理的 MNIST 数据库。

```
wget "https://batchaisamples.blob.core.windows.net/samples/mnist_dataset_original.zip?st=2017-09-29T18%3A29%3A00Z&se=2099-12-31T08%3A00%3A00Z&sp=rl&sv=2016-05-31&sr=b&sig=Qc1RA3zsXIP4oeioXutkL1PXIrHJO0pHJlppS2rID3I%3D" -O mnist_dataset_original.zip
unzip mnist_dataset_original.zip
```

6. 下载 `mnist_replica`。

```
wget "https://raw.githubusercontent.com/Azure/BatchAI/master/recipes/TensorFlow/TensorFlow-GPU-Distributed/mnist_replica.py?token=AcZzrcpJGDHCUzsCyjlWiKVNfBuDdkqwks5Z4dPrwA%3D%3D" -O mnist_replica.py
```

7. 创建一个 Azure 文件共享，在其中上传下载的 MNIST 数据集和 `mnist_replica.py` 文件。

```
az storage share create --name batchaisample
az storage directory create --share-name batchaisample --name mnist_dataset
az storage file upload --share-name batchaisample --source t10k-images-idx3-ubyte.gz --path mnist_dataset
az storage file upload --share-name batchaisample --source t10k-labels-idx1-ubyte.gz --path mnist_dataset
az storage file upload --share-name batchaisample --source train-images-
```

```
idx3-ubyte.gz --path mnist_dataset
az storage file upload --share-name batchaisample --source train-labels-
idx1-ubyte.gz --path mnist_dataset
az storage directory create --share-name batchaisample --name
tensorflow_samples
az storage file upload --share-name batchaisample --source mnist_replica.py
--path tensorflow_samples
```

8. 创建一个集群。对于这个案例，该集群包括两个标准的 NC6 型 GPU 节点，或者 Ubuntu LTS 和 Ubuntu DVSM 型节点。可以使用 Azure CLI 命令创建集群：

对于 Linux 系统来说，命令如下：

```
az batchai cluster create -n nc6 -i UbuntuDSVM -s Standard_NC6 --min 2 --
max 2 --afs-name batchaisample --afs-mount-path external -u $USER -k
~/.ssh/id_rsa.pub
```

对于 Windows 系统来说，命令如下：

```
az batchai cluster create -n nc6 -i UbuntuDSVM -s Standard_NC6 --min 2 --
max 2 --afs-name batchaisample --afs-mount-path external -u <user_name> -p
<password>
```

9. 在 job.json 文件中创建工作参数：

```
{
  "properties": {
    "nodeCount": 2,
    "tensorFlowSettings": {
      "parameterServerCount": 1,
      "workerCount": 2,
      "pythonScriptFilePath": "$AZ_BATCHAI_INPUT_SCRIPT/mnist_replica.py",
      "masterCommandLineArgs": "--job_name=worker --num_gpus=1 --
ps_hosts=$AZ_BATCHAI_PS_HOSTS --worker_hosts=$AZ_BATCHAI_WORKER_HOSTS --
task_index=$AZ_BATCHAI_TASK_INDEX --data_dir=$AZ_BATCHAI_INPUT_DATASET --
output_dir=$AZ_BATCHAI_OUTPUT_MODEL",
      "workerCommandLineArgs": "--job_name=worker --num_gpus=1 --
ps_hosts=$AZ_BATCHAI_PS_HOSTS --worker_hosts=$AZ_BATCHAI_WORKER_HOSTS --
task_index=$AZ_BATCHAI_TASK_INDEX --data_dir=$AZ_BATCHAI_INPUT_DATASET --
output_dir=$AZ_BATCHAI_OUTPUT_MODEL",
      "parameterServerCommandLineArgs": "--job_name=ps --num_gpus=0 --
ps_hosts=$AZ_BATCHAI_PS_HOSTS --worker_hosts=$AZ_BATCHAI_WORKER_HOSTS --
task_index=$AZ_BATCHAI_TASK_INDEX --data_dir=$AZ_BATCHAI_INPUT_DATASET --
output_dir=$AZ_BATCHAI_OUTPUT_MODEL"
    },
    "stdOutErrPathPrefix": "$AZ_BATCHAI_MOUNT_ROOT/external",
    "inputDirectories": [{
      "id": "DATASET",
      "path": "$AZ_BATCHAI_MOUNT_ROOT/external/mnist_dataset"
    }, {
      "id": "SCRIPT",
      "path": "$AZ_BATCHAI_MOUNT_ROOT/external/tensorflow_samples"
    }],
    "outputDirectories": [{
      "id": "MODEL",
      "pathPrefix": "$AZ_BATCHAI_MOUNT_ROOT/external",
      "pathSuffix": "Models"
```

```
    }],
    "containerSettings": {
      "imageSourceRegistry": {
        "image": "tensorflow/tensorflow:1.1.0-gpu"
      }
    }
  }
}
```

10. 使用以下命令创建 Batch AI 作业：

```
az batchai job create -n distibuted_tensorflow --cluster-name nc6 -c
job.json
```

解读分析

Batch AI 自己管理资源，你只需指定作业、输入位置和存储输出的位置。如果在执行作业期间想要查看结果，可以使用以下命令：

```
az batchai job stream-file --job-name myjob --output-directory-id stdouterr
--name stderr.txt
```

作业结束后，使用命令 `az batchai job delete` 和 `az batchai cluster delete` 删除作业和集群。

拓展阅读

上面学习了如何使用 Azure 命令行工具将 Microsoft Azure Batch AI 用于分布式 TensorFlow，也可以使用 Jupyter Notebook 做同样的事情。这将涉及设置 Azure Active Directory，并进行新的应用程序注册。详细信息可从 https://docs.microsoft.com/en-us/azure/azure-resource-manager/resource-group-create-service-principal-portal 获得。

Azure Batch AI 也可以与其他 AI 深度学习库一起工作，建议通过 BatchAI GitHub 了解更多详细信息，网址为 https://github.com/Azure/BatchAI。

12.10 在 Amazon AWS 上运行分布式 TensorFlow

Amazon AWS 提供采用 NVIDIA K8 GPU 的 P2.x 机器。为了能够使用，第一步还需要创建一个 Amazon AWS 账户，如果还没有，可以使用链接 https://portal.aws.amazon.com/billing/signup?nc2=h_ct&redirect_url=https%3A%2F%2Faws.amazon.com%2Fregistration-confirmation#/start 来创建。登录账户后，控制台看起来如下图所示。

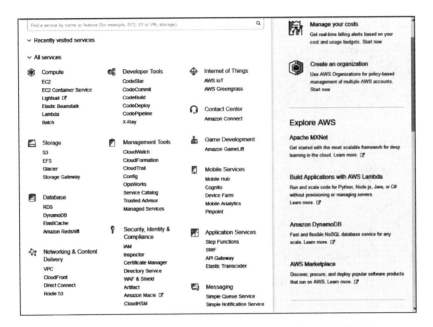

可以看到，Amazon AWS 提供了大量服务，但在这里关注的是使用 Amazon AWS 进行深度学习。

GPU 仅在 P2 实例构建时可用，并且在默认情况下不可用，为了获得该服务，需要通过 AWS support 来提高价格以增加资源，support 位于右上角，一旦进入 support，你会看到一个 Create case 按钮，点击该按钮，并做出以下选择：

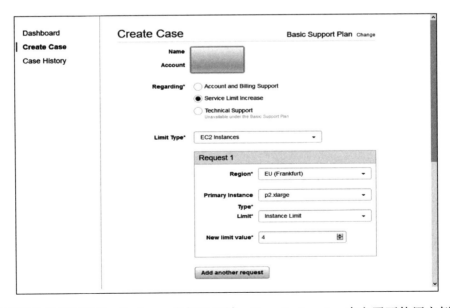

你可以选择任何提供 p2.xlarge 实例的区域。New limit value 决定了可使用实例的最大

数量，请仔细选择该数字，因为该服务不是免费的，每小时大约花费 1 美元。AWS 大约会用 24 小时回复请求。

准备工作

现在已经拥有了 AWS 账户和 p2.xlarge 实例，就可以从 EC2 控制台创建自己的实例。第一步是选择 Machine Image（机器镜像），到目前为止，亚马逊提供预装深度学习库的特殊机器镜像，你可以选择 Ubuntu 或 Linux 版本。接下来，选择 GPU 确定实例类型。

你可以使用默认参数查看和启动（Review and Lanuch）实例，也可以配置设置，选择存储并配置安全组（security group）。配置安全组非常重要，默认情况下，SSH 安全组已经设置，但是如果你要使用 Jupyter Notebook，则需要为端口 8888 添加自定义安全组，选择 source 以备登录实例，其中 source 有三个选项，分别为：Custom、Anywhere 和 My IP。

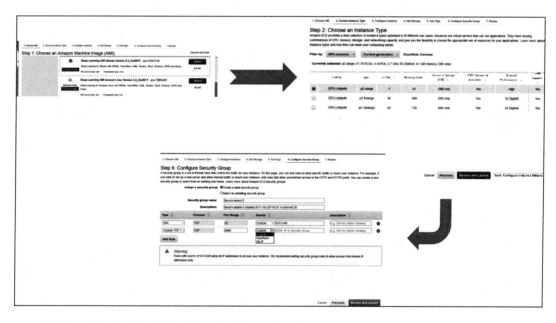

最后，当你启动实例的时候，你会被要求创建一个密钥对（key pair），它允许你登录到指定的实例。创建一个密钥对并下载相应的 .pem 文件，这将在稍后登录时使用。

具体做法

1. 连接到实例，可以通过 ssh 命令或浏览器完成，使用 CLI（命令行界面）。
2. 连接 .pem 文件的可变化模块：

```
chmod 400 <security-keypair-filename>.pem
```

3. 使用下面的命令将 SSH 连接到实例，当要连接时，通过控制台能够看到准确地址。

```
ssh -i " <security-keypair-filename>.pem" ubuntu@ec2-
XXXXXXXXXXXXX.compute-1.amazonaws.com
```

4. 选择的机器实例已经包含了包括 TensorFlow 在内的所有深度学习库，所以不需要安装任何东西：

```
ubuntu@ip-172-31-19-254: ~/src

* Documentation:  https://help.ubuntu.com
* Management:     https://landscape.canonical.com
* Support:        https://ubuntu.com/advantage

 Get cloud support with Ubuntu Advantage Cloud Guest:
   http://www.ubuntu.com/business/services/cloud

4 packages can be updated.
0 updates are security updates.

*** System restart required ***
The programs included with the Ubuntu system are free software;
the exact distribution terms for each program are described in the
individual files in /usr/share/doc/*/copyright.

Ubuntu comes with ABSOLUTELY NO WARRANTY, to the extent permitted by
applicable law.

ubuntu@ip-172-31-19-254:~$ ls
src
ubuntu@ip-172-31-19-254:~$ cd src
ubuntu@ip-172-31-19-254:~/src$ ls
anaconda2        caffe2              caffe_cpu          keras              OpenBLAS              tensorflow_anaconda3    theano
anaconda3        caffe2_anaconda2    caffe_python3      logs               pytorch               tensorflow_cpu          torch
bin              caffe_anaconda2     cntk               mxnet              README.md             tensorflow_python2
caffe            caffe_anaconda3     demos              Nvidia_Cloud_EULA.pdf  tensorflow_anaconda2  tensorflow_python3
ubuntu@ip-172-31-19-254:~/src$
```

5. 每个文件夹都包含一个介绍如何使用相应库的 readme 文件：

```
ubuntu@ip-172-31-19-254: ~/src/tensorflow_anaconda3

emo.apk), [native libs](http://ci.tensorflow.org/view/Nightly/job/nightly-android/lastSuccessfulBuild/artifact/out/native/)
([build history](https://ci.tensorflow.org/view/Nightly/job/nightly-android/))

#### *Try your first TensorFlow program*
```shell
$ python
```
```python
>>> import tensorflow as tf
>>> hello = tf.constant('Hello, TensorFlow!')
>>> sess = tf.Session()
>>> sess.run(hello)
'Hello, TensorFlow!'
>>> a = tf.constant(10)
>>> b = tf.constant(32)
>>> sess.run(a+b)
42
>>>
```

For more information

* [TensorFlow website](https://www.tensorflow.org)
* [TensorFlow whitepaper](http://download.tensorflow.org/paper/whitepaper2015.pdf)
* [TensorFlow Model Zoo](https://github.com/tensorflow/models)
* [TensorFlow MOOC on Udacity](https://www.udacity.com/course/deep-learning--ud730)

The TensorFlow community has created amazing things with TensorFlow, please see the [resources section of tensorflow.org](http
s://www.tensorflow.org/about/#community) for an incomplete list.
ubuntu@ip-172-31-19-254:~/src/tensorflow_anaconda3$
```

解读分析

你可以运行刚才学习的已经创建的实例的相关代码。一旦工作结束，不要忘记退出，并从控制台上停止实例。有关价格和使用的更多细节请访问：`https://aws.amazon.com/documentation/ec2/`。

拓展阅读

AWS 市场上有大量具备预配置库和 API 的 docker 镜像和机器镜像。要启动 jupyter notebook，请在命令行中使用 `<jupyter notebook --ip=0.0.0.0 --no-browser>`，输出如下所示：

```
Copy/paste this URL into your browser when you connect for the first time to login with a token:
http://0.0.0.0:8888/?token=3156e...
```

复制该网址到浏览器中即可开始使用。

此外，使用 AWS CloudFormation 可以简化整个流程。CloudFormation 通过模板创建和配置 Amazon Web Services 资源，能够简化建立分布式深度学习集群的过程，有兴趣的读者请参考：`https://aws.amazon.com/blogs/compute/distributed-deep-learning-made-easy/`。

APPENDIX A
附录 A

利用 AutoML 学会学习（元学习）

深度学习的成功极大地促进了特色工程的工作。传统的机器学习很大程度上取决于特征集合的正确选择，特征集合的正确选择往往比学习算法的选择更重要。而深度学习则改变了这种情况，构建正确的模型仍然非常重要，但是现在的网络对特征的选取并不是那么敏感，而且具备了自动选择真正重要的特征的能力。

随着深度学习的引入，增加了对恰当神经网络结构选择的关注，这意味着研究人员的兴趣逐渐从特征工程转向了网络工程。AutoML（元学习）是一个新兴的研究课题，旨在为给定的学习任务自动选择最有效的神经网络，换句话说，AutoML 展现了一套如何学会有效学习的方法。如机器翻译、图像识别或玩游戏的任务，这些模型通常是由工程师、数据科学家和领域专家组成的团队手工设计的，如果考虑典型的 10 层网络，那么大概存在个候选网络，此时你便能理解传统的模型设计过程非常昂贵、容易出错并且最终可能是次优的原因了。

A.1 基于循环神经网络和强化学习的元学习

解决这个问题的关键思路是建立一个控制器网络，提供概率为 p 的子模型架构，在输入中给定特定网络。对子模型进行训练，并对待解决的特定任务进行评估（例如，子模型获得的精度为 R），评估 R 传回给控制器，而控制器又使用 R 来改进下一个候选架构。给定这么一个框架，可以从候选子模型到控制器进行反馈建模，并作为计算梯度 p 的任务，然后对梯度缩放 R 倍，控制器则可以用循环神经网络实现（见下图）。在下一次迭代中，控制器倾向于对得分为 R 的架构赋予更高的概率，而对得分不好的架构赋予较低的概率。

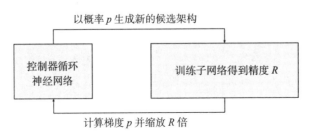

控制器循环神经网络可以对卷积网络进行采样,并且能够预测许多超参数,例如滤波器高度、滤波器宽度、步幅高度、步幅宽度以及一层滤波器的数量,然后进行重复。每个预测由 softmax 分类器执行,然后作为输入馈送到下一个 RNN 时序。下图表达了该思想(来自 *Neural Architecture Search with Reinforcement Learning*, Barret Zoph, Quoc V. Le, https://arxiv.org/abs/1611.01578):

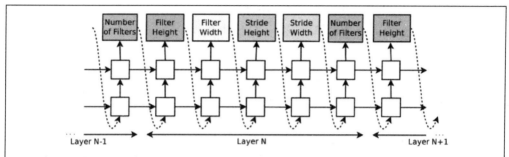

我们的控制器递归神经网络如何采样一个简单的卷积网络。它预测过滤器高度、过渡器宽度、步幅高度、步幅宽度,以及下一层和重复的过滤器数量。每一次预测都由 softmax 分类器执行,然后作为输入馈入下一个时序。

只预测超参数是不够的,其原因在于,在网络中定义一组动作来创建新层次,它必须是最优的。但是这特别困难,因为描述新层的奖励函数很可能不可导,因此不可能用诸如 SGD 的标准技术来对其进行优化。强化学习给出了解决方案,类似第 9 章所描述的,它由策略梯度网络组成。

除此之外,并行训练能够用于优化控制器 RNN 的参数。Quoc Le 和 Barret Zoph 提出采用参数服务器方案,其中参数服务器有 S 个切片,用于存储 K 个控制器副本的共享参数,每个控制器副本都对并行训练的 m 个子架构进行采样,如图所示(来自 *Neural Architecture Search with Reinforcement Learning*, Barret Zoph, Quoc V. Le, https://arxiv.org/abs/1611.01578)。

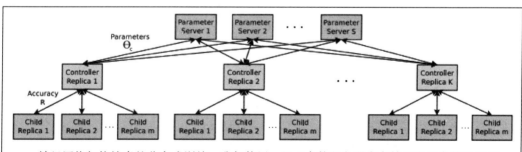

神经网络架构搜索的分布式训练。我们使用一组 S 参数服务器来存储和发送参数到 K 控制器副本。然后每个控制器副本对 m 架构进行采样并行运行多个子模型。记录每个子模型的精确度以计算相对于 θ_c 的梯度,然后返回参数服务器

Quoc 和 Barret 将 AutoML 技术应用到神经架构搜索中，用于处理 Penn Treebank 数据集（https://en.wikipedia.org/wiki/Treebank），该数据集是一个众所周知的自然语言建模的基准。他们的结果改善了目前被认为是最佳的手动设计网络，特别是他们在 Penn Treebank 上达到了 62.4 的测试集困惑度，比之前的最先进模型要好 3.6 倍。类似地，在 CIFAR-10 数据集上（https://www.cs.toronto.edu/~kriz/cifar.html），从头开始，在测试集精确度上，该方法设计的新网络架构能够媲美最好的人工架构。所提出的 CIFAR-10 模型实现了 3.65 的测试错误率，比使用类似架构方案的最先进模型好 0.09%，并且速度要快 1.05 倍。

A.2 元学习 block

论文"Learning Transferable Architectures for Scalable Image Recognition"（Barret Zoph, Vijay Vasudevan, Jonathon Shlens, Quoc V. Le, 2017，https://arxiv.org/abs/1707.07012）提出在小数据集上学习结构构建 block，然后迁移到大数据集上。作者提出在 CIFAR-10 数据集上搜索最优的卷积层（或 cell），然后通过将这个最优 cell 的副本堆叠在一起，把这个 cell 应用到 ImageNet 数据集中，每个副本都有自己的参数。准确地说，所有的卷积网络都是由具有相同结构的卷积层（或 cell）组成的，但权重不同，因此，搜索最佳的卷积架构可简化为寻找最好的 cell 结构，而这种结构更容易推广到其他问题。虽然 cell 不是直接在 ImageNet 上学习的，但也是由最优的 cell 构建的。在 ImageNet 上，在此之前精确度最高的 top-1 达到了 82.7%，top-5 达到了 96.2%，与该论文精确度相同，但是比人工设计的最好架构提高了 1.2%（top-1），FLOPS 则减少了 90 亿次——比现有最好的模型减少了 28%。还有一点需要注意的是，使用 RNN + RL（递归神经网络 + 强化学习）学习的模型正在击败由随机搜索（RS）表示的基线，如下图所示，在 RL 和 RS 确定的 top-5 和 top-25 模型的平均性能上，RL 总是获胜：

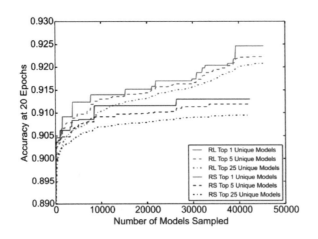

A.3 元学习的新任务

对学习系统进行训练后能够实现大量任务,并测试其学习新任务的能力。这种元学习的一个著名例子是之前讨论的迁移学习,其中网络可以从相对较小的数据集中成功学习新的图像任务。但是却没有语音、语言、文本等非视觉领域的训练方案。

"Model-Agnostic Meta-Learning for Fast Adaptation of Deep Networks"(Chelsea Finn, Pieter Abbeel, SergeyLevine, 2017, https://arxiv.org/abs/1703.03400)提出来一种与模型无关的元学习算法(MAML),它能匹配任何使用梯度下降算法训练的模型,并能应用于各种不同的学习问题,如分类、回归和强化学习等。元学习的目标是在各种学习任务上训练一个模型,因此就可以只使用少量的梯度迭代数来解决新的学习任务。元学习器希望寻求一个初始化,它不仅能适应多个问题,同时还能做到快速适应(少量梯度迭代数)和高效(少量样本)。一个带参数 θ 的函数 f 表示的模型,当适应新任务 T_i 时,模型参数 θ 将变成 $\tilde{\theta_i}$,在利用 MAML 适应新任务时,更新参数 θ_i' 的计算只需少量的梯度迭代。

例如,当使用梯度更新 $\tilde{\theta_i}=\theta-\alpha\nabla_\theta L_{Ti}(f_\theta)$ 时,是任务 T 的损失函数,是元学习参数。MAML 算法描述如下图所示:

```
Algorithm 1 Model-Agnostic Meta-Learning
Require: p(T): distribution over tasks
Require: α, β: step size hyperparameters
 1: randomly initialize θ
 2: while not done do
 3:    Sample batch of tasks T_i ~ p(T)
 4:    for all T_i do
 5:        Evaluate ∇_θ L_{T_i}(f_θ) with respect to K examples
 6:        Compute adapted parameters with gradient descent: θ'_i = θ − α∇_θ L_{T_i}(f_θ)
 7:    end for
 8:    Update θ ← θ − β∇_θ Σ_{T_i~p(T)} L_{T_i}(f_{θ'_i})
 9: end while
```

MAML 在流行的 few-shot 图像分类问题上大大超越了现有的一些方法。few-shot 图像问题旨在依靠一个或很少的几个实例就能学到新的概念,相当具有挑战性。"Human-level concept learning through probabilistic program induction"(Brenden M. Lake, Ruslan Salakhutdinov, Joshua B. Tenenbaum, 2015, https://www.cs.cmu.edu/~rsalakhu/papers/LakeEtAl2015Science.pdf)提出人类通过一张图片就能辨识新式的两轮车,如下图所示:

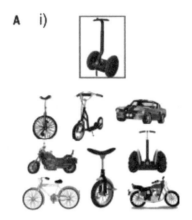

截至2017年年底，AutoML（或元学习）是一个热门研究方向，旨在为指定学习任务自动选择最有效的神经网络。其目标是学习如何有效地自动设计网络，而网络又可以学习指定任务或适应新任务，存在的主要问题是网络设计不能简单地用一个可导的损失函数来描述，因此传统的优化技术不能轻易地用于元学习。针对该问题，已经有了一些解决方案，包括具有控制器循环神经网络（RNN）和强化学习中的奖励策略，以及与模型无关的元学习思想，这两种方法非常强大，但仍然有很大的研究空间。

如果对热门方向感兴趣，*Learning to learn for deep learning* 可以作为你的下一个工作。

- Google提出了将RNN应用到控制器中：*Using Machine Learning to Explore Neural Network Architecture*(Quoc Le & Barret Zoph, 2017, `https://research.googleblog.com/2017/05/using-machine-learning-to-explore.html`)。
- *Neural Architecture Search with Reinforcement Learning*（Barret Zoph, Quoc V. Le, `https://arxiv.org/abs/1611.01578`）是一篇开创性的论文，详细介绍了Google的方法，但是RNN并不是唯一的选择。
- *Large-Scale Evolution of Image Classifiers*（Esteban Real, Sherry Moore, Andrew Selle, Saurabh Saxena, Yutaka Leon Suematsu, Jie Tan, Quoc Le, Alex Kurakin, 2017, `https://arxiv.org/abs/1703.01041`)提出使用演化算法，其中借鉴自遗传算法中的变异操作产生新的候选网络。
- *Learning Transferable Architectures for Scalable Image Recognition*（Barret Zoph, Vijay Vasudevan, Jonathon Shlens, Quoc V. Le, `https://arxiv.org/abs/1707.07012`)提出了在CIFAR上学习cell的思想，然后用于改进ImageNet分类。
- *Building A.I. That Can Build A.I.: Google and others, fighting for a small pool of researchers, are looking for automated ways to deal with a shortage of artificial intelligence experts*(《纽约时报》, `https://www.nytimes.com/2017/11/05/`

technology/machine-learning-artificial-intelligence-ai.
html)。

- *Model-Agnostic Meta-Learning for Fast Adaptation of Deep Networks*(Chelsea Finn, Pieter Abbeel, Sergey Levine, 2017, https://arxiv.org/abs/1703.03400)。
- *Learning to Learn by Gradient Descent by Gradient Descent*(Marcin Andrychowicz, Misha Denil, Sergio Gomez, Matthew W. Hoffman, David Pfau, Tom Schaul, Brendan Shillingford, Nando de Freitas, https://arxiv.org/abs/1606.04474）展示了如何将优化算法的设计作为一个学习问题，使算法自动探索感兴趣问题的结构。在训练任务上，LSMT学习算法要优于手工设计算法，并且能够很好地推广到结构相似的新任务，这个算法的代码可以在GitHub上找到，网址为 https://github.com/deepmind/learning-to-learn。

A.4 孪生神经网络

孪生神经网络是一种特殊类型的网络，由Yann LeCun和他的同事在NIPS 1994(http://www.worldscientific.com/doi/pdf/10.1142/S0218001493000339)提出。其背后的基本思想类似于"连体双胞胎"，网络由两个不同的神经网络组成，两个神经网络共享相同的架构和权重。

下图是孪生架构：

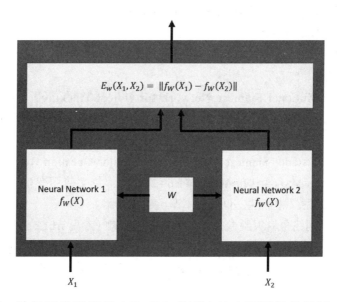

在训练阶段，神经网络训练对（X_1, X_2）的两个输入不同但是相似，例如，X_1=He is smart，X_2=He is a wise man。两个神经网络将产生两个不同的结果，组合网络可以被认为

是标量能量函数,用以测量训练对 (X_1,X_2) 之间的相似性,定义为:
$$E_w(X_1,X_2) = \|f_w(X_1) - f_w(X_2)\|$$

孪生网络的目标是真训练对 (X_1,X_2) 之间的能量要小于其他伪训练对 (X_1,X_2') 之间的能量。通过使用对比损失函数(contrastive loss function)来实现。

在训练阶段,网络提供训练对作为输入并给出训练对真伪的对应标签,$(X_1,X_2',Y)^i$ 是第 i 个训练样本,对比损失函数计算如下:
$$L(W) = \sum_{i=1}^{p} L(W,(X_1,X_2,Y)^i)$$

其中,$L(W,(X_1,X_2,Y)^i) = (1-Y)L_G\left(E_w(X_1,X_2)^i\right) + YL_I\left(E_w(X_1,X_2)^i\right)$,$L_G$ 是只计算真训练对的部分损失函数,L_I 是只计算伪训练对的部分损失函数,P 是训练样本数。真训练对相应标签 Y 为 0,伪训练对相应标签值为 1。部分损失 L_G 和 L_I 的设计应该保证对比损失 $L(W)$ 的最小化将会减小真训练对的能量而增大伪训练对的能量,具体实现时,$L(W)$ 关于 L_G 单调递增,$L(W)$ 关于 L_I 单调递减,部分损失函数可使用余弦函数。

权重使用反向传播算法进行调整。

孪生神经网络的应用

孪生神经网络近年来已经出现了很多应用。首次使用是 LeCun 论文中的签名验证,自此出现了大量的应用,我们将介绍一些最近的应用:

- "Joint Learning of Speaker and Phonetic Similarities with Siamese Networks"(https://pdfs.semanticscholar.org/4ffe/3394628a8a0ffd4cba1a77ea85e197bd4c22.pdf):他们训练一个多输出连体网络,一个作为语音上的相似输出,其他的作为说话者的相似度输出。他们也将工作扩大到包括 Triamese 网络。
- "Fully-Convolutional Siamese Network for Object Tracking"(https://link.springer.com/chapter/10.1007/978-3-319-48881-3_56):论文使用在 ILSVRC15 数据集训练的孪生卷积神经网络检测视频中的目标。
- "Together We stand: Siamese Networks for Similar Question Retrieval"(http://www.aclweb.org/anthology/P16-1036):论文使用孪生网络发现当前和存档问题之间的语义相似性,也使用了孪生卷积网络。

除此之外,孪生网络也对人脸验证/识别进行了探索(参见 https://github.com/harveyslash/Facial-Similarity-with-Siamese-Networks-in-Pytorch),还被应用于问答系统中(参见 https://arxiv.org/pdf/1512.05193v2.pdf)。

实现示例——MNIST

应用示例基于 GitHub 页面:https://github.com/ywpkwon/siamese_tf_mnist,

代码中使用孪生网络将手写 MNIST 数字嵌入二维空间中，其中属于同一类的数字嵌入在一起。代码由三个主要文件组成：
- `run.py`：包含执行训练的基本包装，使用梯度下降算法最小化对比损失。
- `inference.py`：包含定义了 3 层全连接网络的 Siamese 类，代码中两个网络输出之间的相似度使用欧式距离，然后使用部分真训练和部分伪训练对损失计算对比损失。
- `visualize.py`：一个可视化结果。

经过第一个 100000 步训练，结果如下：

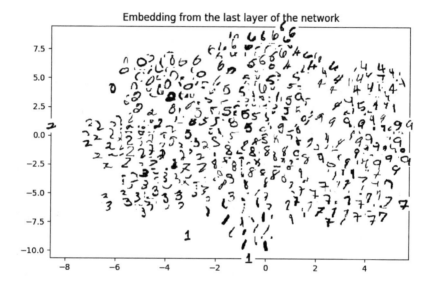

可以看到相同（标记）数字在二维空间中被嵌入到一起。

还有一个有趣的示例，参见 https://github.com/dhwajraj/deep-siamese-text-similarity。

在此示例中，利用 TensorFlow，深度孪生 LTSM 经过训练，使用字符嵌入来捕获短语/句子的相似度。

APPENDIX B

附录 B

TensorFlow 处理器

搜索（RankBrain）、街景、图片和翻译，这些谷歌提供的服务有一个共同点：那就是都使用谷歌的张量处理器（TPU）进行计算。

你可能会思考什么是 TPU，这些服务有什么好处呢？所有这些服务在后台使用最先进的机器学习算法，这些算法需要大量的计算，而 TPU 有助于加速神经网络计算。在 Go 游戏中击败李世石的深度学习项目 AlphaGo 也由 TPU 提供动力，下面就看看 TPU 究竟是什么。

TPU 是由 Google 专门为机器学习而构建的专用集成电路（ASIC），并为 TensorFlow 量身定制。它采用 28nm 工艺制造，主频为 700MHz，功耗为 40W，并被打包成外部扩展加速器，可以放入现有的 SATA 硬盘插槽中。TPU 通过 PCIe Gen 3×16 总线连接到主机 CPU，提供 12.5GB/s 的有效带宽。

到目前为止，第一代 TPU 只能使用已经训练好的模型进行推理。DNN 训练通常需要很长时间，仍然需要在 CPU 和 GPU 上完成训练。2017 年 5 月的博客文章 (`https://www.blog.google/topics/google-cloud/google-cloud-offer-tpus-machine-learning/`) 中宣布的第二代 TPU 既可以用于训练又可以用于推理。

B.1 TPU 组成

本书涵盖的所有深度学习模型，无论是什么样的学习模式，都需要进行三项基本计算：乘法、加法和应用激活函数。

前两个组件是矩阵乘法的一部分：权重矩阵 W 需要与输入矩阵 X 相乘，表示为 W^TX。尽管有 GPU 并行操作，在 CPU 上进行矩阵乘法的计算量依然很大，仍有改进 W^TX 的空间。

TPU 有 65536 个 8 位整数矩阵乘法单元（MXU），最大吞吐量为 92 TOPS。GPU 和 TPU 乘法的主要区别是 GPU 包含浮点乘法器，而 TPU 包含 8 位整数乘法器。TPU 还包含统一缓冲区（UB）——作为寄存器工作的 24 MB 容量的 SRAM，以及激活单元（AU）——包含硬件连接的激活功能。

MXU 使用脉动阵列架构实现，包含由多个运算逻辑单元（ALU）组成的阵列，其中 ALU 以网状拓扑的形式连接到少量的最近邻。每个数据只读一次，但在流经 ALU 阵列时不同操作会使用多次，而不是将其存储到寄存器。TPU 中的 ALU 只能以固定模式执行乘法和加法操作，且 MXU 针对矩阵乘法运算进行了优化，这并不适用于通用计算。

每个 TPU 还有被称作权重寄存器的片外 8-GiB DRAM 池，它有一个四级流水线并执行 CISC 指令。到目前为止，TPU 由六个神经网络组成：两个 MLP、两个 CNN 和两个 LSTM。

TPU 在高级指令的帮助下编程，一些用于编程 TPU 的指令如下：

- `Read_Weights`：从内存中读取权重。
- `Read_Host_Memory`：从内存中读取数据。
- `MatrixMultiply / Convolve`：与数据相乘或卷积并累积结果。
- `Activate`：应用激活功能。
- `Write_Host_Memory`：将结果写入内存。

Google 已经创建了一个 API 堆栈来帮助 TPU 编程，它将来自 TensorFlow 图表的 API 调用转换成 TPU 指令。

B.2　TPU 的优点

TPU 超过 GPU 和 CPU 的首要优势就是性能。Google 将 TPU 性能与 Intel Haswell CPU（服务器级）和 NVIDIA K80 GPU（运行 95% 推理工作负载的基准代码）进行比较，发现 TPU 比 NVIDIA GPU 和 Intel CPU 快 15～30 倍。

第二个重要参数是功耗。降低功耗非常重要，因为它具有双重的能源优势：不仅减少了功耗，而且还能够减小运行过程中的冷却散热成本，从而节省能耗。相比于其他 CPU/GPU 配置，TPU/CPU 每瓦性能提高了 30～80 倍。

TPU 的另一个优势是极简和确定性设计，因为它一次只能执行一项任务。

与 CPU 和 GPU 相比，单线程 TPU 没有任何复杂的微架构特征，而这种复杂的架构将会消耗晶体管和能量来改善平均情况，但却无法改善 99% 的情况，例如，无高速缓存、分支预测、无序执行、多核、推测预取、地址合并、多线程，上下文切换等。极简主义是特定领域处理器的优点。

B.3　访问 TPU

Google 不会直接将 TPU 出售给其他机构，而是通过谷歌云平台提供服务：Cloud TPU Alpha（https://cloud.google.com/tpu/）。Cloud TPU Alpha 将提供高达 180 teraflop 的计算性能和 64 GB 的超高带宽内存，用户能够通过自定义虚拟机连接到这些云端 TPU。

Google 还免费向全世界的机器学习研究人员提供 1000 个云 TPU 的服务器集群，以加快开放式机器学习研究步伐。访问权将以有限的计算时间授权给选定的个人，个人可以使用链接 https://services.google.com/fb/forms/tpusignup/ 注册。参见 Google 博客：

"由于 TensorFlow 研究云的主要目的是使开放机器学习研究社区整体受益，所以成功的申请人需要做以下工作：

通过同行评议的出版物、开源代码、博客文章或其他开放媒体，将他们的 TFRC 支持的研究与世界分享；

与 Google 分享具体的建设性反馈，随着时间的推移帮助改进 TFRC 计划和底层的云 TPU 平台。

想象一下在未来，机器学习研究速度将越来越快，未来将会开发更多的新机器学习模型。"

B.4　TPU 资源

- Norman P. Jouppi 等人 "In-datacenter performance analysis of a tensor processing unit", arXiv preprint arXiv:1704.04760 (2017)。在这篇论文中作者将 TPU 与服务器级 Intel Haswell CPU 和 NVIDIA K80 GPU 进行比较，与 CPU 和 K80 GPU 相比，本文测试了 TPU 的性能。
- Google 博客 https://cloud.google.com/blog/big-data/2017/05/an-in-depth-look-at-googles-first-tensor-processing-unit-tpu 以浅显的语言解释了 TPU 及其工作原理。